Water Supply and Water Scarcity

Water Supply and Water Scarcity

Editors

Vasileios A. Tzanakakis
Nikolaos V. Paranychianakis
Andreas N. Angelakis

MDPI • Basel • Beijing • Wuhan • Barcelona • Belgrade • Manchester • Tokyo • Cluj • Tianjin

Editors
Vasileios A. Tzanakakis Hellenic Mediterranean University
Greece

Nikolaos V. Paranychianakis
Technical University of Crete
Greece

Andreas N. Angelakis
Union of Water Supply and Sewerage Enterprises
Greece

Editorial Office
MDPI
St. Alban-Anlage 66
4052 Basel, Switzerland

This is a reprint of articles from the Special Issue published online in the open access journal *Water* (ISSN 2073-4441) (available at: https://www.mdpi.com/journal/water/special_issues/supply_scarcity).

For citation purposes, cite each article independently as indicated on the article page online and as indicated below:

LastName, A.A.; LastName, B.B.; LastName, C.C. Article Title. *Journal Name* **Year**, *Article Number*, Page Range.

ISBN 978-3-03943-306-3 (Hbk)
ISBN 978-3-03943-307-0 (PDF)

Cover image courtesy of Vasileios Tzanakakis, Nikos Paranychianakis and Andreas N. Angelakis.

Contents

About the Editors vii

Vasileios A. Tzanakakis, Nikolaos V. Paranychianakis and Andreas N. Angelakis
Water Supply and Water Scarcity
Reprinted from: *Water* **2020**, *12*, 2347, doi:10.3390/w12092347 1

Andreas N. Angelakis, Daniele Zaccaria, Jens Krasilnikoff, Miquel Salgot, Mohamed Bazza, Paolo Roccaro, Blanca Jimenez, Arun Kumar, Wang Yinghua, Alper Baba, Jessica Anne Harrison, Andrea Garduno-Jimenez and Elias Fereres
Irrigation of World Agricultural Lands: Evolution through the Millennia
Reprinted from: *Water* **2020**, *12*, 1285, doi:10.3390/w12051285 17

Rodrigo Valdés-Pineda, Pablo García-Chevesich, Juan B. Valdés and Roberto Pizarro-Tapia
The First Drying Lake in Chile: Causes and Recovery Options
Reprinted from: *Water* **2020**, *12*, 290, doi:10.3390/w12010290 67

Amy McNally, Kristine Verdin, Laura Harrison, Augusto Getirana, Jossy Jacob, Shraddhanand Shukla, Kristi Arsenault, Christa Peters-Lidard and James P. Verdin
Acute Water-Scarcity Monitoring for Africa
Reprinted from: *Water* **2019**, *11*, 1968, doi:10.3390/w11101968 79

Maarten V. van de Griend, Luewton L. F. Agostinho, Elmar C. Fuchs, Nigel Dyer and Willibald Loiskandl
Consequences of the Integration of a Hyperbolic Funnel into a Showerhead for Droplets, Jet Break-Up Lengths, and Physical-Chemical Parameters
Reprinted from: *Water* **2019**, *11*, 2446, doi:10.3390/w11122446 95

V. A. Tzanakakis, A. N. Angelakis, N. V. Paranychianakis, Y. G. Dialynas and G. Tchobanoglous
Challenges and Opportunities for Sustainable Management of Water Resources in the Island of Crete, Greece
Reprinted from: *Water* **2020**, *12*, 1538, doi:10.3390/w12061538 111

Tatiana Borisova, Matthew Cutillo, Kate Beggs and Krystle Hoenstine
Addressing the Scarcity of Traditional Water Sources through Investments in Alternative Water Supplies: Case Study from Florida
Reprinted from: *Water* **2020**, *12*, 2089, doi:10.3390/w12082089 147

John J. Erickson, Yamileth C. Quintero and Kara L. Nelson
Characterizing Supply Variability and Operational Challenges in an Intermittent Water Distribution Network
Reprinted from: *Water* **2020**, *12*, 2143, doi:10.3390/w12082143 169

Stavros Yannopoulos, Ioanna Giannopoulou and Mina Kaiafa-Saropoulou
Investigation of the Current Situation and Prospects for the Development of Rainwater Harvesting as a Tool to Confront Water Scarcity Worldwide
Reprinted from: *Water* **2019**, *11*, 2168, doi:10.3390/w11102168 189

Jéssica Kuntz Maykot and Enedir Ghisi
Assessment of A Rainwater Harvesting System in A Multi-Storey Residential Building in Brazil
Reprinted from: *Water* **2020**, *12*, 546, doi:10.3390/w12020546 205

Jéssica Kuntz Maykot and Enedir Ghisi
Correction: Maykot, J.K. and Ghisi, E. Assessment of A Rainwater Harvesting System in A Multi-Storey Residential Building in Brazil. *Water* 2020, *12*, 546
Reprinted from: *Water* **2020**, *12*, 1482, doi:10.3390/w12051482 . . . . **227**

Debjani Sihi, Biswanath Dari, Zhengjuan Yan, Dinesh Kumar Sharma, Himanshu Pathak, Om Prakash Sharma and Lata Nain
Assessment of Water Quality in Indo-Gangetic Plain of South-Eastern Asia under Organic vs. Conventional Rice Farming
Reprinted from: *Water* **2020**, *12*, 960, doi:10.3390/w12040960 . . . . **229**

Guangyi Deng, Xiaohan Yao, Haibo Jiang, Yingyue Cao, Yang Wen, Wenjia Wang, She Zhao and Chunguang He
Study on the Ecological Operation and Watershed Management of Urban Rivers in Northern China
Reprinted from: *Water* **2020**, *12*, 914, doi:10.3390/w12030914 . . . . **239**

Oscar Molina, Thi Thanh Luong and Christian Bernhofer
Projected Changes in the Water Budget for Eastern Colombia Due to Climate Change
Reprinted from: *Water* **2020**, *12*, 65, doi:10.3390/w12010065 . . . . **253**

About the Editors

Vasileios A. Tzanakakis is Assistant Professor in the Department of Agriculture of Hellenic Mediterranean University. He has received a bachelor's, master's and Ph.D. degrees from Agricultural University of Athens, Greece, Department of Natural Resources Management and Agricultural Engineering. He has received scholarship for post-doc research in Norwegian University of Life Sciences, in Norway, by State Scholarships Foundation (IKY)—EEA Grands from 2014 to 2016, and has been granted by Fulbright Foundation Greece to contact research, in USA, in Oregon State University (2018). Also, he has participated in research projects in Greece; in Region of Crete, in Technical University of Crete, in Hellenic Agricultural Organization (HAO)-"Demeter", and, recently, in Hellenic Mediterranean University. Research work of Vasileios has been published in peer-reviewed international journals, book chapters and conference' proceedings. He also has been guest editor in journals' special issues. His current research interests are soil (bio) chemistry and fertility, soil and water resources management, and reuse and climate change mitigation and adaptation practices in agriculture.

Nikolaos V. Paranychianakis received his diploma in 1994 and his Ph.D Thesis in 2001 from the Department of Natural Resources and Agricultural Engineering of the Agricultural University of Athens. He was employed as a post-doc researcher in the Department of Biology of the University of Crete from 2004 to 2006. Nikolaos associated with the School of Environmental Engineering in 2008 and leads the Lab of Agricultural Engineering. Nikolaos has participated in many EU-founded and national projects dealing with water-use efficiency, effluent reuse in agriculture, natural wastewater treatment systems, plant response to abiotic stressors and biogeochemical cycles of nutrients. He has published three book chapters as well as 40 journal articles. Nikolaos has been guest editor in two journal issues. His current research activities focus on the mechanisms regulating the biogeochemistry of C and N in natural and manmade ecosystems, and most specifically on their ecology, biochemistry, and molecular biology, their response to environmental stimulus and their modeling.

Andreas N. Angelakis was a Water Resources Engineer at the HAO-Demeter, Agricultural Research Institution of Crete, Iraklion and Technical Consultant of Hellenic Union of Municipal Enterprises for Water Supply and Sewerage, Larissa, Greece. He graduated from Agricultural Univ. of Athens, Greece (BS in Agronomy), UC Riverside, USA (MS in Soil Sci.), and UC Davis, USA (BS in Civil Eng., MS in Water Sci., and Ph. D. in Soil Physics). He is the author/co-author of over 500 publications mostly in the scientific fields of environmental engineering; Aquatic wastewater management systems; small systems of water and wastewater management; treated wastewater and reclamation and reuse; and water and wastewater technologies in ancient civilizations. He has over 5500 SCH citations and an i10-index of 95. He has participated, mostly as invited speaker by the organizing/scientific committee in more than 120 international symposia and/or conferences in the last 25 years. He is Editor, Associated Editor, and/or a member of the Editorial Boards of several scientific journals. He is an active member of the European Academy of Sciences and Arts and an honorary professor of Hubei University, Wuhan, China. Moreover, he is an IWA (International Water Association) distinguished fellow and an honorary member of the IWA and of the Hellenic Water Association (HWA), as well as the IWA Strategic Council. He is the president of the IWA SG on water and wastewater in ancient civilizations. In addition, he is the past president of EurEau (Federation of European Water and Wastewater Services) and the EurEau WG on Water Reuse. He was recently awarded by the Hellenic Committee of the International Association of Hydrogeology.

Editorial

Water Supply and Water Scarcity

Vasileios A. Tzanakakis [1,*], Nikolaos V. Paranychianakis [2] and Andreas N. Angelakis [3,4]

1. Department of Agriculture, School of Agricultural Science, Hellenic Mediterranean University, Iraklion, 71410 Crete, Greece
2. School of Environmental Engineering, Technical University of Crete, 73100 Chania, Greece; niko.paranychianakis@enveng.tuc.gr
3. Hellenic Agricultural Organization (HAO)-Demeter, Agricultural Research Institution of Crete, 71300 Iraklion, Greece; angelak@edeya.gr
4. Union of Hellenic Water Supply and Sewerage Operators, 41222 Larissa, Greece

* Correspondence: vtzanakakis@hmu.gr

Received: 3 August 2020; Accepted: 19 August 2020; Published: 21 August 2020

Abstract: This paper provides an overview of the Special Issue on water supply and water scarcity. The papers selected for publication include review papers on water history, on water management issues under water scarcity regimes, on rainwater harvesting, on water quality and degradation, and on climatic variability impacts on water resources. Overall, the issue underscores the need for a revised water management, especially in areas with demographic change and climate vulnerability towards sustainable and secure water supply. Moreover, general guidelines and possible solutions, such as the adoption of advanced technological solutions and practices that improve water use efficiency and the use of alternative (non-conventional) water resources are highlighted and discussed to address growing environmental and health issues and to reduce the emerging conflicts among water users.

Keywords: water management; water scarcity regime; water reuse; water use efficiency; rain harvesting; desalination

1. Prolegomena

Water scarcity refers to the lack of fresh water resources to meet water demand. Thomas S. Eliot (1888–1965) reported that "Drought is the death of the earth". The disruption of agriculture and social order by intense and prolonged droughts, called megadroughts, appears to have dictated the cultural time horizons of several civilizations [1]. In prehistoric times, for example the Hittite Empire, the Egyptians of the Pharaohs, and other civilizations collapsed due to the prevalence of intense and prolonged droughts that occurred in their lowlands [2]. Later, the Mayas; Salinas Puebloans; and the Khmer Empire, also known as the Angkor civilization, collapsed from the impacts of megadroughts [1,3–5]. In more recent history, the Dust Bowl (1930–1936) was the driest and hottest drought that hit the USA with significant and long-lasting effects in land productivity and society [6]. In line, intense droughts have hit Europe, the USA, and Australia in recent years, with significant socio-economic, environmental, and ecological impacts [7–9]. Despite the significant improvements in relevant infrastructure, updated water management plans and technological solutions improving water use efficiency (WUE), water scarcity remains a major concern in several parts of the world, listed as one of the largest global risks over the next decade [10]. Millions of families around the world remain vulnerable to water scarcity or do not yet have access to clean and adequate drinking water. More specifically:

(a) Over 2 billion people are living in regions experiencing high water stress and this number is expected to increase.

(b) Over 1 billion people do not have access to clean and safe drinking water.
(c) About 3.4 million people die each year due to the use of contaminated water.
(d) Millions of women and children spend several hours each day collecting water from an average distance of 6 km.
(e) At any given time, half of the world's hospital beds are occupied by patients suffering from diseases associated with lack of access to clean water.

Water scarcity imposes strong constraints in terms of social integrity and economic development. The primary sector that is affected is agriculture, accounting for more than 80% of the total water use [11]. Domestic use also follows an increasing trend over the years due to population growth, living standards requirements, and increasing temperature. These human alterations in the natural hydrologic cycle in conjunction with global warning will cause strong shifts in water availability and demands, and will intensify conflicts between users, outlining the need for updating the existing water governance plans to meet future demands and to ensure sustainable use of water. The improvements will require water resource planning at a finer spatial scale than the basic hydrologic unit (watershed) and give greater emphasis on water recycling, improved WUE by users, and real-time monitoring of water reserves and demands.

Considering the uneven spatial and temporal distribution of water resources and increased water demand, it is necessary to investigate the exploitation of alternative water sources, e.g., recycled water, brackish water, and rainwater [12,13], in order to close the gap between offer and demand [14]. Water recycling, particularly in agriculture, provides comparative advantages since it increases water availability for other activities (domestic and industrial use), reducing the competition between users and preventing overexploitation and degradation of natural water bodies. This perspective seems to be developing in many countries around the world [13,15–17].

The above-mentioned challenges of the water sector in a changing world underline the need for updating the existing water governance frameworks, policies, and applied management strategies to provide incentives and generate opportunities for sustainable use of water resources. Such measures are (a) the need to re-examine all potential sources including non-conventional sources, (b) development of region-wide water resource management programs, and (c) implementation of voluntary and mandatory water conservation measures.

This Special Issue on water supply and water scarcity addresses some of the above aspects, emphasizing on the current knowledge, future trends, and challenges in the water sector under water scarcity. More specifically, this special issue advances our existing knowledge on water resource management on five disciplines, focusing on (a) evolution of hydro-technologies through the centuries, (b) water management issues under water scarcity regimes, (c) rainwater harvesting (RWH), (d) quality of water resources, and on (e) climatic changes and/or variability impacts on water resources.

2. The Main Contribution of This Special Issue

The articles included in this issue cover a wide spectrum of thematology. The 12 papers published are grouped into 5 categories: (a) one paper deals with the evolution of irrigation technologies, (b) six studies focus on water management issues under water scarcity, (c) two papers investigate rainwater harvesting (RWH), (d) two papers deal with water quality and degradation of water resources, and (e) one paper addresses the chimeric changes impacts on water resources.

Angelakis et al. [18] review the evolution of irrigation practices through the millennia, considering archeological evidence from remnants and the relevant literature. Compiled knowledge indicates the development of sophisticated irrigation and water storage systems since the prehistoric times to adapt to water scarcity. Examples are provided from the Bronze Age civilizations (Minoans, Egyptians, and Indus valley), pre-Columbian societies, those grown in historic times (Chinese, Hellenic, and Roman), late-Columbian societies (Aztecs and Incas), Byzantines, Ottomans, and Arabs [19]. In ancient Egypt, for example, farmers took advantage of the periodic flooding of the Nile River to increase crop yields by putting out seeds in soils that had been recently covered and fertilized with floodwater and silt deposits. In arid and semi-arid regions, farmers used perennial springs and seasonal runoff under conditions completely different from the rivers of Mesopotamia, Egypt, India, and the first dynasties in China. The implications and impacts of irrigation on modern management of water resources, as well as on irrigated agriculture, are also discussed and the major challenges are outlined. An important finding from the study is that ancient practices could be adapted to cope with the present challenges in agricultural production and environmental protection.

2.1. Water Management under Water Scarcity Regimes

Preservation of ecological flow and natural water bodies remains a high priority under water scarcity conditions. Effective restoration and management plans for water can lead to significant benefits to the economy, society, and environment. Such a case is the historical Aculeo Lagoon, which is one of the largest natural bodies of water in central Chile [20]. The lake, from 2012 to 2018, was progressively dried as consequence of intense droughts in the surrounding area, causing imbalances between water reserves and withdraws. In the study, the modelling (MODFLOW) simulations confirmed the water imbalances between lake inflows and outflows, attributable to (i) high groundwater demands; (ii) drying of the lagoon's natural and/or man-made stream tributaries; and (iii) decreases in precipitation that affected water capture, storage, and natural drainage, resulting in the lake drying up. To address the problem, the study proposed the implementation of a monitoring and recovery plan (MRP) based on the simulation, considering the combination of three feasible options: (i) the recovery of natural tributaries, (ii) reductions of groundwater pumping, and (iii) feasibility analysis of water importation alternatives either from groundwater or nearby basins. Moreover, the authors argued that the restoration of the Aculeo Lagoon will require supporting actions, such as investments (USD 10 million) in infrastructure for water transfer into the lagoon and training of the involved stakeholders.

Acute and chronic water scarcity affects 4 billion people worldwide, a number that is likely to climb in the upcoming years due to population growth [21]. McNally et al. [22] investigated the development and implementation of a monthly acute water scarcity monitoring system on the basis of hydrologic data gathered from the Famine Early Warning System Network (FEWS NET) and the Land Data Assimilation System (FLDAS), as well as population data from WorldPop. The system computes the annual water availability per capita and yields updated maps of acute water scarcity at monthly intervals by using the Falkenmark classifications and departures from the long-term mean classification. The maps, designed to serve FEWS NET objectives, highlight the acute water scarcity events and provide up-to-date and interpretable information to decision-makers. Further improvements could include the applicability of the approach to lower spatial scales, improved coverage of the populations living in marginal areas (the Sahara Desert, Eastern Kenya, the Kalahari Desert), and addressing the uncertainties stemming from hydrologic or land surface modeling.

The study of van de Griend et al. [23] deals with the indoor use of water, examining the bathing technology. More specifically, the study showed that the inclusion of a hyperbolic vortex in a showerhead can increase the flow rate compared to a showerhead without a vortex for a given discharge without reducing the nozzle diameter. This was achieved by air bubbles introduced from the central part of the nozzle matrix in the sprayed liquid, causing higher liquid velocities and break-up length in the peripheral nozzles. The study argues that a vortex showerhead could save up to 14% of the water compared to conventional showerheads. Additionally, they detected an increase in pH and a reduction of the redox potential compared to conventional technology, indicating an increased degassing of CO_2 and an increased intake of O_2.

The Mediterranean region is among the regions that will be affected by climate fluctuations. Tzanakakis et al. [17] reviewed the availability of water resources and water uses in the island of Crete, highlighting the current and future challenges and opportunities for water management. In the island, despite the high theoretical water potential, there are areas under water scarcity, particularly in the southeastern part of the island, related to local soil-climatic conditions and the imbalances between water availability and demand. Important challenges highlighted by the study are the over-exploitation of groundwater, over-consumption mainly in the agricultural sector, mismanagement at the local level, low overall water use efficiencies, limited use of non-conventional water sources, lack of modern mechanisms of control and monitoring, and inadequate cooperation among stakeholders. The study proposes the improvement of the current water governance framework encouraging the implementation of an integrated and flexible water management plan, accounting for local social and economic specificities to allow for the successful adaptation to changing climatic conditions and to increasing water needs [17]. Moreover, it proposes the exploitation of alternative water sources (recycled water and brackish water); however, further work should be done on legislative framework to promote water reuse, particularly in agriculture, while ensuring the product safety and marketability. Finally, to alleviate the pressure on groundwater resources, the authors propose the adoption of cost-effective technological advances that improve water use efficiency in fields (efficient irrigation methods, crop adaptations, reduced soil tillage, and improvement of soil health).

Expenditure forecasting should be an integral part of long-term water resource management [24,25]. Borisova et al. (2020) estimated the expenditures required to develop alternative water supplies (e.g., reclaimed water, brackish groundwater, surface water storage, and stormwater) in the state of Florida, USA, to cope with the increasing needs for water, mostly driven by the constantly growing population, as well as to protect water resources from over-exploitation. The projections were based on estimations of previous projects using scenarios relying on such commonly used water sources. It was estimated that the state total investments needed to meet future water demands could reach USD 2 billion in the next 20 years, with the project implementation cost being dependent on project capacity, type, implementation status, and implementation region. The authors propose the expansion of stormwater use and the adoption of water conservation practices (defined as practices reducing wasteful and inefficient water uses) as more cost-effective options.

Urban water supply requires improved administration and operation of the domestic water distribution networks. Decision making processes should rely on reliable data that describe system operation, such as flows, potential failures, losses, and/or other problems, in order to addresses all issues properly and in a timely manner. Erickson et al. (2020) provided a detailed and long-term description of the water supply patterns in four areas in Arraiján, Panama, characterized by an intermittent water distribution network, identifying concurrently the challenges and opportunities for the current and future network management. The authors proposed an improved monitoring scheme for the water network that is based on the pressure and flow accounts, which could be helpful for longer-term planning and for the prioritization of system improvements. On a larger scale, they proposed reduction of water losses along with the increase in distribution storage capacity as a proper means to mitigate the adverse effects of the potential operational failures. Finally, the authors highlighted the need for investments in monitoring and data analysis to improve the potential and reliability of the intermittent water supply.

2.2. *Rainwater Harvesting (RWH)*

Rainwater harvesting (RWH) is a sustainable water management practice that has been adopted since the ancient times to augment water-potable and non-potable supplies in water-limited areas. Following a decline in the development of RWH systems in the last century, a renewed interest has emerged since the second half of the 20th century, driven mainly by rising water demands due to growing population, urbanization, climate variability, and by food security [26].

Yannopoulos et al. [27] provide a concise overview of the historical evolution of RWH systems, their current status, and the need for incentives for spreading RWH practice worldwide, particularly in water-limited countries. The compiled information indicates a renewed concern for RWH systems on a global basis, either as a standalone or combined with conventional technologies to confront water scarcity. They successfully state, "*Worldwide, rainwater harvesting has retrieved its importance as a valuable water resource, alternative or supplementary, in conjunction with more conventional water supply technologies. If rainwater harvesting is practiced more widely, many water shortages, actual or potential, can be alleviated*".

They also underline the need for more research, investments, and public awareness on the importance of RWH; economic incentives (subsidies and tax exemptions); and the development and enactment of pertinent regulations to meet the full potential of RWH systems as a complementary water supply technology, not only in rural areas but in urbanized areas as well.

Kuntz and Chisi [28] investigated the economic feasibility and user satisfaction in RWH systems in a residential building in Florianópolis (Brazil) by using a questionnaire survey. The economic feasibility analysis considered different rainwater demands, residents' habits, user satisfaction, and the importance of potable water savings. The findings of the study documented the economic feasibility of RWH systems in residential buildings for the middle and upper socioeconomic class. Showers had the greatest share (54.2%) of water consumption, followed by washing machines (21.3%), kitchen tap (9.3%), toilet flushing (9.2%) (the most economically feasible), and washbasins (2.6%). Overall, residents were satisfied with the perspective of a RWH system, indicating its high potential not only in reduction of the potable water consumption but also as a new marketing strategy for the private sector.

2.3. Quality of Water Resources

Water pollution is a critical issue in intensively managed agricultural areas derived from over-application of nutrients and pesticides and the adoption of non-sustainable field management practices [29]. Diffusing pollution from agricultural watersheds may cause severe problems in ecosystem functioning, quality of water resources, biodiversity, and human health [30].

Sihi et al. (2020) investigated the impacts of different farming systems (organic vs. conventional) of basmati rice on water quality during the rainy season at the Kaithal area, India. Drinking water quality and additional parameters were monitored and evaluated, including nitrates, total dissolved solids (TDS), soil salinity (as electrical conductivity EC), sodium adsorption ratio (SAR), and pH. Most parameters were kept below the regulated thresholds, except nitrates, for which an almost twofold increase was found in conventional fields compared to the organic fields, indicating potential risks for the drinking water supplies. This finding has profound implications for decision-makers in terms of managing nutrients and protecting water quality in agricultural areas more effectively.

The rapid rates of urbanization and industrialization have resulted in increased risks for the ecological degradation of rivers and, thereby, of the derived services [31]. This problem is widespread in China, particularly in the water-limited regions of the northern of part of the country [32]. Thus far, a key role to address the problem is the proper reservoir operation, which can restore the damaged river environment. Deng et al. [33] investigated the urban section of the Yitong River in Changchun, northern China, providing estimations of the ecological water demands and the reservoir operation. A reservoir operation scheme was proposed to restore the ecological quality of the river in its urban section, considering the existing limitations in the process of such operation schemes, including clarification of department responsibilities, updated regulations, strengthening service management, and encouraging public participation. The effect of proposed scheme on water quality and natural habitat of the river was evaluated by simulations with MIKE 11 a one-dimensional hydraulics-water quality model and the Physical Habitat Simulation Model (PHABSIM), indicating improvements in the ecological quality of the urban section of Yitong River.

2.4. Climate Change Impacts on Water Resources

There is a growing body of literature investigating the impacts of climate change on the availability of water resources at either the global and continental or country level. However, the pertinent simulations at local scales are still subjected to additional challenges arising from the downscaling procedure and from uncertainties [34,35]. Molina et al. [36] investigated the effect of climate change on the water budget in eastern Colombia, which currently experiences water scarcity due to increasing water demands for food production, industry, and domestic use. Using the model BROOK90 and historical and projected meteorological data, the authors provide information for potential changes in the water balance in four different regions characterized by distinct climatic conditions.

Projections were performed via a statistical regional downscaling procedure in which two climate change scenarios (RCP2.6 and RCP8.5) were simulated by two global climate models (CanESM2 and IPSL-CM5A-MR). The projections showed clear reduction in stream flow and changes in temporal distribution of water balance components and in hydrological regimes. Moreover, the projected changes in evapotranspiration, stored water, and soil moisture were found to be dependent on soil and land use characteristics and the climate scenario.

3. Challenges and Opportunities for in Improving Water Supply

Water has vital role for sustaining life on earth by regulating ecosystem functioning, preserving environmental quality, and supporting human health and welfare [37,38]. To date, the sustainable use of water resources faces grand challenges arising from population growth, fragile economic context, increased water demand, need to ensure food security, quality deterioration, and ageing infrastructures [19]. These challenges are not addressed effectively in the existing water governance plans, underlining the need for developing more sophisticated water management schemes adapted to changing conditions and requirements. Current insights in water resource management stress the importance of integrated, multi-sectoral, and (inter) national management plans, considering background specialties of the areas and motivating all the involved actors from governmental services and agencies, the private sector, the academy, and the public [17,39,40].

3.1. Growing Population and Urbanization

The world population will approach 10 billion in the next 30 years, and a significant proportion of this growth will take place in developing countries. Today, more than half of the world population is living in urban areas, particularly in highly dense cities; by 2050, more than two-thirds of the population will live in urban areas (Figure 1) [41]. On the basis of the facts that the available fresh water supplies on earth will remain the same, being unevenly distributed, these urban areas will become water-stressed [42,43], enhancing the conflicts among users, particularly among the urban, agricultural, and industrial sectors. At the same time, significant impacts are expected on availability and quality of water resources [44,45], quality of soil resources [46], potential water demand increases [47–49], flood intensity and frequency [50], and ecosystem functioning and derived services (e.g., food security) [51–54] impacts that are tightly inter-connected and influenced by background climate and terranean (land use/cover, geomorphology, and hydrologic) characteristics of the areas as well as by human activities. These factors should carefully be considered in the long-term planning of water resources [33,49,54]. Considering the above, it is urgent to improve water resource management by considering the option of alternative water supplies as well as by adopting strategies and measures promising improved water use efficiency by the potential users. Both options are discussed below in detail.

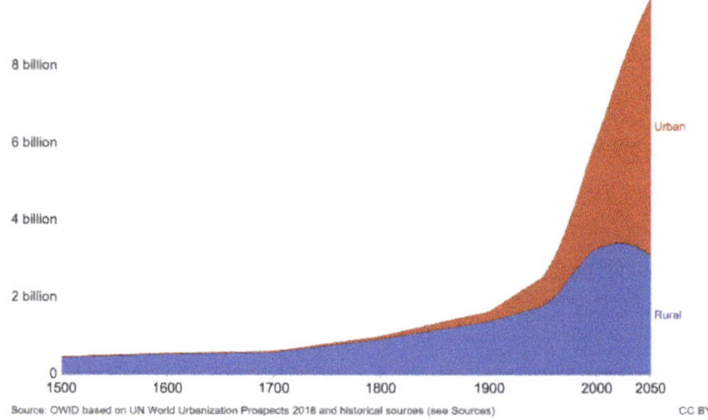

Figure 1. Urban and rural population projected to 2050 [41].

3.2. Climate Change (and/or Variability)

Climate change has already begun to affect water resources worldwide, through warming, shifts in precipitation patterns, and occurrence of extreme weather events (droughts, heat waves, floods) [55,56]. These impacts are not uniformly distributed, but they show strong spatial and temporal variations following climate variation [49,57]. For instance, in the Mediterranean basin, the pace of warming has been significantly greater than the global mean [58], which will likely lead to significant changes in water resources availability and water demands to cope with the higher frequency of droughts [59].

Pertinent studies reveal either intensification of the global hydrological cycle, i.e., increases in both evaporation and precipitation fluxes, or alterations from intensification to de-intensification with corresponding fluctuations on precipitation and evaporation patterns and an overly decreasing trend of global humidity [57,60,61]. Despite the general agreement regarding the climate model projections at global and regional scales [62,63], downscaling of these projections at scales that allow planning and effective management of water resources (e.g., watershed scale) still remain a methodological challenge [64]. Even at larger scales, uncertainty of global climate models and global hydrological models remains large [65].

Apart from the availability of water resources, it still remains highly uncertain as to how climate change will affect water use, particularly in the agricultural and domestic sector. Early studies' observations have shown increases in the WUE of agricultural crops [66], but this positive effect of elevated CO_2 may be eliminated under intense droughts due to the greater leaf area [67]. The effect of climate change on irrigation needs depends on climate change scenario and irrigation method [68]. A 9% increase in evaporative losses was reported that, however, was nearly counteracted by a reduction of non-evaporative losses. Moreover, projected increases in the aridity index [69] raise questions about the future of rainfed agriculture. High water deficits will require additional volumes of water to be allocated to rainfed agriculture to maintain its economic viability and the development of rural areas; however, ensuring additional water supplies for agriculture remains highly uncertain under the existing water management plans, policies, and existing infrastructure.

The situation is also comparable in the domestic sector. Accurate estimation of future demands requires information on the relationships between temperature and water consumption. Xenochristou et al. [70], using a combination of smart water metering data, household characteristics, and socio-economic data, developed such relationships, which could potentially be used for the planning of water use in the domestic sector. These relationships were complex and showed seasonal and weekly variations as well as strong dependencies on socio-economic status and household characteristics (e.g., presence or absence of gardens). More studies are needed to allow accurate estimation of domestic water needs in different climatic backgrounds.

3.3. Improving Water Use Efficiency

A major issue in water resources management is the reduction of water losses and the improvement of WUE. This issue is becoming more challenging nowadays due to population growth, need for economic recovery, and climate change. The main targets for improving WUE are the agricultural and domestic use, which account for the majority of water use (>95%) worldwide. Significant gains, in certain cases, can also be achieved in the industrial sector.

Although only marginally covered in this issue (Tzanakakis et al., 2020), significant water savings can be achieved through methodological and technological innovations in the agricultural sector. New methods of evapotranspiration (ET) estimation (eddy covariance towels), deficit irrigation, smart technologies for soil moisture monitoring, and user-friendly software can substantially improve WUE in the agricultural sector [71]. The currently used methods of ET do not account for the effect of deficit irrigation, resulting in overestimation up to 30% of the irrigation demands [72]. Investments in infrastructure and particularly in the maintenance of irrigation networks will result in significant water savings. Considering the fact that agriculture is the largest water user with a share up to 75% in (semi)arid climates, small improvements in WUE can result in significant gains in water availability, allowing for the adaptation of the sector to climate change and the ameliorating of the conflicts between users. In addition, the pressure on surface and ground water bodies will be alleviated decreasing the risk of ecological degradation as in the case of lake Chile [20].

Regarding the domestic water use, household-centered measures have the potential to result in significant reduction (30%) of drinking water consumption with small effort and without limitation of comfort, or even more (50%) if they are combined with effluent reuse and RWH [73]. Significant savings in domestic use can be achieved by decreasing the application of potable water for landscape irrigation [39]. In addition, technological innovations, updated policy frameworks, and market-based solutions that increase water supply and decrease demand will be needed to meet the challenge of sustainable domestic water supply. More specifically, provision of incentives and a flexible regulatory framework for promoting the adoption of appropriate technological solutions and tools, applicable at different levels of organization (e.g., households, urban infrastructures in conjunction with the effective motivation and education of end-users) will result in positive results. Taken together, optimizing WUE across different sectors poses as an important measure to mitigate or prevent water overconsumption and to improve water balances at either the regional or global level; the importance of WUE of terrestrial ecosystems on global water and carbon cycles is already under consideration [74,75].

3.4. Alternative (Non-Conventional) Water Resources

3.4.1. Wastewater Reuse

Water recycling has been proven to be a reliable solution to increase water availability in many areas of the world, especially in those suffering from water scarcity, serving agricultural, urban (non-potable and potable), industrial, and environmental needs and at the same time protecting human health and the environment [76–79]. Water reuse is also recognized as an adaptation solution to climate change [80,81], compatible with the concept of circular economy that is highly promoted in developed countries [15,82–84]. However, water reuse still lacks widespread implementation in several areas of the world, among them the EU [16], due to often strict regulations, social-economic issues, lack of awareness for the potential benefits, and economic constrains arising from value chain adaptation needs and product marketability [13,17,83]. Currently wastewater treatment plants (WWTPs), particularly in the developed world, are upgraded with new, more reliable, and more energy-sustainable processes, providing high-quality effluent and meeting the requirements, even for unregulated yet emerging pollutants and agents (pharmaceuticals, antibiotic resistance) [85,86] and decreasing significantly (or even eliminating) risks for public health and the environment [87,88]. Promoting water recycling for various beneficial uses instead of discharge to surface water bodies provides significant advantages to control pollution, preserve the spreading of antibiotic genes and emerging pollutants, maintain biodiversity, and improve the adaptation and resilience of urban and rural communities to climate change. There is a need, however, to effectively include the recycled water within the water management plans. The example of Spain is in the right direction, wherein a database with information about the treatment processes treatment costs, recycled water quality and volumes, and reuse activities in every autonomous community allows for the effective use of water [89].

3.4.2. Rainwater Harvesting

Rainwater harvesting is an alternative water supply that can have a significant contribution in meeting future water demands and to maintain/improve the quality of water resources. It has been proven to be a cost-effective solution in urban and sub-urban areas [27,28,90–93]. In recent years, there is a growing interest in both developing and developed countries (the EU, the USA, the United Kingdom, Japan, South Korea, Australia, and Africa) for RWH systems [27], driven mainly by their cost efficiency and potential benefits to different sectors of the economy, environment, water resources, and human health [94–96]. A comparative advantage of RWH is its flexibility for adaptation in various types of collection systems., ranging from small private-owned and managed structures to large-scale structures (multi-stores, schools, stadiums, airports and others), as well as to storm water collection systems from urban, suburban, industrial, and rural areas [27,95,97]. Further developments in the domains of technology, research, and education [27,98]; urban and water planning [91,95]; policy and legislative framework [27,91,95,99]; economic viability [100,101]; and public health risks [98] are still required.

3.4.3. Desalination

Desalination poses an important alternative water source, at least locally, to meet the water needs of the growing population in urban areas. It has been also used in intensified agriculture activities in order to cope with water pollution and it is a viable adaptation practice, particularly for urban areas to climate change [102,103]. Current advances in the field (membranes, decreasing costs of operation, lower energy) [104–109] have expanded desalination applications worldwide [110]. However, the potential environmental impacts of desalination processes is still a great challenge that imposes constrains in the expansion of the systems, particularly for developing countries [103,108]. Besides optimization of the current desalination technologies, promising alternatives include the use of small-scale desalination plants, combined use of seawater with brackish water where feasible, and energy recovery during desalination processing [17,108]. Such options may boost the adoption of desalination systems in the near future, corroborated by the ever-increasing needs for alternative water sources and the updates in applied technologies.

In summary, promoting the use of alternative water sources and effectively integrating these resources to existing management plans of (conventional) water resources will undoubtedly result in significant benefits (social, economic) and helping to cope with the challenges of the changing word (global warming, population growth, food security, environmental protection, public health).

3.5. Preserving Water Quality

Protecting the quality of water is a principal goal for humanity in the 21st century in order to preserve water availability and sustain life on the planet [111]. Despite the current knowledge of potential sources of pollution of water resources and adverse effects on the environment, biodiversity, and human health [30], critical issues still remain to be resolved with relation to inefficient policies, economic pressures, and competitions; lacking knowledge about the fate of emerging pollutants and contaminants and their impacts on biodiversity and humans; and knowledge transfer issues across policymakers, the research sector, and stakeholders [17,112,113]. Thus far, the emphasis has been placed on chemical pollution, particularly on inorganic and organic pollutants [30], and waterborne diseases (pathogens) [88,114,115], particularly in developing countries, arising from spreading of contaminants from various water sources [116]. In these areas, it is urgent to improve sanitation by applying early warning and prevention at source to mitigate pollution of freshwater resources [117,118].

The growing demands for alternative water resources to cope with the water scarcity phenomena, whether for agricultural and local irrigation or for industrial use and drinking water, underscore the need to implement safeguards. Technological insights in water treatment as well as updating existing reuse criteria for different types of reuse to integrate emerging pollutants should be of high priority [13,88]. For developing countries to improve sanitation and supply the population with safe drinking water, it is necessary to develop and implement cost-effective sanitation systems and household-centered sanitation, especially in rural areas [30]. Finally, it is necessary to develop technological tools and new approaches to identify and address the challenges arising from a growing population and climate change. The latter is likely to cause changes in pollutant/contaminant spreading patterns and in the frequency of infectious disease outbreaks due to intensified rainfall and flooding events [30,119].

4. Epilogue

As century continues, increased freshwater resources will be required in many parts of the world to meet the growing needs of the population and the uncertainties and consequences of climatic variability. At the same time, more efforts will be needed to identify and address the challenges that arise, including new threats to the quality of resources and ecosystems, as well as emerging needs to adapt and mitigate detrimental factors, especially in the regions under water scarcity. This Special Issue, "Water Supply and Water Scarcity", by considering historical and current research data, evaluates, discusses, and highlights most these challenges, providing in parallel general guidelines and possible solutions to improve water management. The main messages of the issue can be summarized as follows:

(a) There is an urgent need to review water management, particularly in areas with demographic changes and vulnerability to climatic conditions, in order to ensure sustainable and safe water supply. Implications by climate fluctuations should be carefully evaluated, covering a wide range of human activity and environment. Water management should address the emerging conflicts between water users by providing primary options and alternatives in distribution and use of water resources while protecting the sustainability of water resources.
(b) The adoption of advanced technological solutions and practices that improve water use efficiency by users should be a primary goal for water management to reduce water loss, support the sustainability of water resources, and increase the economic profitability of water.
(c) Increasing the use of alternative (non-conventional) water resources is an important option for the protection of water resources, especially in areas that are already experiencing water scarcity. In developed countries, new developments in WWTP technology provide the basis for expanding water reuse and reducing the competition between water users. Rainwater harvesting and storage, and desalination, especially of brackish water, also remain strong alternatives, depending on local conditions.
(d) Degradation of water quality should be of a global concern due to its impact on human health, biodiversity, and the sustainability of ecosystems. For developing countries, sanitation and supplying the population with safe drinking water should be of high priority, being supported by cost-effective and household-centered sanitation systems. The potential impacts on water resource quality by growing demands and climate change and/or variability arising from changes in spreading pattern of contaminants should be considered and evaluated.

Author Contributions: V.A.T. contributed to the project idea development, prepared the manuscript (reviewing and editing of it), produced data collection of the main text, and supervised the research work; N.V.P. reviewed, revised, and edited the manuscript; A.N.A. had the original idea and prepared the original draft of the manuscript. All authors have read and agreed to the published version of the manuscript.

Funding: This research received no external funding.

Acknowledgments: Authors are grateful to Hailey Wu for her collaboration.

Conflicts of Interest: The authors declare no conflict of interest.

References

1. Stahle, D.W. Anthropogenic megadrought. *Science* **2020**, *368*, 238–239. [CrossRef]
2. Smith, K. *Photochemical and Photobiological Reviews: Volume 1*; Springer Science & Business Media: Berlin, Germany, 2013; Volume 1.
3. Haug, G.H.; Günther, D.; Peterson, L.C.; Sigman, D.M.; Hughen, K.A.; Aeschlimann, B. Climate and the collapse of Maya civilization. *Science* **2003**, *299*, 1731–1735. [CrossRef]
4. Gill, R.B. *The Great Maya Droughts: Water, Life, and Death*; University of New Mexico (US) Press: Albuquerque, NM, USA, 2001.
5. Lovgren, S. How Water Built and Destroyed This Powerful Empire. Available online: http://news.nationalgeographic.com/2017/04/angkor-wat-civilization-collapsed-floods-drought-climate-change/ (accessed on 20 June 2020).
6. Hornbeck, R. The Enduring Impact of the American Dust Bowl: Short- and Long-Run Adjustments to Environmental Catastrophe. *Am. Econ. Rev.* **2012**, *102*, 1477–1507. [CrossRef]
7. Dai, A.; Trenberth, K.E.; Karl, T.R. Global variations in droughts and wet spells: 1900–1995. *Geophys. Res. Lett.* **1998**, *25*, 3367–3370. [CrossRef]

8. Ciais, P.; Reichstein, M.; Viovy, N.; Granier, A.; Ogée, J.; Allard, V.; Aubinet, M.; Buchmann, N.; Bernhofer, C.; Carrara, A. Europe-wide reduction in primary productivity caused by the heat and drought in 2003. *Nature* **2005**, *437*, 529–533. [CrossRef]
9. Whetton, P.; Fowler, A.; Haylock, M.; Pittock, A. Implications of climate change due to the enhanced greenhouse effect on floods and droughts in Australia. *Clim. Chang.* **1993**, *25*, 289–317. [CrossRef]
10. WEF. World Economic Forum. Global Risks Report 2019. Available online: https://www.weforum.org/reports/the-global-risks-report-2019 (accessed on 25 March 2019).
11. Dalezios, N.R.; Angelakis, A.N.; Eslamian, S. Water scarcity management: Part 1: Methodological framework. *Int. J. Glob. Environ. Issues* **2018**, *17*, 1–40. [CrossRef]
12. Tzanakakis, V.; Koo-Oshima, S.; Haddad, M.; Apostolidis, N.; Angelakis, A.; Angelakis, A.; Rose, J. The history of land application and hydroponic systems for wastewater treatment and reuse. In *Evolution of Sanitation and Wastewater Technologies through the Centuries*; IWA Publishing: London, UK, 2014; p. 457.
13. Paranychianakis, N.V.; Salgot, M.; Snyder, S.A.; Angelakis, A.N. Water Reuse in EU States: Necessity for Uniform Criteria to Mitigate Human and Environmental Risks. *Crit. Rev. Environ. Sci. Technol.* **2015**, *45*, 1409–1468. [CrossRef]
14. Salgot, M.; Oron, G.; Cirelli, G.L.; Dalezios, N.R.; Díaz, A.; Angelakis, A.N. *Criteria for Wastewater Treatment and Reuse Under Water Scarcity*; CRC Press: Boca Raton, FL, USA, 2016.
15. Voulvoulis, N. Water reuse from a circular economy perspective and potential risks from an unregulated approach. *Curr. Opin. Environ. Sci. Health* **2018**, *2*, 32–45. [CrossRef]
16. Menegaki, A.N.; Hanley, N.; Tsagarakis, K.P. The social acceptability and valuation of recycled water in Crete: A study of consumers' and farmers' attitudes. *Ecol. Econ.* **2007**, *62*, 7–18. [CrossRef]
17. Tzanakakis, V.; Angelakis, A.; Paranychianakis, N.; Dialynas, Y.; Tchobanoglous, G. Challenges and Opportunities for Sustainable Management of Water Resources in the Island of Crete, Greece. *Water* **2020**, *12*, 1538. [CrossRef]
18. Angelakιs, A.N.; Zaccaria, D.; Krasilnikoff, J.; Salgot, M.; Bazza, M.; Roccaro, P.; Jimenez, B.; Kumar, A.; Yinghua, W.; Baba, A. Irrigation of World Agricultural Lands: Evolution through the Millennia. *Water* **2020**, *12*, 1285. [CrossRef]
19. Angelakis, A.; Voudouris, K.; Tchobanoglous, G. Evolution of water supplies in the Hellenic world focusing on water treatment and modern parallels. *Water Supply* **2020**, *20*, 773–786. [CrossRef]
20. Valdés-Pineda, R.; García-Chevesich, P.; Valdés, J.B.; Pizarro-Tapia, R. The First Drying Lake in Chile: Causes and Recovery Options. *Water* **2020**, *12*, 290. [CrossRef]
21. Mekonnen, M.M.; Hoekstra, A.Y. Four billion people facing severe water scarcity. *Sci. Adv.* **2016**, *2*, e1500323. [CrossRef] [PubMed]
22. McNally, A.; Verdin, K.; Harrison, L.; Getirana, A.; Jacob, J.; Shukla, S.; Arsenault, K.; Peters-Lidard, C.; Verdin, J.P. Acute Water-Scarcity Monitoring for Africa. *Water* **2019**, *11*, 1968. [CrossRef]
23. van de Griend, M.V.; Agostinho, L.L.F.; Fuchs, E.C.; Dyer, N.; Loiskandl, W. Consequences of the Integration of a Hyperbolic Funnel into a Showerhead for Droplets, Jet Break-Up Lengths, and Physical-Chemical Parameters. *Water* **2019**, *11*, 2446. [CrossRef]
24. Zeff, H.B.; Kasprzyk, J.R.; Herman, J.D.; Reed, P.M.; Characklis, G.W. Navigating financial and supply reliability tradeoffs in regional drought management portfolios. *Water Resour. Res.* **2014**, *50*, 4906–4923. [CrossRef]
25. Mitchell, G. Demand forecasting as a tool for sustainable water resource management. *Int. J. Sustain. Dev. World Ecol.* **1999**, *6*, 231–241. [CrossRef]
26. König, K.W.; Sperfeld, D. Rainwater Harvesting—A global issue matures. *Fachver. Betr. Regenwassernutzung Dispon.* **2007**, *25*, 2015.
27. Yannopoulos, S.; Giannopoulou, I.; Kaiafa-Saropoulou, M. Investigation of the current situation and prospects for the development of rainwater harvesting as a tool to confront water scarcity worldwide. *Water* **2019**, *11*, 2168. [CrossRef]
28. Kuntz Maykot, J.; Ghisi, E. Assessment of A Rainwater Harvesting System in A Multi-Storey Residential Building in Brazil. *Water* **2020**, *12*, 546. [CrossRef]
29. Evans, A.E.; Mateo-Sagasta, J.; Qadir, M.; Boelee, E.; Ippolito, A. Agricultural water pollution: Key knowledge gaps and research needs. *Curr. Opin. Environ. Sustain.* **2019**, *36*, 20–27. [CrossRef]

30. Schwarzenbach, R.P.; Egli, T.; Hofstetter, T.B.; Von Gunten, U.; Wehrli, B. Global water pollution and human health. *Ann. Rev. Environ. Resour.* **2010**, *35*, 109–136. [CrossRef]
31. Dunham, J.B.; Angermeier, P.L.; Crausbay, S.D.; Cravens, A.E.; Gosnell, H.; McEvoy, J.; Moritz, M.A.; Raheem, N.; Sanford, T. Rivers are social—Ecological systems: Time to integrate human dimensions into riverscape ecology and management. *Wiley Interdiscip. Rev. Water* **2018**, *5*, e1291. [CrossRef]
32. Jia, W.; Dong, Z.; Duan, C.; Ni, X.; Zhu, Z. Ecological reservoir operation based on DFM and improved PA-DDS algorithm: A case study in Jinsha river, China. *Hum. Ecol. Risk Assess. Int. J.* **2019**, *26*, 1723–1741. [CrossRef]
33. Deng, G.; Yao, X.; Jiang, H.; Cao, Y.; Wen, Y.; Wang, W.; Zhao, S.; He, C. Study on the Ecological Operation and Watershed Management of Urban Rivers in Northern China. *Water* **2020**, *12*, 914. [CrossRef]
34. Maraun, D. Bias correcting climate change simulations-a critical review. *Curr. Clim. Chang. Rep.* **2016**, *2*, 211–220. [CrossRef]
35. Smitha, P.; Narasimhan, B.; Sudheer, K.; Annamalai, H. An improved bias correction method of daily rainfall data using a sliding window technique for climate change impact assessment. *J. Hydrol.* **2018**, *556*, 100–118. [CrossRef]
36. Molina, O.; Luong, T.T.; Bernhofer, C. Projected Changes in the Water Budget for Eastern Colombia Due to Climate Change. *Water* **2019**, *12*, 65. [CrossRef]
37. Sivapalan, M.; Konar, M.; Srinivasan, V.; Chhatre, A.; Wutich, A.; Scott, C.A.; Wescoat, J.L.; Rodríguez-Iturbe, I. Socio-hydrology: Use-inspired water sustainability science for the Anthropocene. *Earth's Future* **2014**, *2*, 225–230. [CrossRef]
38. Orlove, B.; Caton, S.C. Water Sustainability: Anthropological Approaches and Prospects. *Ann. Rev. Anthr.* **2010**, *39*, 401–415. [CrossRef]
39. MacDonald, G.M. Water, climate change, and sustainability in the southwest. *Proc. Natl. Acad. Sci. USA* **2010**, *107*, 21256–21262. [CrossRef] [PubMed]
40. Kumar, M.; Deka, J.P.; Kumari, O. Development of Water Resilience Strategies in the context of climate change, and rapid urbanization: A discussion on vulnerability mitigation. *Groundw. Sustain. Dev.* **2020**, *10*, 100308. [CrossRef]
41. Ritchie, H.; Roser, M. *Urbanization*; Our World in Data. 2018. Available online: https://ourworldindata.org/how-urban-is-the-world (accessed on 20 June 2020).
42. Niva, V.; Cai, J.; Taka, M.; Kummu, M.; Varis, O. China's sustainable water-energy-food nexus by 2030: Impacts of urbanization on sectoral water demand. *J. Clean. Prod.* **2020**, *251*, 119755. [CrossRef]
43. Flörke, M.; Schneider, C.; McDonald, R.I. Water competition between cities and agriculture driven by climate change and urban growth. *Nat. Sustain.* **2018**, *1*, 51–58. [CrossRef]
44. McGrane, S.J. Impacts of urbanisation on hydrological and water quality dynamics, and urban water management: A review. *Hydrol. Sci. J.* **2016**, *61*, 2295–2311. [CrossRef]
45. Tam, V.T.; Nga, T.T.V. Assessment of urbanization impact on groundwater resources in Hanoi, Vietnam. *J. Environ. Manag.* **2018**, *227*, 107–116. [CrossRef]
46. Cui, Y.; Xiao, X. Temporal consistency between gross primary production and solar-induced chlorophyll fluorescence in the ten most populous megacity areas over years. *Sci. Rep.* **2017**, *7*, 14963. [CrossRef]
47. Hao, L.; Huang, X.; Qin, M.; Liu, Y.; Li, W.; Sun, G. Ecohydrological Processes Explain Urban Dry Island Effects in a Wet Region, Southern China. *Water Resour. Res.* **2018**, *54*, 6757–6771. [CrossRef]
48. Mi, Z.; Guan, D.; Liu, Z.; Liu, J.; Viguié, V.; Fromer, N.; Wang, Y. Cities: The core of climate change mitigation. *J. Clean. Prod.* **2019**, *207*, 582–589. [CrossRef]
49. Singh, R.; Biswal, B. Assessing the Impact of Climate Change on Water Resources: The Challenge Posed by a Multitude of Options. In *Hydrology in a Changing World: Challenges in Modeling*; Singh, S.K., Dhanya, C.T., Eds.; Springer International Publishing: Cham, Switzerland, 2019; pp. 185–204. [CrossRef]
50. Angelakis, A.N.; Antoniou, G.; Voudouris, K.; Kazakis, N.; Dalezios, N.; Dercas, N. History of floods in Greece: Causes and measures for protection. *Nat. Hazards* **2020**, *101*, 833–852. [CrossRef]
51. Grimm, N.B.; Faeth, S.H.; Golubiewski, N.E.; Redman, C.L.; Wu, J.; Bai, X.; Briggs, J.M. Global Change and the Ecology of Cities. *Science* **2008**, *319*, 756–760. [CrossRef]
52. Song, W.; Deng, X.; Yuan, Y.; Wang, Z.; Li, Z. Impacts of land-use change on valued ecosystem service in rapidly urbanized North China Plain. *Ecol. Model.* **2015**, *318*, 245–253. [CrossRef]

53. Sun, G.; Li, C.; Hao, L.; Mack, E.; Boggs, J.; McNulty, S.; Caldwell, P.; Sanchez, G.; Meentemeyer, R. *Effects of Urbanization on Water Yield, Ecosystem Productivity, and Micro-Climate: Case studies in the United States and China*; American Geophysical Union: Washington, DC, USA, 2019; p. 1. [CrossRef]
54. Li, C.; Sun, G.; Cohen, E.; Zhang, Y.; Xiao, J.; McNulty, S.G.; Meentemeyer, R.K. Modeling the impacts of urbanization on watershed-scale gross primary productivity and tradeoffs with water yield across the conterminous United States. *J. Hydrol.* **2020**, *583*, 124581. [CrossRef]
55. AghaKouchak, A.; Chiang, F.; Huning, L.S.; Love, C.A.; Mallakpour, I.; Mazdiyasni, O.; Moftakhari, H.; Papalexiou, S.M.; Ragno, E.; Sadegh, M. Climate Extremes and Compound Hazards in a Warming World. *Ann. Rev. Earth Planet. Sci.* **2020**, *48*, 519–548. [CrossRef]
56. Bell, J.E.; Brown, C.L.; Conlon, K.; Herring, S.; Kunkel, K.E.; Lawrimore, J.; Luber, G.; Schreck, C.; Smith, A.; Uejio, C. Changes in extreme events and the potential impacts on human health. *J. Air Waste Manag. Assoc.* **2018**, *68*, 265–287. [CrossRef]
57. Koutsoyiannis, D. Revisiting global hydrological cycle: Is it intensifying? *Hydrol. Earth Syst. Sci.* **2020**, *24*, 3899–3932. [CrossRef]
58. Cramer, W.; Guiot, J.; Fader, M.; Garrabou, J.; Gattuso, J.-P.; Iglesias, A.; Lange, M.A.; Lionello, P.; Llasat, M.C.; Paz, S.; et al. Climate change and interconnected risks to sustainable development in the Mediterranean. *Nat. Clim. Chang.* **2018**, *8*, 972–980. [CrossRef]
59. Dai, A. Increasing drought under global warming in observations and models. *Nat. Clim. Chang.* **2013**, *3*, 171. [CrossRef]
60. Durack, P.J.; Wijffels, S.E.; Matear, R.J. Ocean Salinities Reveal Strong Global Water Cycle Intensification During 1950 to 2000. *Science* **2012**, *336*, 455–458. [CrossRef] [PubMed]
61. Huntington, T.G. Evidence for intensification of the global water cycle: Review and synthesis. *J. Hydrol.* **2006**, *319*, 83–95. [CrossRef]
62. Fowler, H.J.; Wilby, R.L. Detecting changes in seasonal precipitation extremes using regional climate model projections: Implications for managing fluvial flood risk. *Water Resour. Res.* **2010**, *46*. [CrossRef]
63. Hausfather, Z.; Drake, H.F.; Abbott, T.; Schmidt, G.A. Evaluating the Performance of Past Climate Model Projections. *Geophys. Res. Lett.* **2020**, *47*, e2019GL085378. [CrossRef]
64. Abatzoglou, J.T.; Brown, T.J. A comparison of statistical downscaling methods suited for wildfire applications. *Int. J. Climatol.* **2012**, *32*, 772–780. [CrossRef]
65. Wada, Y.; Wisser, D.; Eisner, S.; Flörke, M.; Gerten, D.; Haddeland, I.; Hanasaki, N.; Masaki, Y.; Portmann, F.T.; Stacke, T. Multimodel projections and uncertainties of irrigation water demand under climate change. *Geophys. Res. Lett.* **2013**, *40*, 4626–4632. [CrossRef]
66. Bernacchi, C.J.; Kimball, B.A.; Quarles, D.R.; Long, S.P.; Ort, D.R. Decreases in stomatal conductance of soybean under open-air elevation of [CO2] are closely coupled with decreases in ecosystem evapotranspiration. *Plant Physiol.* **2007**, *143*, 134–144. [CrossRef]
67. Gray, S.B.; Dermody, O.; Klein, S.P.; Locke, A.M.; McGrath, J.M.; Paul, R.E.; Rosenthal, D.M.; Ruiz-Vera, U.M.; Siebers, M.H.; Strellner, R.; et al. Intensifying drought eliminates the expected benefits of elevated carbon dioxide for soybean. *Nat. Plants* **2016**, *2*, 16132. [CrossRef]
68. Malek, K.; Adam, J.C.; Stöckle, C.O.; Peters, R.T. Climate change reduces water availability for agriculture by decreasing non-evaporative irrigation losses. *J. Hydrol.* **2018**, *561*, 444–460. [CrossRef]
69. Berdugo, M.; Delgado-Baquerizo, M.; Soliveres, S.; Hernandez-Clemente, R.; Zhao, Y.; Gaitan, J.J.; Gross, N.; Saiz, H.; Maire, V.; Lehman, A.; et al. Global ecosystem thresholds driven by aridity. *Science* **2020**, *367*, 787–790. [CrossRef]
70. Xenochristou, M.; Kapelan, Z.; Hutton, C. Using Smart Demand-Metering Data and Customer Characteristics to Investigate Influence of Weather on Water Consumption in the UK. *J. Water Resour. Plan. Manag.* **2020**, *146*, 04019073. [CrossRef]
71. Fisher, J.B.; Melton, F.; Middleton, E.; Hain, C.; Anderson, M.; Allen, R.; McCabe, M.F.; Hook, S.; Baldocchi, D.; Townsend, P.A.; et al. The future of evapotranspiration: Global requirements for ecosystem functioning, carbon and climate feedbacks, agricultural management, and water resources. *Water Resour. Res.* **2017**, *53*, 2618–2626. [CrossRef]
72. Kustas, W.P.; Anderson, M.C.; Alfieri, J.G.; Knipper, K.; Torres-Rua, A.; Parry, C.K.; Nieto, H.; Agam, N.; White, W.A.; Gao, F.; et al. The Grape Remote Sensing Atmospheric Profile and Evapotranspiration Experiment. *Bull. Am. Meteorol. Soc.* **2018**, *99*, 1791–1812. [CrossRef]

73. Schuetze, T.; Santiago-Fandiño, V. Quantitative assessment of water use efficiency in urban and domestic buildings. *Water* **2013**, *5*, 1172–1193. [CrossRef]
74. Tang, X.; Li, H.; Desai, A.R.; Nagy, Z.; Luo, J.; Kolb, T.E.; Olioso, A.; Xu, X.; Yao, L.; Kutsch, W.; et al. How is water-use efficiency of terrestrial ecosystems distributed and changing on Earth? *Sci. Rep.* **2014**, *4*, 7483. [CrossRef]
75. Seibt, U.; Rajabi, A.; Griffiths, H.; Berry, J.A. Carbon isotopes and water use efficiency: Sense and sensitivity. *Oecologia* **2008**, *155*, 441. [CrossRef]
76. Asano, T.; Burton, F.; Leverenz, H.; Tsuchihashi, R.; Tchobanoglous, G. *Water Reuse-Issues, Technologies, and Applications*; McGrow-Hill: New York, NY, USA, 2007.
77. Angelakis, A.; Gikas, P. Water reuse: Overview of current practices and trends in the world with emphasis on EU states. *Water Util. J.* **2014**, *8*, e78.
78. Paranychianakis, N.V.; Angelakis, A.N.; Leverenz, H.; Tchobanoglous, G. Treatment of Wastewater With Slow Rate Systems: A Review of Treatment Processes and Plant Functions. *Crit. Rev. Environ. Sci. Technol.* **2006**, *36*, 187–259. [CrossRef]
79. Ghernaout, D. Increasing trends towards drinking water reclamation from treated wastewater. *World J. Appl. Chem.* **2018**, *3*, 1–9. [CrossRef]
80. Gude, V.G. Desalination and water reuse to address global water scarcity. *Rev. Environ. Sci. Bio/Technol.* **2017**, *16*, 591–609. [CrossRef]
81. Morote, Á.-F.; Olcina, J.; Hernández, M. The use of non-conventional water resources as a means of adaptation to drought and climate change in Semi-Arid Regions: South-Eastern Spain. *Water* **2019**, *11*, 93. [CrossRef]
82. EC-COM. European Commission. From the Commission to the European Parliament, the Council, the European Economic and Social Committee and the Committee of the Regions on the Implementation of the Circular Economy Action Plan (190 Final). Available online: https://ec.europa.eu/environment/circular-economy/index_en.htm (accessed on 20 June 2020).
83. Sgroi, M.; Vagliasindi, F.G.A.; Roccaro, P. Feasibility, sustainability and circular economy concepts in water reuse. *Curr. Opin. Environ. Sci. Health* **2018**, *2*, 2020–2025. [CrossRef]
84. Shen, K.-W.; Li, L.; Wang, J.-Q. Circular economy model for recycling waste resources under government participation: A case study in industrial waste water circulation in China. *Technol. Econ. Dev. Econ.* **2020**, *26*, 21–47. [CrossRef]
85. Pazda, M.; Kumirska, J.; Stepnowski, P.; Mulkiewicz, E. Antibiotic resistance genes identified in wastewater treatment plant systems—A review. *Sci. Total Environ.* **2019**, *697*, 134023. [CrossRef]
86. Zerva, I.; Alexandropoulou, I.; Panopoulou, M.; Melidis, P.; Ntougias, S. Antibiotic Resistance Genes Dynamics at the Different Stages of the Biological Process in a Full-Scale Wastewater Treatment Plant. *Proceedings* **2018**, *2*, 650. [CrossRef]
87. Sabri, N.A.; Schmitt, H.; Van der Zaan, B.; Gerritsen, H.W.; Zuidema, T.; Rijnaarts, H.H.M.; Langenhoff, A.A.M. Prevalence of antibiotics and antibiotic resistance genes in a wastewater effluent-receiving river in the Netherlands. *J. Environ. Chem. Eng.* **2020**, *8*, 102245. [CrossRef]
88. Adegoke, A.A.; Amoah, I.D.; Stenström, T.A.; Verbyla, M.E.; Mihelcic, J.R. Epidemiological Evidence and Health Risks Associated With Agricultural Reuse of Partially Treated and Untreated Wastewater: A Review. *Front. Public Health* **2018**, *6*, 337. [CrossRef]
89. Iglesias, R.; Ortega, E.; Batanero, G.; Quintas, L. Water reuse in Spain: Data overview and costs estimation of suitable treatment trains. *Desalination* **2010**, *263*, 1–10. [CrossRef]
90. Kotsifakis, K.; Kourtis, I.; Feloni, E.; Baltas, E. *Assessment of Rain Harvesting and RES Desalination for Meeting Water Needs in an Island in Greece*; Springer: Cham, Switzerland, 2020; pp. 59–62.
91. GhaffarianHoseini, A.; Tookey, J.; GhaffarianHoseini, A.; Yusoff, S.M.; Hassan, N.B. State of the art of rainwater harvesting systems towards promoting green built environments: A review. *Desalination Water Treat.* **2016**, *57*, 95–104. [CrossRef]
92. Campisano, A.; Butler, D.; Ward, S.; Burns, M.J.; Friedler, E.; DeBusk, K.; Fisher-Jeffes, L.N.; Ghisi, E.; Rahman, A.; Furumai, H. Urban rainwater harvesting systems: Research, implementation and future perspectives. *Water Res.* **2017**, *115*, 195–209. [CrossRef]
93. Yosef, B.A.; Asmamaw, D.K. Rainwater harvesting: An option for dry land agriculture in arid and semi-arid Ethiopia. *Int. J. Water Resour. Environ. Eng.* **2015**, *7*, 17–28.

94. Elgert, L.; Austin, P.; Picchione, K. Improving water security through rainwater harvesting: A case from Guatemala and the potential for expanding coverage. *Int. J. Water Resour. Dev.* **2016**, *32*, 765–780. [CrossRef]
95. Suleiman, L.; Olofsson, B.; Saurí, D.; Palau-Rof, L. A breakthrough in urban rain-harvesting schemes through planning for urban greening: Case studies from Stockholm and Barcelona. *Urban For. Urban Green.* **2020**, *51*, 126678. [CrossRef]
96. Saurí, D.; Palau-Rof, L. Urban drainage in Barcelona: From hazard to resource? *Water Altern.* **2017**, *10*, 475–492.
97. Demetropoulou, L.; Lilli, M.A.; Petousi, I.; Nikolaou, T.; Fountoulakis, M.; Kritsotakis, M.; Panakoulia, S.; Giannakis, G.V.; Manios, T.; Nikolaidis, N.P. Innovative methodology for the prioritization of the Program of Measures for integrated water resources management of the Region of Crete, Greece. *Sci. Total Environ.* **2019**, *672*, 61–70. [CrossRef] [PubMed]
98. Gwenzi, W.; Dunjana, N.; Pisa, C.; Tauro, T.; Nyamadzawo, G. Water quality and public health risks associated with roof rainwater harvesting systems for potable supply: Review and perspectives. *Sustain. Water Qual. Ecol.* **2015**, *6*, 107–118. [CrossRef]
99. Lee, K.E.; Mokhtar, M.; Hanafiah, M.M.; Halim, A.A.; Badusah, J. Rainwater harvesting as an alternative water resource in Malaysia: Potential, policies and development. *J. Clean. Prod.* **2016**, *126*, 218–222. [CrossRef]
100. Devkota, J.; Schlachter, H.; Apul, D. Life cycle based evaluation of harvested rainwater use in toilets and for irrigation. *J. Clean. Prod.* **2015**, *95*, 311–321. [CrossRef]
101. Morales-Pinzón, T.; Rieradevall, J.; Gasol, C.M.; Gabarrell, X. Modelling for economic cost and environmental analysis of rainwater harvesting systems. *J. Clean. Prod.* **2015**, *87*, 613–626. [CrossRef]
102. Garcia-Rodriguez, L. Seawater desalination driven by renewable energies: A review. *Desalination* **2002**, *143*, 103–113. [CrossRef]
103. Gude, V.G. Desalination and sustainability—An appraisal and current perspective. *Water Res.* **2016**, *89*, 87–106. [CrossRef] [PubMed]
104. Fritzmann, C.; Löwenberg, J.; Wintgens, T.; Melin, T. State-of-the-art of reverse osmosis desalination. *Desalination* **2007**, *216*, 1–76. [CrossRef]
105. Ang, W.L.; Wahab Mohammad, A.; Johnson, D.; Hilal, N. Forward osmosis research trends in desalination and wastewater treatment: A review of research trends over the past decade. *J. Water Process. Eng.* **2019**, *31*, 100886. [CrossRef]
106. Khawaji, A.D.; Kutubkhanah, I.K.; Wie, J.-M. Advances in seawater desalination technologies. *Desalination* **2008**, *221*, 47–69. [CrossRef]
107. Shannon, M.A.; Bohn, P.W.; Elimelech, M.; Georgiadis, J.G.; Marinas, B.J.; Mayes, A.M. Science and technology for water purification in the coming decades. In *Nanoscience and Technology: A Collection of Reviews from Nature Journals*; World Scientific: Singapore, 2010; pp. 337–346.
108. Voutchkov, N. Energy use for membrane seawater desalination—Current status and trends. *Desalination* **2018**, *431*, 2–14. [CrossRef]
109. Oatley-Radcliffe, D.L.; Walters, M.; Ainscough, T.J.; Williams, P.M.; Mohammad, A.W.; Hilal, N. Nanofiltration membranes and processes: A review of research trends over the past decade. *J. Water Process. Eng.* **2017**, *19*, 164–171. [CrossRef]
110. Global Water Intelligence. *Desalination Yearbook 2016–2017, Water Desalination Report*; Global Water Intelligence: Oxford, UK, 2016.
111. World Water Assessment Programme. *The United Nations World Water Development Report 3: Water in a Changing World (Two-Volume Set)*; Earthscan: London, UK, 2009.
112. Smith, L.; Inman, A.; Lai, X.; Zhang, H.; Fanqiao, M.; Jianbin, Z.; Burke, S.; Rahn, C.; Siciliano, G.; Haygarth, P.M. Mitigation of diffuse water pollution from agriculture in England and China, and the scope for policy transfer. *Land Use Policy* **2017**, *61*, 208–219. [CrossRef]
113. Olmstead, S.M. The Economics of Water Quality. *Rev. Environ. Econ. Policy* **2009**, *4*, 44–62. [CrossRef]
114. Fenwick, A. Waterborne infectious diseases—Could they be consigned to history? *Science* **2006**, *313*, 1077–1081. [CrossRef]
115. Craun, G.F. *Waterborne Diseases in the USA*; CRC Press: Boca Raton, FL, USA, 2018.
116. Malik, A.; Yasar, A.; Tabinda, A.; Abubakar, M. Water-borne diseases, cost of illness and willingness to pay for diseases interventions in rural communities of developing countries. *Iran. J. Public Health* **2012**, *41*, 39.

117. Lu, Y.; Song, S.; Wang, R.; Liu, Z.; Meng, J.; Sweetman, A.J.; Jenkins, A.; Ferrier, R.C.; Li, H.; Luo, W.; et al. Impacts of soil and water pollution on food safety and health risks in China. *Environ. Int.* **2015**, *77*, 5–15. [CrossRef] [PubMed]
118. Royston, M.G. *Pollution Prevention Pays*; Elsevier: Amsterdam, The Netherlands, 2013.
119. Myers, S.S.; Patz, J.A. Emerging threats to human health from global environmental change. *Ann. Rev. Environ. Resour.* **2009**, *34*, 223–252. [CrossRef]

Review

Irrigation of World Agricultural Lands: Evolution through the Millennia

Andreas N. Angelakis [1], Daniele Zaccaria [2,*], Jens Krasilnikoff [3], Miquel Salgot [4], Mohamed Bazza [5], Paolo Roccaro [6], Blanca Jimenez [7], Arun Kumar [8], Wang Yinghua [9], Alper Baba [10], Jessica Anne Harrison [11], Andrea Garduno-Jimenez [12] and Elias Fereres [13]

1 HAO-Demeter, Agricultural Research Institution of Crete, 71300 Iraklion and Union of Hellenic Water Supply and Sewerage Operators, 41222 Larissa, Greece; angelak@edeya.gr
2 Department of Land, Air, and Water Resources, University of California, California, CA 95064, USA
3 School of Culture and Society, Department of History and Classical Studies, Aarhus University, 8000 Aarhus, Denmark; hisjk@cas.au.dk
4 Soil Science Unit, Facultat de Farmàcia, Universitat de Barcelona, 08007 Barcelona, Spain; salgot@ub.edu
5 Formerly at Land and Water Division, Food and Agriculture Organization of the United Nations-FAO, 00153 Rome, Italy; m.bazza55@gmail.com
6 Department of Civil and Environmental Engineering, University of Catania, 2 I-95131 Catania, Italy; proccaro@dica.unict.it
7 The Comisión Nacional del Agua in Mexico City, Del. Coyoacán, México 04340, Mexico; blanca.jimenez@conagua.gob.mx
8 Department of Civil Engineering, Indian Institute of Technology, Delhi 110016, India; arunku@civil.iitd.ac.in
9 Department of Water Conservancy History, China Institute of Water Resources and Hydropower Research, Beijing 100048, China; wangyh@iwhr.com
10 Izmir Institute of Technology, Engineering Faculty, Department of Civil Engineering, Urla/İzmir 35430, Turkey; alperbaba@iyte.edu.tr
11 Independent Scholar, Baton Rouge, Louisiana, LA 70802, USA; jharri32@gmail.com
12 Food Water Waste Research Group, The University of Nottingham, Nottingham NG7 2RD, UK; Andrea.GardunoJimenez@nottingham.ac.uk
13 IAS-CSIC and University of Cordoba, 14004 Córdoba, Spain; ag1fecae@uco.es
* Correspondence: dzaccaria@ucdavis.edu

Received: 28 March 2020; Accepted: 24 April 2020; Published: 1 May 2020

Abstract: Many agricultural production areas worldwide are characterized by high variability of water supply conditions, or simply lack of water, creating a dependence on irrigation since Neolithic times. The aim of this paper is to provide an overview of the evolution of irrigation of agricultural lands worldwide, based on bibliographical research focusing on ancient water management techniques and ingenious irrigation practices and their associated land management practices. In ancient Egypt, regular flooding by the Nile River meant that early agriculture probably consisted of planting seeds in soils that had been recently covered and fertilized with floodwater and silt deposits. On the other hand, in arid and semi-arid regions farmers made use of perennial springs and seasonal runoff under circumstances altogether different from the river civilizations of Mesopotamia, Egypt, India, and early dynasties in China. We review irrigation practices in all major irrigation regions through the centuries. Emphasis is given to the Bronze Age civilizations (Minoans, Egyptians, and Indus valley), pre-Columbian, civilizations from the historic times (e.g., Chinese, Hellenic, and Roman), late-Columbians (e.g., Aztecs and Incas) and Byzantines, as well as to Ottomans and Arabs. The implications and impacts of irrigation techniques on modern management of water resources, as well as on irrigated agriculture, are also considered and discussed. Finally, some current major agricultural water management challenges are outlined, concluding that ancient practices could be adapted to cope with present challenges in irrigated agriculture for increasing productivity and sustainability.

Keywords: irrigation practices; Aztecs; bronze age; Byzantine times; Chinese dynasties; Egyptians; Harappans; Hellenic civilizations; Incas; medieval times; Mayas; Mesopotamia; Minoans; modern times; Ottoman times; Romans

1. Prolegomena

Historians and archaeologists believe that the Fertile Crescent [1] is the cradle of agriculture, written language, and monotheist religions. The area comprising the territories formerly called Mesopotamia (the origin of Mesopotamia's name is from the Greek words meso and potamos, i.e., the land between (meso) the Euphrates and Tigris rivers (potamos) and their tributaries), Assyria and Phoenicia, together with Lower and Upper Egypt, witnessed the domestication of crops by mankind some 10,000 years ago. It could be hypothesized that the birth of irrigation also took place there, soon after the start of agriculture. Given the long periods without rainfall in the Fertile Crescent, one could imagine that the first farmers diverted water to the dry lands as soon as they saw an opportunity to do so. In arid and semi-arid climates, irrigation is needed to sustain agricultural production because of insufficient or unevenly distributed rainfall during the crop-growing season and therefore water storage facilities must be constructed to buffer water demand and supply during the irrigation periods [2].

The initial irrigation activity must have been the simple diversion of water onto nearby cropped lands, which may have been done with bare hands (the diversion of water from streams into cropped land with bare hands is still practiced today as witnessed by the authors in several traditional societies) or with primitive tools used at the time. The use of earthen ridges and digging canals to convey water to land located further away from the water source must have followed some time later. Flood spreading—also called spate irrigation—as is still practiced today in several traditional rural communities, is probably the closest practice to the origin of irrigation.

Throughout history, the rise and decline of the civilizations that thrived in the Fertile Crescent, was closely dependent on harnessing water for agricultural production, which provided economic prosperity, social stability, and military power. Drought episodes of long duration often caused the replacement of ruling dynasties by new ones. It was also in this part of the world that the negative impacts of irrigation first appeared. Salt accumulation in the soil, stemming from lack of, inadequate drainage, or usage of poor-quality waters led to soil salinization and declining agricultural production. Agricultural practices require hydraulic and reclamation works such as flood protection, land levelling, and drainage.

It is well known that the development of agriculture in the Bronze Age allowed the subsistence of larger populations. There is also evidence that during the first millennium BC (Before Christ) agricultural productivity in Europe reached for the first-time levels capable of sustaining highly stratified societies [3]. According to [4], the beginnings of agricultural developments that led to the creation of Classical civilizations should be traced to the early Chinese dynasties, ancient Egyptians, Minoan and Mycenaean states of the actual Greece and to Indus valley civilizations. Additionally, Mayans and other pre-Columbian civilizations paralleled similar developments in Central and South America. Angelakis et al. [5] indicate that Egyptians and other neighboring civilizations exchanged knowledge with Minoans, Myceneans, Archaic and Classical Greeks. Thereafter civilizations, mainly Romans, inherited prehistoric technologies and developed them further, mainly by scaling up and implementing them both in urban and rural areas.

The use of drainage and irrigation practices are probably nearly as old as farming, although the earliest recorded examples of drainage and irrigation systems date to the Classical Greek and Imperial Roman periods [6,7]. However, archaeological studies revealed that the oldest irrigated agriculture in the Near East Region along the Nile riverbanks appeared in Egypt some 5000 years BC [8,9].

This paper describes the historical evolution of irrigation and discusses major achievements in terms of practices and technologies through the millennia. The most important and well-known civilizations worldwide are considered. It also attempts to provide insights into ancient irrigation technologies and management, highlighting features related to sustainability and adaptation to the environment, and its role in converting "marginal" lands into regular fields suitable for cultivation. Finally, the paper compares different technological developments among civilizations. These technologies are the underpinning of modern achievements in water science and are one proof that "the past is the key for the future".

This study is organized as follows: (a) Section 1 is an introductory one; (b) Sections 2 and 3 refer to the irrigation in pre-historic and historical times; (c) Section 4 refers to early Chinese dynasties; (d) Section 5 refers to late pre-Columbian civilizations; (e) Section 6 refers to medieval times; (g) Section 7 refers to late Chinese dynasties; (f) Section 8 refers to modern times; (h) Section 9 discusses the main agricultural water management challenges and future trends; and (i) Section 10 includes the epilogue and the concluding remarks.

The conceptual diagram of Figure 1 depicts the structure of this review paper.

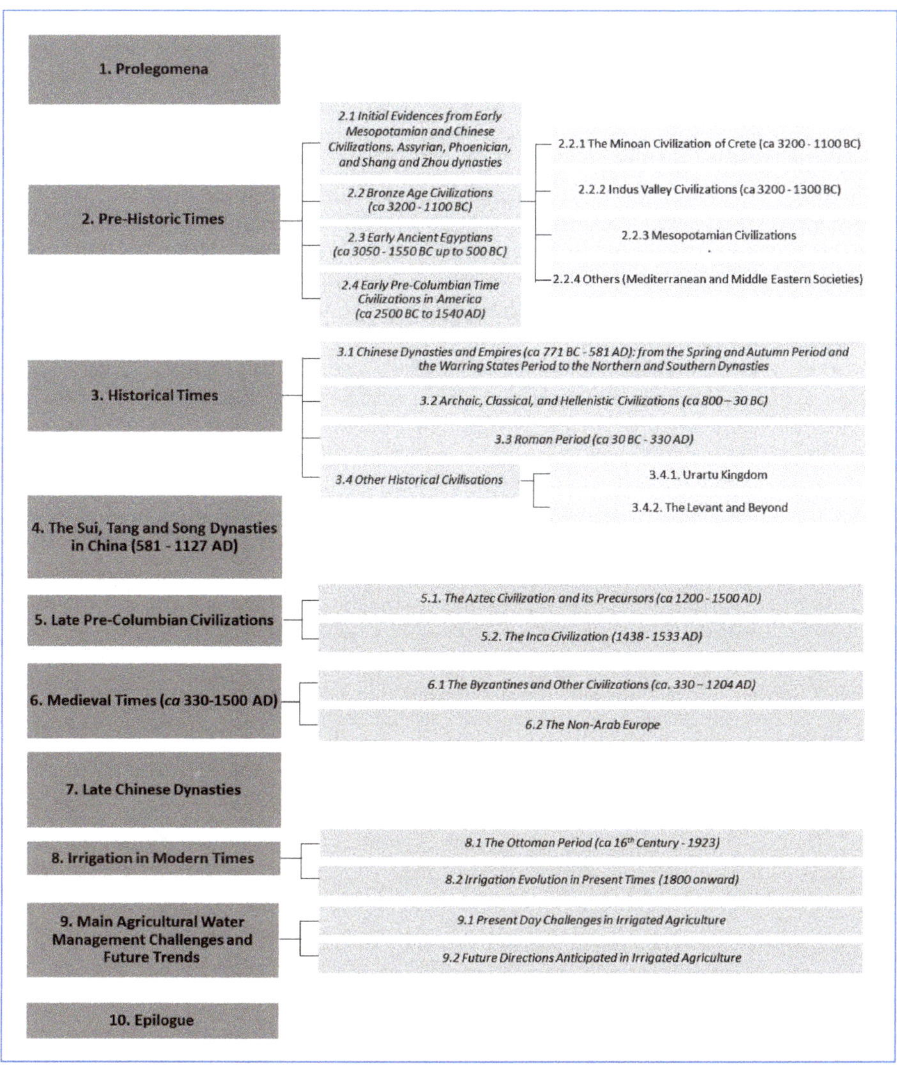

Figure 1. Structure of the present review paper.

2. Pre-Historic Times

2.1. Initial Evidences from Early Mesopotamian and Chinese Civilizations. Assyrian, Phoenician, and Shang and Zhou Dynasties

The very early phases of irrigation development are difficult to document today. As indicated above, the initial irrigation practices at that time probably consisted of a simple diversion of water onto fields adjacent to watercourses. Physical remnants of such practices are not likely to last for more than decades after ceasing operation. The existing archaeological evidence of irrigation relates only to that irrigation infrastructure built for decades to a century of use. For instance, there is evidence that the oldest irrigated agriculture in the Fertile Crescent was practiced in Egypt some 5000 years BC and consisted of large flat basins built for diverting floods to grow winter crops along the Nile riverbanks [10]. At the peak of floods, water was naturally diverted into these basins where it remained for 40 to 60 days [9]. The system was later refined and gave rise to basin irrigation. More advanced systems comprising earthen banks, basins of various sizes, regulated sluices to direct floodwater, and other structures have been reported in the area [11].

Techniques similar to those used in Egypt, with adaptations to the Tigris and Euphrates rivers, were used in Mesopotamia *ca* 3000–5000 years BC [12]. In Mesopotamia, a canal connected to and crossing through the ancient Semitic city of Mari was used for irrigation [13,14].

Early farming and stock breeding communities appeared in China in the mid Neolithic period about 7000 years ago. Primitive paddy fields emerged in the mid and lower reaches of the Yangtze River, where abundant water resources and flat terrains could be found. Humans began to plant rice in the lowland areas near rivers and lakes, which resulted in the emergence of the primitive irrigation and drainage engineering practices in the present Hemudu, Yuyao County, Chekiang Province [15]. In the mid-late Neolithic period, humans began to settle in areas far away from the rivers, which led to the excavation of wells for irrigation.

According to archaeological discoveries, the earliest water wells in China were excavated about 6000 years ago also in Hemudu [16]. However, in this period wells for irrigation were mostly confined to courtyards and gardens.

Later, the Xia, Shang, and Zhou dynasties (*ca* 2100–771 BC) developed in the lower reaches of the Yellow River. The Shang dynasty—the first Chinese dynasty to leave historical records—is thought to have ruled from about 1600 to 1046 BC. However, one must distinguish Shang as an archaeological term from Shang as a dynastic one [10]. In this region, summer and autumn are typically heavy rainy seasons, so it became necessary to dig canals to drain off excess water from farmlands and reduce flooding. According to legends, excavation of farmland canals began taming water in the times of Yu the Great (*ca* 2123–2025 BC). Based on archaeological research, the symbols for farmland and canal ('田巛') appeared in the oracle of the Shang dynasty, and referred to a canal near a plot of farmland [17]. Farmland canals became an integral part of a well system in the Western Zhou dynasty (*ca* 1045–771 BC), leading to the well-field irrigation canals system of Jing Tian Gou Xue. Under this system, a piece of land was divided into nine '井'-shaped plots with a storage well in the middle, while the other eight plots were devoted to agricultural fields and surrounded by irrigation canals. When the dry spring and winter periods came, farmers diverted water from the canals and/or wells to the fields. Farmers who lived near the city of Haojing, the capital of the Western Zhou dynasty, diverted water from Biaochi, a storage pond connected to the rice paddocks with irrigation canals. In this period, the number and size of irrigation areas were very small.

2.2. Bronze Age Civilizations (ca 3200–1100 BC)

2.2.1. The Minoan Civilization of Crete

One example of an early civilization that focused on irrigation is the Minoan. The Minoans selected their first settlements based on defense, food and water conditions. During the Minoan era (*ca* 3200–1100 BC) the development of agriculture in Crete and other islands in South-Eastern Greece was necessary to support a rapidly growing population. Agriculture played an important role in the development of water management. Cereals, wine and olive oil were the three main agricultural products in Crete and throughout the Mediterranean prehistory. Sophisticated techniques were applied to increase the amount of food produced and secure its quality with the introduction of new plant species and terraced agriculture [18]. Although not documented in the archaeological records, irrigated agricultural lands should have developed in prehistoric Crete, especially in cereal and grapes growing areas [19] not only to increase but also to stabilize production.

In the Neopalatial period (*ca* 1750–1450 BC), the practice of irrigation and drainage of agricultural lands became very important. The most famous irrigation systems of that time are called linies (linea = straight line), identified in the Lasithi Plateau in eastern Crete that were irrigated during the Minoan era [20]. This conclusion follows from findings in the Neolithic and Minoan settlements in Kastellos, Plati, and the Kronion's sacred caves in Trapeza and at Idaion Andro in the Lasithi Plateau. Numerous drainage and irrigation channels intersected groves, creating a very attractive view. These techniques are thought to have been transferred later by Minyans, another prehistoric people of Greece like Minoans, to Orchomenos in central Greece [20]. In fact, traces of these techniques were observed during the reclamation project of Lake Kopais in 1886 [2].

In Minoan Crete, environmental and socio-economic conditions forced the development of pioneering actions for implementing water technology in agriculture and water management in general. Such creative and innovative technologies can be traced in different areas of the world, at different times and for several activities, like in the terraced agriculture in pre-Columbian Peru [19].

In the Minoan palaces and settlements, management of water resources varied according to local conditions, which were determined mainly by climate, rainfall, groundwater availability and soils [5]. In settlements of the eastern Crete (e.g., Zakros, Palaikastro, and Komos) water needs were mainly met with groundwater. So far, several wells have been discovered in Palaikastro with depths ranging from 10 to 15 m [21]. There are also indications that the Minoans invented the shaduf or shadoof—the simple hand operated water-lifting device still used in India, Egypt, and some other countries—during the Meso-Minoan period (*ca* 2150–1600 BC) in the eastern part of the island. Similar devices were in use in Mesopotamia as early as at the time of King Sargon of Akkad of the Sumerian city-states in the *ca* 23rd and 22nd centuries BC [22].

In the valley of Choiromandres, located on the southeastern edge of the Zakros area at the eastern end of Crete, Greece, an integrated management system covering an area of around 7.5 hm^2 has been discovered [23]. The valley consists of a rocky ravine with high slope from east to west. During the second millennium BC, the residents attempted to regulate the flow of streams through a system of two dams in series. In addition to using water for irrigation, the dams were built for the protection of arable land from soil erosion caused by the storm water runoff flowing into the river after heavy rainfall. The first attempt was made during the Propalatial period (*ca* 1900–1750 BC), with the dams built using megaliths during the Neopalatial period (*ca* 1750–1450 BC). The largest of them has a length of 27 m (at the top) and a height today of 3.10 m, while the thickness of the base is greater than that of the crest. In the east side, a channel at the rock surface existed that probably served as a spillway. The upper part of the dam was rebuilt in the late Classical and Hellenistic periods. Hydro-technologies including irrigation systems developed by Minoans and Mycenaeans were transferred to neighboring civilizations such as Egyptians, Etruscans, and Dorians and later transferred by them to Archaic and Classical Greece, with which they had "built bridges" [5].

Along the watercourse downstream of the dam, retaining walls were built to support other walls parallel or perpendicular to the riverbed in order to achieve containment and proper flow channeling (Figure 2). At the lower end of the ravine, a permeable barrier was constructed, enabling control of irrigation water in the downstream terraces.

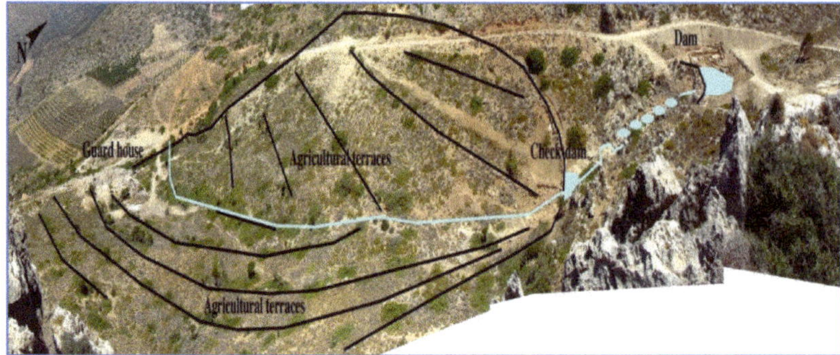

Figure 2. The main dam, the check dam, and the irrigated agricultural terraces in Choiromandres (modified from [23]).

Irrigation projects at Choiromandres remained in use until the end of the Classical period or the beginning of the Hellenistic period. This water management system can be considered as an example of public works originally constructed in the Bronze Age. This case could inspire to investigate other archaeological sites in Crete, contributing to a better understanding of the Bronze Age practices related to the management of water resources [24]. Other irrigation projects of that period are known from Pitsidia and Pseira, which are in Southeastern Crete [25].

General movement of populations from higher elevations and marginal zones with little arable land to landscapes of low alluvial terraces occurred during the late Protopalatial and the early Neopalatial period [26]. These areas provided better access to water supplies, whether groundwater or lower reaches of riverbeds, and they were better suited to the intensive cultivation of cereals, legumes, olives and grapes, as long as they were efficiently irrigated. Such developments are recorded at Malia and Pseira, where Neopalatial dams allowed more intensive exploitation of the limited agricultural potential of the island in response to increasing population [27].

Some other water management systems have been discovered in Crete. The outputs of the central sewerage and drainage systems in palaces and cities at Knossos, Phaistos, Zakros, and Malia appear to be similar. However, the palaces of Knossos and Zakros disposed rainwater and wastewater into the torrent Kairatos and the sea, whereas the palace of Phaistos had collection facilities and diversion of raw runoff into farmland. Angelakis and Spiridakis [28] described how similar techniques of sewerage and storm water runoff management were used in other cities and palaces of Minoan Crete.

2.2.2. Indus Valley Civilizations (*ca* 3200–1300 BC)

The Indus culture, one of the world's great civilizations, was based on two river systems, the Indus and the Ghaggar-Hakra [29]. To feed the inhabitants of its cities, this civilization relied on farmers of the Indus valley who produced peas, sesame seed and cotton, among other main crops. These farmers cultivated large fields using wooden ploughs pulled by oxen, as witnessed by model ploughs—possibly toys—that archaeologists have found in the area. They also domesticated wild animals for harvesting crops at their farms, and utilized their power for cultivation [30].

Irrigation systems of ancient India were infused from those developed earlier in Egypt and Mesopotamia. Violett [31] describes these early Indus valley cultures as the great hydraulic civilization of Harappa, which developed around the 3rd and 2nd millennium BC.

Farmers relied on water from the Indus and the Ghaggar-Hakra rivers for agriculture purposes, where multi-cropping systems were practiced. Channels connected to these rivers conveyed water to agriculture fields, which used "flooding-based irrigation" [29,32]. Seeds were sown after floods, which made the soil rich in nutrients and allowed growing different crops in winter and summer. During cooler wet winter periods, "Rabi" crops were grown and harvested in spring, whereas "Kharif" crops were cultivated and harvested in India, Pakistan, and Bangladesh during the subcontinent's monsoon season.

According to Wright et al. [29], the source of water used for irrigation and the method of its application for irrigating crops differed among the sites located in different agro-ecological zones. These differences can be seen in three subdivisions of the Harappan civilization, that is, sites such as Harappa, Mohenjo-Daro, and Lothal. In a Harappa site, on the Ravi River, people used summer flooding, summer oxbow lakes and summer and winter monsoons for agriculture, whereas in Mohenjo-Daro, located on the Indus River, farmers relied on summer sheet flooding, winter rains and hillside dikes. In a Lothal site, located on the Bhagavo River, monsoon season rains supported agriculture.

The Indus Valley Civilization (*ca* 2600 BC), in present-day Pakistan, also had early canal irrigation systems. Large-scale agriculture was practiced in the area and an extensive network of canals was used for irrigation. Sophisticated irrigation and storage systems of surface tanks and underground reservoirs were developed [33], like at Girnar *ca* 3000 BC [34]. These farmers were probably among the earliest to take water from underground wells, in addition to surface river water. The shaduf, known as denkli or paecottah in India, was used in the Indus valley at the time for pumping water and irrigating agricultural land. Besides, some of the pictures on toys from the Indus Valley civilization indicate that there was a proper system of water supply to different houses and places with underground pipelines or aqueducts (Figure 3). Women were responsible for the water supply to the different places in town.

Figure 3. Underground pipe/aqueduct at Mohenjo Daro, Pakistan (modified from [32]).

2.2.3. Mesopotamian Civilizations

The city-state of Babylon, located along the Euphrates River, was founded ca. 2300 B.C. by the ancient Akkadian-speaking people of southern Mesopotamia, which became a major military power under Amorite King Hammurabi, who ruled from *ca* 1792 to 1750 BC. The regulation of water usage followed the initial irrigation developments in that region. The first known written rules related to irrigation date back to the era of that Babylonian King who wrote a code of law based on previous Sumerian rules. Additionally, the Hammurabi law code introduced three key concepts that created the foundation for collective irrigation management [35]. The key concepts were: (a) proportional distribution, whereby a grower receives water in proportion to the amount of farmed land; (b) definition of an individual farmer's responsibility towards the community, by safeguarding the canal sections on his property, accepting community-shared rules such as water rotations and liability for damages caused to neighbors owing to negligence or malice; and, (c) water apportionment and policy of irrigation arrangements being the collective responsibility of beneficiary farmers [36,37]. These concepts constituted the foundations of irrigation development in the region, and although they were abandoned for long periods in different areas, today they still represent excellent key principles for collective irrigation management. The goals were to ensure farmers' participation in the construction and management of infrastructures and achieve an equitable distribution of water to avoid conflicts. It is disputed among legal historians whether early Greek law was inspired by these early Mesopotamian examples [35].

The scale of irrigation in the Indus Valley was also larger than that in Egypt and Mesopotamia, but also more active due to the different characteristics of the rivers involved. Eridu is believed to be the first place that grew into a city as the result of irrigated agriculture development [9]. The region flourished later, during the period of the Babylonian dynasty, especially under the rule of Hammourabi. The well-known Hanging Gardens of Babylon—one of the Seven Wonders of the World—are believed to have been built during the Neo-Babylonian dynasty, under the king Nebuchadnezzar (*ca* 604–562 BC).

2.2.4. Others (Mediterranean and Middle Eastern Societies)

Other Mediterranean and Middle Eastern societies of the region also developed elaborate techniques for small-scale water collection, storage and conservation in the highlands. Rock-walled bench terraces and rainfall water diversion have been used in the present Lebanon nearly 3000 years ago. Similarly, Yemen is well known for its ancient terraces that facilitated the successful cultivation of crops on steep terrains. Archaeological evidence in the Negev have shown that arid-zone farming/irrigation techniques of the type known for North Africa (e.g., wadi valleys terracing) started developing as a viable alternative to exploiting the arid environment since the Bronze Age [38]. Examples of traditional rainwater harvesting, and soil and water conservation works known as "Jessours" have been developed in the southern part of Tunisia (Figure 4). They consist of systems with earthen dikes collecting runoff and spillways in areas where annual rainfall is less than 250 mm.

Figure 4. Jessours in southern Tunisia for growing tree crops (adapted from [39]; photo of A. Bahri).

Farmers used to build earthen dams (known as tabias) across the valley floors to trap run-off water and silt. Small terraces collecting this water were created for growing crops and fruit trees that would not survive otherwise in such arid climate. They were based on hydro-solidarity: farmers would help each other rebuild the spillways when they were destroyed by heavy rains and flash floods. Studies have been conducted at the INRGREF [39] on these early examples of water regulation and management works.

Another clever system of water collection and conservation in agriculture, believed to have originated in North Africa in the past, is "pot-watering" or "jar irrigation". It consists of burying a water-filled clay jar near a tree seedling to allow slow moisture diffusion to the plant roots. Such system is still used today to grow trees for halting the progression of sand dunes in southern Morocco. The jar helps capture and maintain moisture in the upper soil layers. All these desert agricultural systems represent general evidence of a well-established level of development by 1000 BC, which reached a rather high degree of sophistication in development by the so-called "Israelite III Period" (*ca* 850–600 BC) [38]. The use of soil-embedded porous jars is one of the oldest localized, high-frequency (or continuous) irrigation methods. Although the exact origin and antiquity of the method cannot be established with certainty, numerous reports have attested to its use by traditional farmers throughout North Africa and the Near East [40].

2.3. Early Ancient Egyptians (ca 3050–1550 BC up to 500 BC)

The ancient Egyptians developed irrigation systems to use water of the Nile River for a variety of purposes. Notably, the irrigation granted greater control over the agricultural practices [41]. Despite the fact that irrigation was crucial to their agricultural success, there were no statewide regulations on water control; rather, local farmers were responsible for irrigation. They practiced basin irrigation when water levels of the Nile were sufficiently high during flooding. The earliest and most famous reference to irrigation in Egypt was found on the mace head of the Scorpion King, which was dated to about 3100 BC [42].

The first major irrigation project in Egypt was built about 3050 BC by the First Egyptian dynasty. From around 2100 BC, several ingenious systems for irrigation were in use, including a 20 km long canal to divert Nile floodwaters to a lake. The Sadd el-Kafara was a masonry embankment dam, one of the oldest dams in the region that was built during the Third or Fourth dynasty, i.e., between *ca* 2686 and 2498 BC [43,44], but was eroded by floods before being used. Across the Red Sea, the oldest dam on the Arabian Peninsula, Marib Dam in today's Yemen, was built *ca* 500–600 BC.

The Egyptian "shaduf" (Figure 5) and the water wheel, or "noria" or "sania" are probably among the earliest devices for lifting water from the Nile River for irrigation and domestic uses [12]. A modified version of the shaduf, called locally Diou or Dlou, was developed in North Africa and the Arabian Peninsula at the beginning of the *ca* 12th century BC [45]. It was refined later on with the introduction of a pulley and animal traction for lifting water from deep wells.

Figure 5. Egyptian Shaduf (adapted from [12]).

The noria or Egyptian Wheel, thought to be the first vertical water wheel, was invented in Egypt around the *ca* 4th century BC [46]. However, its diffusion is typically associated with the Arab civilization, being the animal-powered noria considered as the higher symbol of the Islamic imprint upon irrigation technology. The hydraulic wheel was first built in Fez, Morocco, in the 13th century [47], then spread to other parts of North Africa.

A different version of the noria is the Persian Wheel (Figure 6), whose date of invention is not well known. It consists of an endless series of pots of unequal weight turned over two pulleys [48], and it is therefore classified as a pump rather than a water wheel. This device, in its different versions, represents the ancestor of water pumps and modern hydropower systems. The system used for lifting water to irrigate the Hanging Gardens of Babylon remains a mystery, although Greek historians describe it as consisting of something similar to an Archimedes' screw or chain pumps, each consisting of two large wheels powered by slaves.

Figure 6. Noria in Hama, Syria [48].

The ancient Egyptians developed a system for monitoring the water flow in several places along the Nile River. The system (nilometer) consisted of marking the level of water and comparing it with those of previous years, thus allowing predictions with some accuracy of the following year's high marks. At least 20 "nilometers" were spaced along the river, and the maximum level of each year's flood was recorded in the palace and temple archives [49]. The early version of the system consisted of marked flights of stairs and has been used for thousands of years. The knowledge of Nile's flow level helped managing irrigation schemes.

2.4. Early Pre-Columbian Time Civilizations in America (ca 2500 BC to 1540 AD)

The Prehispanic Maya lived and farmed in a region with few permanent water bodies that were affected by distinct rainy and dry seasons (map presented in Figure 7). Maya farmers had to predict the start of the rainy season and the time for planting their crops, so that seedling could be sustained by the rains. Seasonal water bodies, such as 'aguadas' (rain-fed depressions), 'bajos' (wetlands), and swamps provided natural water resources that disappeared by the end of the dry season after supplying the needed water. As settlements grew, the Maya began to develop new ways of providing adequate water supplies. Unlike other civilizations, the Maya did not use real direct irrigation, but instead constructed rather elaborate water management systems to capture, store, and redistribute water for supporting both urban life and large-scale agriculture.

While most of Maya's water management systems were constructed to cope with water scarcity, in the wetlands of northern Belize the Maya built ditches and canals that recovered arable land from the karst seasonal wetlands, i.e., the bajos [50].

Figure 7. Map of the Maya area (modified from [51]).

Prehispanic Maya farmers practiced swidden agriculture (slash and burn). They cleared forest areas by burning prior to planting multi-crop fields, and then fertilized the system by letting the cinders remain in the field. Those areas were named "milpas" and were planted with both annual and perennial crops varying from corn to fruit trees. These sites were farmed for a few years, and then allowed to fallow to recover fertility and regrow forest vegetation before starting a new burning cycle. This agricultural system better maintained the fertility of the land, limited the susceptibility of crops to pest and diseases, and required little input from water management systems. Population expansion caused these systems to be unsustainable [52], as the fallow cycles were shortened beyond reasonable periods to maintain production. Nevertheless, milpa agriculture continues being practiced today by some traditional small farmers in Central America, with limited revenues. By the Late Preclassic period (*ca* 300 BC–250 AD), Maya farmers modified the natural water bodies to create reservoirs and support growing communities. The largest Preclassic Maya centers, such as El Mirador, were built in the swampy Petén of North Guatemala, where naturally abundant water provided an ideal basis for constructing reservoirs capable of supporting cities [53]. Many of the great water management systems in the height of the Late Classic period (*ca* 600–900 AD) were built on the foundation of Preclassic reservoirs, modifying the natural 'aguadas' and 'bajos'. This is especially true at Tikal, where extensive study work has been done to understand the characteristics of the reservoirs system.

The natural water reservoirs were expanded, lined with plaster, and maintained with a system of plants and microbial flora that kept stored water potable throughout the dry season [54]. At its height, Tikal was home to as many as 100,000 inhabitants, whose needs for drinking and irrigation water were met by a system of ten reservoirs, at least four of which laid within the central precinct of the city. The reservoirs of Tikal stored runoff from rainfall that was flowing over the plastered ground, channeled into a series of reservoirs that fed first the elite residences, then those of lower-status families and finally ran towards agricultural fields in the outskirts of the city. Though the construction of

Tikal's reservoirs (Figure 8) began in the Late Preclassic period, the construction program was greatly expanded in the Classic period to meet the needs of a rapidly growing urban population [52,55]. Later, despite Tikal being abandoned for nearly a millennium, the reservoirs continue storing and regulating water throughout the year.

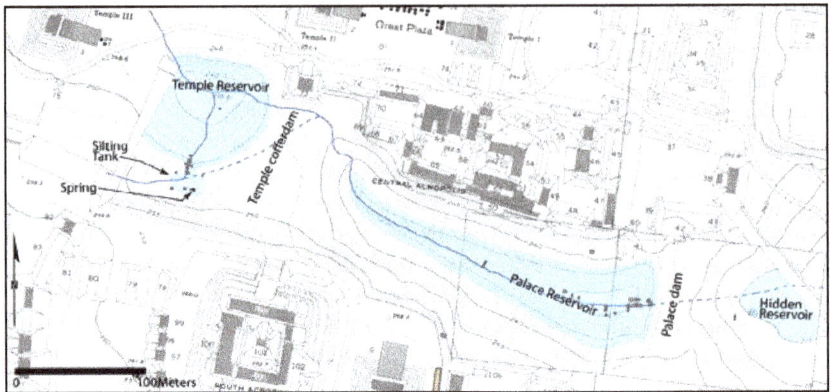

Figure 8. Map of Tikal's central reservoirs (adapted from [56]).

Unlike the Southern Lowlands, the Northern Lowlands of the Yucatán Peninsula have far fewer water bodies. In order to create reservoirs and store water, the northern Maya constructed 'chultunes', small storage tanks excavated into the bedrock. The chultunes were used by both hinterland farming households and larger centers, where the storage pits complemented the reservoir systems to supply larger urban populations with adequate water throughout the year [57].

In the wetlands of Northern Belize, the Maya faced a different problem, i.e., the overabundance of water. Maya farmers built a series of raised beds surrounded by canals, which enabled them to farm in swampy lands that provided fertile sediments and water for their crops, and aquatic animals to supplement their diet [58]. This system is reminiscent of the Aztec chinampas, though on a smaller scale.

Despite these elaborate water management systems, the Classic Maya, like all human societies, faced serious environmental challenges. A series of multiyear droughts, known from speleothems and sediment cores, hit Central America between *ca* 500 and 1000 AD, with the greatest occurring during the 9th century AD [59]. The most severe droughts coincided with dramatic political changes, as Maya cities were abandoned and the elite fell from power. After the fall of major urban centers in what is often referred to as the Maya Collapse, the majority of water management systems were abandoned. Nevertheless, hinterland communities remained in place, and smaller reservoirs and other systems continued to be used by the Postclassic (*ca* 900–1521 AD) Maya. Spanish colonizers attempted to subvert indigenous practices and enforce European agricultural traditions; however, many traditional agricultural practices continued to today.

3. Historical Times

3.1. Chinese Dynasties and Empires (ca 771 BC–581 AD): from the Spring and Autumn Period and the Warring States Period to the Northern and Southern Dynasties

As a country built on the basis of agriculture, China has made a great deal of admirable achievements in agricultural related technologies, as indicated for the earlier periods of Chinese civilization. Irrigation, as a critical part of agriculture, witnessed some of the most important inventions during that time. Generally, all governing dynasties in Chinese history spent a great deal of manpower, material resources and money into various water conveyance projects. However, most of these projects were built in the major agricultural regions, and could not cover the high altitude areas or those located far away from water sources. Subsequently, the Chinese searched for water distribution equipment and devised the waterwheel to meet the need to extend irrigation throughout the entire country [60].

After the capital of the Zhou dynasty moved to Luoyang on Henan Province in 771 BC, China entered the Spring and Autumn Period (*ca* 770–476 BC) with the imperial power declining and the country torn apart by vassals. Along with the transformation of social system and the development of social productivity in this age, the ancient well-field system, described before, collapsed, thus the well-field irrigation canal system was abandoned. In the Spring and Autumn Period a new style of irrigation system suitable for the hilly areas was created, i.e., constructing several storage ponds at the sites of natural depressions, and connecting the ponds in series through excavated canals, storing surface run-off, and regulating water resources through these ponds for more efficient use of water. It was called melon-on-the-vine irrigation because its canals looked like vines and its storage ponds looked like melons on the vine. The earliest irrigation system of this type was in Quebei, located in Shou County, Anhui Province, and came into operation about 2500 years ago.

In the Warring States Period (*ca* 457–221 BC), while vassal states vied for supremacy, irrigation received universal attention in order to make their states rich and build military power, leading to the construction of large irrigation projects, which had additional functions. Three of these projects are famous: the first is the Zhengguo canal diverting water from the Jinghe river, built by the Qin dynasty and located in the middle of the Yellow River basin; the second is the Dujiang weir, located in the Minjiang River, tributary of the Yangtze River, also built by the state of Qin; the third consists of 12 canals diverting water from the Zhanghe River, tributary of the Haihe River, which was built by the Wei dynasty. The capital of Qin state, Xi'an, was located in the Yellow river basin as previously indicated. The construction of the Zhengguo canal made the state of Qin powerful and prosperous, ultimately leading to annexation of the Qin state by other vassal states. In addition, mostly depending on irrigation by the Zhengguo canal, the Yellow Rriver basin became the first economic center of China [61]. Afterwards, building of the Dujiang weir made the state of Qin smoothly unify the South of Five Ridges area, and made the Chengdu Plain known as the "land of abundance" because people there were able to control floods and droughts diminishing hunger in the region.

During the Qin and Han dynasties (*ca* 221 BC–220 AD), with the capital still located in Xi'an, development of irrigation in the Yellow River basin continued receiving high attention. Many irrigation projects, such as Baigong, Liufu, Bai and Longshou and other canals, were constructed during the Western Han dynasty. The Central Shaanxi Plain became the richest region in China, with its wealth representing 60% of the entire country. Meanwhile, the storage pond irrigation systems had greatly developed in the Central Shaanxi Plain of the Hanshui River basin, and in the Nanyang and Runan areas of the Huaihe River basin. Moreover, irrigation appeared in the Ningxia Hui Autonomous Region and the Hetao area of the Inner Mongolia, with the implementation of measures directed at cultivating wilderness and guarding the frontier in north-west region.

With Luoyang as the capital of the Eastern Han dynasty, the economic center of China moved southwards. During this period, the number of storage-pond irrigation systems in the Huaihe River basin had increased substantially. Several storage-pond irrigation systems were built along the main rivers in Nanyang area, such as the Tuanhe, Tanghe, and Baihe rivers, and built along the main rivers in Runan area such as the Rushui River. Among these storage ponds, the Liumenbei built along the Tuanhe River, and the Hongxibei built at the Ruhe River were the most famous. The use of these storage-pond irrigation systems led the areas of Southeast Henan and West Anhui to richness and prosperity. Millet had once been the major cereal crop in the north, but wheat gradually grew in importance. Rice, imported from the south, was extended to the dry soil of the north. The soybean, in a number of varieties, proved to be one of the most important crops, and Chinese farmers gradually developed a sort of intensive agriculture. Soil was improved by adding manure and other soil amendments. Planting fields in carefully regulated rows replaced the fallow system. Great importance was placed on plowing and seeding at the proper time, especially in the fine-grained loess soil of Northern China. Fields were weeded frequently throughout the growing season. Farmers also knew the value of rotating crops to preserve the fertility of the soil, and soybeans were often part of the rotation. Irrigation became necessary as population pressure forced cropland expansion, and construction of irrigation works in many states beginning in the late Chunqiu period. These projects were built to drain swampy areas, leach alkaline soil and replace it with fertile topsoil, and, in the south and in the Sichuan Basin, to carry water into the rice paddies. The irrigation systems unearthed by archaeologists indicate that these were small-scale works presumably carried out by state or local authorities [10].

In the period from the end of Eastern Han dynasty to Sui dynasty (*ca* 220–581 AD), the previously united country had been divided and the regime changed frequently. The longest government lasted about 100 years, while the short ones could only last 10 or 20 years. Because of incessant wars, a large population of Chinese Central Plains moved southward with their agricultural technologies. Areas in the south of Yangtze River were mountainous but suitable for building ponds, thus storage-pond irrigation systems were popularly built there. Examples are the Chengongtang Pond in Yangzhou, Lianhu Lake in Danyang of Jiangsu Province; Jianhu Lake in Shaoxing, and Nanhu Lake in Yuhang of Chekiang Province [60].

3.2. *Archaic, Classical, and Hellenistic Civilizations (ca 800–30 BC)*

Past and present unpredictable precipitation patterns and geological variations of mainland Greece and the Aegean islands remain the "raison d'être" for the erratic nature of hydrology in the regions where the Greeks developed city-states. Since early dates, the existing evidence suggests developments of diversified and changing strategies for water management, including drainage and irrigation. With the emergence of the polis, the city-state in the Archaic age (*ca* 800–500 BC), the state formation involved controversial questions on how communities should manage and distribute water resources.

The division of Greece into a "dry" southwestern part and a "wet" northwestern part serves as a general and valid description of the regions' hydrology [62]. Early epic poetry—Homer and Hesiod—alluded sporadically to the complexities of cultivation and water management; however, these were not performed in a practical manner until the reforms of the Athenian magistrate (Archon) Solon in 594 BC. From that, there is a substantial evidence for state-induced legal action to secure "neighbors" access to water. If their distance to public water supply was more than "four furlongs", citizens were entitled to "fill a five-gallon jar twice a day" (Plutarch. Solon, 23.5). Clearly, this limited volume suggests the fulfillment of basic requirements for households and gardens rather than sustainment of large-scale irrigated farmland.

However, Solon's law makes it clear that Archaic and Classical Attica was a region of diverse hydrology, alluding to the fact that the present Greek peninsula experienced profound variations in rainfall—past and present [63,64]. In addition, frequent food crises triggered by local droughts or by the disruption of grain imports, due to overseas calamities, were recurring threats to the city-state's food supplies [65].

Crouch [66] documented how urban and rural developments from the Archaic through the Hellenistic period were intrinsically associated with water management and drainage. Moreover, he elucidated the dependency of urban and rural developments on the presence of karst areas, suggesting that many cities of the ancient world depended upon this geological structure for their water supply [66,67]. Crouch [66] also pointed to the importance of wastewater disposal, and indicated how drainage techniques and technology derived from agriculture could also be applied to clean urban infrastructure.

The pivotal question of when and where Greek societies developed irrigation techniques is however debatable. Hitherto scholars have convincingly argued that the kepos —the garden—developed from the archaic age onwards as a confined cultivated location where "small-scale" irrigation was practiced [7,67]. The epigraphic evidence from Classical and Hellenistic Athens and elsewhere frequently lists lease contracts of gardens as well as other forms of cultivated plots' management.

Theophrastus (*ca* 371–287 BC) relates how gardeners cultivated a variety of plants, including the so-called potherbs, vegetables, but also flowers and other herbs. Thus, intrinsically, the 'kepos' potentially held wider range of crops than the modern concept of the kitchen garden [7,68–71]. Undoubtedly, in the Classical and Hellenistic periods, the ubiquitous kepos represented a location where manure, water and manpower were concentrated and it is borne out of the literature and epigraphic evidence that it was often located at or near the household (oikos).

It has been speculated, however, that a noticeable development surged in the Classical period where regular kepos-landscapes emerged [72]. According to Plutarch (*ca* 1st century BC), a larger area of Athens was named "The Gardens" and later tradition indicates that Boeotian Thebes was renowned for its gardens before it was destroyed by Alexander in 335 BC (Heracleides Creticus, 1.1-2 in [73]). The fundamental question is to determine whether large-scale irrigation was implemented to aid cultivation of garden systems or field crops. Possible locations of large-scale irrigation facilities in the Classical and Hellenistic periods were in Athens, Kopais, Eretria, Delos and Gortys (in Crete) and Metaponto and Herakleia (Greek colonies in the present-day Italy).

The material evidence suggests that the Minyan culture developed water works, irrigation and drainage in the 2nd millennium BC to manage the effects of seasonal rainfall at the greater polje of the Lake Kopais [74–76]. Also, Minoan Crete probably saw an intensification of irrigation schemes at the Messara plain [20]. The question however is whether this early development left a lasting impact on the Hellenic water management for the following centuries.

The major challenge is to understand how large-scale irrigation and drainage may have developed in the Archaic and Classical periods to support regular field cultivation. First, available facts on the ground seem indicating development of such a scheme in the rugged "highlands" around 5th century BC Delos. Whether this development emerged in *ca* 4th century BC Southern Attica remains to be seen [64,77]. In effect, this example points to collection and storage of water by controlling the torrential autumn and winter rains, which made it possible to perform limited irrigation of large cultivated areas, predominantly terraced agricultural lands [7,64]. Terraces demanded intensive labor to be built and maintained, but at the same time, their construction improved soil depth and quality. This was done partly by removing the stones from the land, which were then used for building terraces and field retaining walls [70,78,79].

Finally, the material and written evidence indicates the construction of large-scale irrigation systems at Metaponto of Magna Graecia and Gortyn, Crete, and a large drainage project at Eretria on Euboea in the late Classical-early Hellenistic period has been interpreted as a combined drainage and irrigation system [7,80]. The famed water works at Lake Kopais, allegedly clocked and malfunctioning, called the attention of Alexander the Great (Strabo IX 2.18; in [81]). Ostensibly, Alexander launched a project for its restoration but evidently, the project was never completed until the restoration and drainage of the lake in the 19th century.

3.3. Roman Period (ca 30 BC–330 AD)

Cereals were the most important part of the Roman diet and were grown during the rainy season, but required sufficient water to achieve adequate yields. Their production in Mediterranean environments was dependent on water availability through rain or irrigation, and was often associated with the fallow practices, where fields were left to rest in rotation. This agricultural practice was well known by Greeks and Romans [82], and such rotation was also applied to the cultivation of legumes, as reported by Pliny the elder (Nat. Hist., XVIII, 91). The crop yield varied with region and also with rainfall fluctuations. For these reasons, and also to ensure crop safety, irrigation was implemented when possible. For instance, Varro reports that in some areas of Etruria a single seed, with sufficient water, could make up to fifteen times its equivalent [83]. There, irrigation was an important element for agricultural practices during the dry summers. Drainage schemes for large areas, involving also runoff water reclamation, flood control, and navigation, were typical of the Hellenistic period. In Italy, the Etruscans used mainly underground galleries, while Romans preferred simple ditches or channels [84].

Romans did not add to science as Greeks did; however, they contributed tremendously to the practical application and deployment of engineering techniques [85]. Water was channeled from its sources, conveyed long ways, and distributed for irrigation of arable fields, orchards, vineyards, pastures, and gardens [86]. However, this method was rarely used for the wheat fields. Irrigation channels could be made of wood, clay, stone, or excavated into the ground. Romans also invented the Roman concrete (opus caementitium), which allowed the construction of long canals, very large bridges and long tunnels in soft rock [87]. For instance, these works were used for eleven aqueducts, which had a total length of about 350 km that ensured the supply of water to ancient Rome. Some of these channels are still working and used nowadays to supply the fountains in the city.

In the Italian peninsula, there was a widespread use of drains for irrigation, especially in orchards and pasture crops, more rarely for vineyards and olive orchards. In Central and Northern Italy drains were used for irrigation of crops and they often marked the limits of centuriation; i.e., the process or act of dividing land into centuries or equal areas undertaken by the Romans and known as Roman grid [88]. Emissaries of lakes, such as Lake Albano (Figure 9), were also used for irrigation and this and other water infrastructures were built by the legions (Figure 10). Among the eleven aqueducts, the Old Aqua Anio and the Aqua Alsietina were used only for irrigation and for feeding the Nymphaea because of their low water quality.

Figure 9. Lake Albano, Italy (Photo by M. Salgot).

(a) **(b)**

Figure 10. Cistern built by the II Legio Partica": (**a**) access, in the Albano municipality, Italy and (**b**) view of the interior (Photo by M. Salgot).

Gardens were closely linked to irrigated agriculture [89]. Pomarium (apple tree garden) was a common feature of every agricultural property, irrespective of its size. It included several fruit trees such as almonds, hazelnuts, nuts, apples, pears, plums, figs, and quinces; with the later addition of other species coming from the East (e.g., cherries and peaches). The newly introduced species were planted interspersed among the existing ones and grafting was frequent. Columella (V, 10, 6) stressed the importance of this practice, which enabled farmers to improve the quality of native species through the introduction of stronger or most valuable varieties [90].

The Romans exported their skills as engineers not only throughout Italy but also all across the Mediterranean basin, including Gaul and Iberia, and especially along the African coast. Indeed, during the Roman Empire the cultivation of grains was extended even to deserts. In Libya, large areas, now abandoned, were enclosed by terraced walls along the edges of the hills, which, at the end of the wadis, collected the silt from floodwaters and the scarce runoff. The land bounded by these terraces was fertilized and the deposited moist layer often led to very high crop yields. These techniques of "dry or rain farming", although traditionally Saharan, were undoubtedly enhanced by the Romans. Generally, in North Africa and in the Eastern regions, Romans established permanent irrigation systems, which allowed for agricultural exploitation of otherwise arid and unproductive territories [91].

Research and examination of aerial photos make it possible to recognize the exceptional breadth of areas served by irrigation systems, in some cases related to the centuriation. Water from natural or man-made watercourses and other sources was collected in canals, with parallel paths, concentric or fan in shape. In some cases, the basins were constructed either isolated or arranged in series along the slopes, resulting in terraces, as in the case of the impressive facility of Oued Ogrib, in Algeria. The canals could be excavated into the ground or contained on the sides of the land, possibly by using retaining walls made of stones and gravel; they were generally small in size, however there are examples of larger size (e.g., in the plain of Caesarea (now Israel), a 27.5 m wide and 0.46 m deep canal. In other Regions of the Roman Empire (e.g., Egypt and Syria), Romans also improved the irrigation systems, using the existing structures in some cases and in others building new canals (e.g., Trajan's works). Furthermore, they introduced technical and administrative changes such as tax incentives, operating rules, and cadastral divisions linked to irrigation systems [90].

The existing regional differences in geo-morphology and river water availability had a significant impact on the distribution of irrigation techniques across the Near East in the Roman period. The distribution of irrigation canals and qanats was mutually exclusive, as illustrated in Figure 11 [92]. Irrigation technologies were inextricably tied up in the complex and shifting relationships between technology, social factors and environmental challenges. Sharing the existing water availability between urban centers and agricultural fields illustrates the fact that agricultural concerns coexisted with broader social and environmental challenges, and the unpredictability of water supply in the region played a role in tightening these relationships [92].

Finally, when the Sassanian Shah Shapur I defeated the Roman emperor Valerian (*ca* 250), he is said to have ordered the captive Roman soldiers to build a large bridge and dam stretching over 500 m. Lying deep in Persian territory, the structure exhibits typical Roman building techniques, and became the most eastern Roman bridge and Roman dam [93]. Its dual-purpose design exerted a profound influence on later Iranian civil engineering and was instrumental in developing Sassanid water irrigation and management techniques.

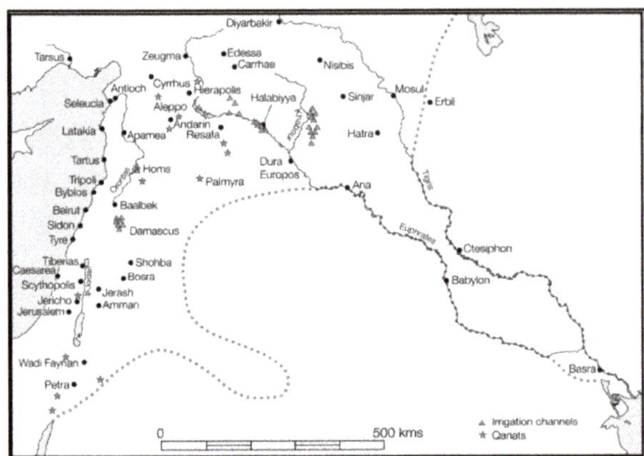

Figure 11. Map of the near east in the Roman period, showing the distribution of irrigation channels (triangles) and qanats (stars) (prepared by Alison Wilkins [92]).

3.4. *Other Historical Civilisations*

3.4.1. Urartu Kingdom

Advanced water management emerged in Anatolia during the Urartu Kingdom, which rose to power *ca* mid-9th century BC. Belli [94] performed a systematic study in eastern Anatolia to identify Urartian dams, reservoirs and irrigation systems, with particular focus on the agricultural infrastructure of the Van Lake basin. In the Van basin and its surrounding areas, located in the western part of Anatolia, irrigation canals and dams dating from the time of the Urartian Kingdom (*ca* 840–590 BC) can be still observed. This Kingdom was located in the highlands of Eastern Anatolia, the Ararat plain of Armenia, and in the present northwestern Iran. This region is located on an active tectonic zone, but earthquakes have not affected its irrigation systems. There are very few places in the world with a water supply and conveyance infrastructure that has been in operation for *ca* 2700–2800 years. That is why the Urartian Kingdom is named the greatest Hydraulic Civilization of Anatolia and Asia [94,95]. The tradition of building water works began with the Hittites and Assyrians, which used to live in Urartian/Urartu and thrived during the medieval and Ottoman times in Anatolia. The Urartian thus represent an important connection in the development of agricultural irrigation and dam construction in Anatolia up to the present day. The Urartu dams built nearly three thousand years ago on streams and small rivers are the forerunners of the modern dams on the Euphrates and Tigris rivers in Eastern Anatolia today [94,95].

One of the most interesting sites relating to the development of Urartian water facilities is Lake Aygır in northwestern Van. Here, terracotta (clay) water pipes and stone channels were laid down to convey water out of the artificial lake [96].

The irrigation systems set up by the Urartians made use of surface water from one of two existing systems. The Menua or Shamram Canal, built by King Menua around 800 BC conveyed water for downstream irrigation (Figure 12). To our knowledge, this canal is the only irrigation canal in the world that has been in continuous use for nearly 2800 years. The second is irrigation supplied by water from Lake Keşiş (Lake Rusa) and the smaller dam lakes in the vicinity. Lake Keşiş, which was dammed by Rusa II (680–654 BC) and lies east of Van at an altitude of 2890 m, stored sufficient water for meeting the needs of the new kingdom capital, Rusahinili, and its surrounding areas to last through the summer months. About 20×10^6 m^3 of water could be stored in the still-existing dam.

Figure 12. View of ruins from Menua canal [97], which still conveys water from Lake Keşiş [94].

3.4.2. The Levant and Beyond

Water diversion devices for irrigation and other purposes have been used in various other locations in the Fertile Crescent. For instance, in Yemen, the oldest dam, constructed in masonry over a length of about 500 m, was built in *ca* 600–500 BC for flood protection and irrigation [10].

North of Mesopotamia, in western Persia and eastern Turkey, farmers began construction of elaborate systems for extracting groundwater for irrigation about 2500 years ago [98]. The system, called "qanat", consists of a series of wells for extraction of material, 20 to 30 m apart, connected at their bottoms by a tunnel with controlled slope, in order to tap groundwater without lifting devices. The technology extended with Persian rule during the period *ca* 550–331 BC, from the Indus to the Nile River basins [10]. Further expansion took place in the Mediterranean basin, Egypt, Afghanistan, the Silk Road, oases settlements of central Asia, and Chinese Turkistan [99]. During the Roman-Byzantine era, large numbers of qanats were constructed in Syria and Jordan. The Romans also used qanats as subterranean parts of aqueducts, forming complex water transportation systems, in Tunisia and Turkey. Another major expansion of the technology in North Africa, Cyprus and Sicily took place during the early Islamic period.

The early developments of irrigation in Egypt and Mesopotamia extended first to North Africa and the Mediterranean under the Carthaginians (Phoenicians), during the *ca* 9th century BC, and then to the south under the Greco-Roman Empire. The latter focused on harnessing existing water sources and rainfall water collection, with structures such as blocks of masonry dams, reservoirs and cisterns for rainwater, canals and aqueducts, and dams made of dry stone to divert water for irrigation.

Solomon's Pools in Bethlehem, Palestine, are three large catchment reservoirs, each of around 160,000 m^3, built with stone and masonry between *ca* 1000 and 30 BC in two stages. Ancient aqueducts in the area collected water from springs and conveyed it to the pools. Some of the springs with channels are still used for irrigation today.

4. The Sui, Tang, and Song Dynasties in China (581–1127 AD)

The Tang dynasty, established on the basis of the Sui dynasty that reunified China, represented a period of brilliant ancient Chinese civilization. Before the An Lushan-Shi Siming rebellion in 755 AD, with the unity and prosperity of the country and regime stability, irrigation in China underwent further development, especially in the Yellow River basin. Along with the spread of rebellion, the Tang dynasty experienced a process from prosperity to decline. Not long after that, several separatist regimes were set up by force of arms. It was the same situation in the following Five Dynasties and Ten Kingdoms. There was a long-term confrontation with Khitan, Dangxiang, and Nuchen in the Northern Song dynasty, and the emperors of the Southern Song dynasty were all contending to exercise sovereignty over only a part of the country. Incessant war spread all over the middle and lower reaches of the Yellow River in this period, but the regime was relatively steady in South China because of the effective defense of the Yangtze River, which was viewed as a natural moat, enabling further economic development demands in South China. Thus, the pond-canal polder system in the Taihu Lake and the Yangtze River basin, and projects limiting seawater intrusion and restoring freshwater in the southeast coast emerged and gradually grew as required.

The pond-canal polder system in the Taihu Lake basin was built during the Tang dynasty and was more advanced than the original fragmented water reclamation system [100]. Natural rivers in the basin were used to develop the ponds and canals, with the excavated soil used to build the dike, so that farmlands were enclosed. Inside the polder, locks were installed on the dike to enable irrigation and drainage. With the building of Taihu Dike and the seawalls, the pond-canal polder system was gradually improved. However, shortsighted reclamation in the Song dynasty destroyed the original irrigation and drainage canal systems, leading to large polders being divided into small pieces. Polders in the lower reach of Yangtze River were developed rapidly during the Tang and Song dynasties. Because of the fairly large hydraulic head of this region, their dikes were very high.

Projects for preventing seawater intrusion and storing freshwater in the southeast coast played an important role in the rapid development of local economy. The infrastructures were mainly consisting of the flash dam and the canal. The flash dam was used for warding off tides and storing freshwater from the upper reach. The canal was used for diverting stored freshwater and conveyance to farmland. Among these projects, two are most famous: one is Tashan weir, located along the Yinjiang River of Yin County, in the Chekiang Province; the other is Mulanbei, located along the Mulan River of Putian County, in the Fujian Province. With all these water conservation projects, the Taihu Lake basin and the Yangtze River basin gradually replaced the Yellow River basin as the economic center of China during the Tang and Song dynasties. Bamboo pipes were utilized as inverted siphon during the Tang dynasty [101], and then used for irrigation. These pipes are illustrated in Figure 13. The proverb "when there is a bumper crop in Taihu Lake basin, the country will be free from hunger" vividly reflected this change.

Figure 13. Bamboo pipes.

5. Late Pre-Columbian Civilizations

5.1. *The Aztec Civilization and its Precursors (ca 1200–1500 AD)*

Recent awareness of the need to address water and food security as a joint issue for achieving sustainable development [102] runs alongside the need to recover nutrients from treated wastewater re-used for irrigation. Recovery of nutrients as fertilizers during agricultural irrigation is being promoted as a recent discovery in some developed countries. However, these goals were already achieved by ancient agricultural techniques such as the one of chinampas. The chinampas were a crop-growing method used in the Valley of Mexico before Aztec times [103] and extensively developed by the Aztec from *ca* 1350 to 1500 AD [104]. The chinampas are a UNESCO world heritage center since 1987. They are a model of sustainable agriculture because of their high productivity, use of waste as fertilizer, beneficial impacts on local fauna, responsible use of natural resources [103], and recovery of nutrients.

Around the 12th and 13th centuries, several civilizations settled in the Valley of Mexico, in an area of around 9600 km^2. This Valley is located in the center of present Mexico at an altitude of 2430 m a.s.l. and is surrounded by mountains. The climate is temperate humid in the south (1200 mm year^{-1} of rain) and temperate dry in the center and the north (800 mm year^{-1} of rain). The rainy season is well defined and lasts from May to November. These are the conditions for which irrigation is necessary to produce food all year round.

The Aztec people (*ca* 1200–1500 AD) was the most important of those civilizations and was the head of an empire comprising around 400 towns [105]. In 1519, before the Spanish conquest, the population of the Valley was very large, estimated to be 1.3 to 3 million inhabitants [106]. This population presented a high demand for food and water, a situation forcefully linked to the development of innovative solutions for both these supplies.

During the Aztec times, the valley was covered by several lakes—some of them were saline—and wetlands, marshlands, springs, and perennial and permanent rivers [107] were also present. The Aztec developed several hydraulic solutions to separate saline water from freshwater, control lake levels and floods, supply water for municipal use, as well as consciously feed water for agriculture [108]. A map of the lakes of Mexico Valley at the Aztec time showing the area occupied by the chinampas is presented in Figure 14. Because of the particular geography of the area, the Aztecs developed a complex and efficient management of the water [109].

Figure 14. Lakes of Mexico Valley at the Aztec time, with areas occupied by the chinampas (adapted from [107]).

One of their main achievements was the mentioned chinampas. The study of the chinampas has been conducted through techniques such as aerial photographs [103,104], archeology, investigation of historical documents, and more recently computer programs that integrate geographical and archeological data [109–112].

The chinampas occupied at least 1200 km^2, i.e., 120,000 hm^2 [104], and were enough to produce the food to feed the entire population of the Valley. Thus, just prior to the Spanish conquest in 1521, the chinampas supported about 10–18 persons per hectare. The most productive chinampa area of the Valley, located near the Xochimilco-Chalco Lake, was used to feed the high social classes in Tenochtitlán [113].

The chinampas (Figure 15) consists of small manmade islands for growing crops, placed on lakes or marshes, 40–100 m long and 2–9 m wide in strictly rectilinear and parallel plots surrounded by canals [112]. The soil of the chinampas is continuously renewed by adding sediments and mud from the bottom of the waterways over the previous surface to maintain their height and fertility. It has been estimated that 10,000 hm^2 of chinampa fields could provide at least half a million people with their staple food [113]. There are some inaccuracies regarding the details of the original use and productivity of chinampas (Figure 14). This is due to the bias of historical sources and the difficulty of conducting archeological investigations on the Valley of Mexico, given that it is now the urban area of Mexico City [104]. However, it is still possible to get a picture of the role the chinampas played in the Aztec civilization. Chinampas are considered the most productive and ecologically sustainable form of agriculture in pre-Hispanic Mesoamerica, as they recycled nutrients and increased biodiversity [112].

(a) (b)

Figure 15. Chinampas: (**a**) View of current plots and (**b**) view of an entire chinampa (http://trabajosnakos22.blogspot.fr/).

The site of a new chinampa was selected using a long pole or based on the remains of an old one. The foundations were built by anchoring strong reeds to the roots of native cypress trees [113]. Afterwards, mud was piled atop the reeds as a base and then alternate layers of chopped algae, tule, and mud were laid making the bottom of the chinampa very porous, thus allowing water flow and permitting capillary action for watering the crops. The topmost layer was made with mud from the bottom of the canals, manure and human waste to make it high in nutrients [111,112,114]. The height chinampas were raised above the water is estimated to be between 0.2 and 0.7 m.

Seeds were not planted directly in the chinampas; instead, they were grown on rafts, back yards or 'almácigas'. Almácigas were customized environments consisting of low terraces at the water body's edge. They were perpetually moist and filled with ultra-nutritious sediments scooped from the bottom of the canals. When seedlings were mature enough, the healthiest were selected and planted on the chinampas by cutting a cube of the soil around them and placing them on a spot preconditioned with canal mud and water plants on the chinampa [115]. The soil of the chinampas was kept rich in nutrients by periodically spreading manure, water algae, human excrements and mud from the surrounding canals.

There are several reasons for the high productivity of chinampas. First, the practice of growing seedlings and then transplanting only the healthy ones to the chinampas led to higher yields since the space and resources were not wasted on unhealthy plants, and the crop cycle was shortened. Secondly, the use of local organic matter to fertilize the plants meant that the crops were nourished from richer soil [104].

The chinampas were also efficient at using water. Since the chinampas are narrow strips of land perpetually surrounded by water, irrigation was not necessary. Consequently, time and energy were saved and the cultivation was possible all-year round, including the dry season. It is also important to highlight that chinampas allowed plants to be moist at root level (capillary rise), therefore avoiding the water loss occurring when watering plants from the top. In addition, chinampas promoted the growth of local fauna, being the patches of land in the lake served as resting places or as habitats for several animals; and promoted the growth of populations of migratory birds, fish and axolotl [111].

Nowadays it is still possible to visit 1000–2000 hm^2 of chinampas at the UNESCO world heritage center (established in 1987) on what remains of the ancient Xochimilco Lake located in the southern axis of Mexico City's urban sprawl. These chinampas are used for the cultivation of corn, vegetables and flowers commercialized in Mexico City and use the same ancient procedures for production.

5.2. The Inca Civilization (1438–1533 AD)

The Inca or Inka (Quechua: Tawantinsuyu) was the largest empire in pre-Columbian America. The administrative, political and military center of that empire was located in Cusco (or Cuzco in Spanish) in the modern-day Peru. The Inca civilization arose from the highlands of Peru sometime in the early 13th century, but the Inca Empire was short-lived (1438–1533 AD). The Incas used a variety of methods, from conquests to peaceful assimilation, to incorporate a large portion of western South America, centered on the Andean mountain ranges.

Andean farmers faced the problem of two dramatically different environmental regimes: the coastal desert that receives minimal rainfall, and the highlands that receive substantial rain that drains through river valleys and finally passes into the coastal plain. In order to control, divert, and redistribute river water, ancient Andean peoples built a series of canals that grew in scale as population densities rose and a more sophisticated arrangement was needed to support larger and more complex societies. The canals began in the valleys but soon extended into the coastal desert, greatly increasing the areas of arable land suitable for farming.

Fields in the upper reaches of river valleys had the greatest access to water, while the fields at the ends of the canals had water only during the rainy season. Netherly [116] classifies Andean farmlands according to their access to water and productivity. In her system, Category 1 land may have permanent access to water, whether through irrigation or natural water abundance. Category 2 land is irrigated and produces two crops annually, whereas Category 3 land is flooded annually, like the farms along the Nile River, and produces a single annual crop. Finally, Category 4 land is irrigated only during the part of the year when water is most abundant, and produces a single crop each year. Each of these land types expands the naturally arable land, so that Andean agricultural farmers were able to move from farming only the riverbanks to successfully producing an annual crop in the desert at the edges of the canal systems.

By the rise of the Inca Empire, these canal systems were controlled by elites at several levels, ranging from valley-wide distribution to a corporate groups' control of local water allotments. The Inca took an existing system and added a second level of political control, instituting a moiety system that doubled the levels of socio-political control over this elaborate system. The Incan imperial hegemony operated by choosing two new local authority figures from the conquered indigenous elite to administer tribute and supervise the allotment of water and maintenance of the canals. While this system appears to disrupt local socio-political organization, it adheres to Andean practice. The new moiety division drew on an existing Andean framework, whereby the world is a series of opposing but complementary halves [117]. What we know of this system comes primarily from post-Conquest Spanish chronicles, but it seems that the existing Andean system continued largely undisturbed into the historic period, with local elites at different levels of political power controlling each step of the water allotment process.

The system governing the canals can be classified as one of the three types: canals controlled by one polity, canals controlled by multiple polities, and canals that connected two valleys [116].

6. Medieval Times (*ca* 330–1500 AD)

The concept of medieval times is basically applied to the civilizations related to Europe and the Mediterranean, which implies that the rest of the world (present America, China, and India) is somewhat independent from this definition. In this limited part of the world, water-related history is more known than elsewhere due to the presence of historians who have been finding, recovering, elaborating and publishing documents on the subject.

Concerning irrigation, since it developed in places where growing crops required supplementary water in addition to rainfall, there is an additional geographic limitation: the arid and semiarid climates are the only ones included in the work and will be a center of the attention for this section. As such, Northern Europe is excluded because irrigation was mainly not needed there. This difference was evident, even in the type of plough used (soil differences were also important) and the animal force (ox instead of horses) as explained by [118]. The development of agriculture was related with the need to supply food to an increasing population with the same amount of land. Barnebeck [118] reported that economy appears to relate to agriculture in the moment when farmers were supposed to supply food to a certain number of persons not living on the agricultural lands, and this is when agricultural water management matters.

Populations did not increase linearly during the medieval ages in Europe because of recurring local, regional or continental wars and epidemics, including the bubonic plague or Black Death that ravaged Europe on several occasions. Malaria also caused many problems in relation to water management [119].

The irrigation practices during those times were an evolution of previous techniques implemented by Greeks, Romans, and Arabs. While in the north of the Mediterranean Basin medieval times was a dark period with little technological progress, the Arab domination of the basin expanded irrigation technologies from the east to the west of Southern Europe [120]. Some limited technology survived in the north however, mainly due to the knowledge maintained in monasteries of religious orders. The crusaders also helped to transfer Arabic science and technology to the Christian world. In fact, due to their expansion throughout the Mediterranean, the ancient irrigation systems were implemented in the dry lands of North Africa and in the arid part of the Iberian Peninsula, as well as in the south of the present-day Italy and the large Mediterranean islands.

During a few centuries, several Italian city-states dominated the Mediterranean Sea, sometimes fighting, other times collaborating. These cities aimed at extending their possessions and establish rule in other countries or cities [66]. Among them, Venice and Genoa established commercial routes and ruled over foreign territories, e.g., parts of Crete, Sardinia, south of Italy, and Sicily. The commercial exchanges and domination favored technology transfer including irrigation. Apart from "high-technology" facilities, small solutions were maintained in the entire Mediterranean area, e.g., the cisterns, still in use in many places. "Subperiods" can be defined in medieval times, in relation with the dominating states or towns, as indicated in the next two sections.

6.1. The Byzantines and Other Civilizations (ca 330–1204 AD)

Since the Eastern Roman Empire, also known as Byzanthium, continued the Roman civilization, the technologies in use for water management did not change in this part of the Mediterranean basin, where the remnants of water infrastructures using Roman or older techniques are common. Some of the related features remain visible, like the public baths or hammams, maintained and utilized by the Arabs throughout the centuries, and afterwards backed again by the Turkish domination of the eastern part of the Mediterranean in later historical periods. Remnants of irrigation systems (rain farms), cisterns and other features from Nabatean and Byzantine periods can still be seen in several places, e.g., in En-Avdat, near the Ben-Gurion University in the Negev in present-day Israel [121–123].

The Arab conquests of the 8th century and the later initiated a great era of agricultural revival, which resulted in an intensification of irrigation practices throughout the Islamic world. Technologically, the civilization of Islam was a synthesizing one, just as the prior Roman Empire. The Arabs may have invented little by themselves, but they preserved, refined, developed, and intensified the technological practices of the ancient world. The spread of Hellenic scientific ideas to the West of the Mediterranean through the Arabic language is well known, and a parallel process can be described in the preservation and extension of ancient Iranian, Babylonian, and Nabataean agricultural techniques [124]. The westward spread of irrigation technology across the Mediterranean followed a pattern that might also serve to illustrate the diffusion of eastern concepts of water distribution and measurement [120].

With the spread of the Islamic Empire westward in the Mediterranean Basin, agricultural and irrigation methods and techniques were brought into conquered areas. The rulers of Al-Andalus (Andalucía, Spain) and many of their followers were of Syrian origin, and the climate, terrain and hydraulic conditions in parts of southern and eastern Spain and North Africa resemble those of Syria. Therefore, it is hardly surprising that the irrigation methods—technical and administrative—in Valencia (Spain) closely resemble the methods applied in the Ghuta of Damascus [120].

There is a unanimous consensus among historians that the present Spanish irrigation systems of Valencia, Murcia and Andalucía are of Muslim origin [125], as well as those in the South of Portugal. In 1960, a celebration was held in Valencia commemorating the "Millennium of the Waters", which expressed public recognition of the establishment of the irrigation system, and specifically of the Tribunal of Waters (Tribunal de les Aigües) during the reign of "Abd al-Rahman III". It should be noticed that the Tribunal of Waters is a UNESCO world immaterial heritage since 2009. Water distribution and irrigation in the Balearic Islands also have Arab origins as is the case for Palma de Mallorca (Madina Mayurqa) where many 'sequias' (siquias or small canals) still remain [126].

As an example, it is worth noting that there are many remains of historical irrigation systems of Muslim heritage still in operation in Spain, as is the case in the watershed of the Poqueira river, in the high Alpujarra, Southeast Spain. The area of the watershed is 9000 hm^2 and its elevation varies from 400 m, where the Poqueira River joins the Guadalfeo River, to 3479 m at the peak Mulhacén, the highest summit in the Iberian Peninsula [127]. The irrigation canals, known as acequias, were excavated; however, the history of the acequias is uncertain and to some extent speculative. Documents from the Early Modern Age improved our understanding of the layout and organization of the acequias network. The longest acequia is the Acequia Nueva (Figure 16), with a total length of 10045 m, a conveying capacity of 459 l s^{-1}, and an irrigable area of 827 ha. The smaller acequia is the Acequia Cachariche, with less than 5 km of length and a conveyance capacity of 243 l s^{-1}, as indicated by [128].

Figure 16. Views of the Acequia Nueva, watershed of the Poqueira River (adapted from [128]).

6.2. The Non-Arab Europe

The Arab armies were halted in their northward expansion in the early 8th century by the army of the Franks, near Poitiers, and later on, they returned to the south and ruled an important part of the Iberian Peninsula for more than seven centuries. While in the Christian Western Europe science and knowledge were mainly sheltered and survived in Monasteries, the rest of Europe entered a time of technical obscurity. For the monastic orders to survive, it was extremely important to have water to establish a monastery and keep it active. The requisite of sufficient water availability was so strict that the monastery should be rebuilt in a different location if the original site did not have sufficient or reliable water supply, or if it was exposed to flooding hazards. For example, the bylaws of the Benedictine order in its Chapter 66 recommended that the monasteries of the Cister (a branch of Benedictines) must have sufficient water supplies to be independent from external resources. Thus, the majority of the monasteries of the order were located in valleys, where usually plenty of water was available and the water level was high enough to allow direct access and use of the resource. Water was so important that the church could even be rebuilt in another location in the precinct in order to avoid interference with water management. An additional requisite was the preservation of buildings from possible flooding, and specific defenses against flood water and drainage channels were usually installed. If needed, riverbeds could be diverted to protect monastery infrastructures [129].

By the end of the medieval period, several Mediterranean states started developing as independent town-states and to guarantee food supplies, they re-developed, recovered or constructed brand-new irrigation and drainage infrastructures to control seasonal floods and cope with droughts. In this respect, great advances were also made. It should be noted however, that in medieval records there is no clear indication of distinction between irrigation systems and drainage; the two purposes were related and, where possible, combined. However, the conditions were not equally favorable everywhere [130].

Thereafter, references in Piedmont about partnerships for the maintenance of irrigation works were found. Shortly after, the first record of permanent water meadows or marcite appears in 1138 in the present-day Italy on the estates of the Cistercian abbey of Chiaravalle near Milan (Figure 17). South of the Po River, in Emilia, where water supply was dependent on the irregular Apennine streams, irrigation was more limited and subjected to state control. However, here too, by 1330, in the territory of Parma, Modena and Bologna, it had begun to encroach on farm and meadowland. With irrigation and drainage, dykes, canals and ditches, medieval farmers and engineers were preparing changes in Northern Italy, far exceeding anything achieved in Etruscan or Roman antiquity [131].

Figure 17. Chiaravalle abbey, near Milan, Italy (photo by M. Salgot).

In the hills and highlands particularly, irrigation was restricted, as it still is today, to certain privileged areas where easy access to markets encouraged limited seasonal irrigation of market-gardens and orchards. Irrigation practices are found of this form in present Italy by the 13th and 14th centuries near several Tuscan towns and on the slopes of coastal Liguria, as well as in the upper Adige valley. In the arid Val d'Aosta there may also have been some irrigation of cereals [131]. In other places of the Mediterranean, where towns were developing an industry (e.g., textile), water was brought from different distances to feed the population, industry and agriculture, sometimes with huge problems of coexistence of the different uses. This was the case, for example, in the medieval Barcelona, Spain.

7. Late Chinese Dynasties

A united country and steady regimes during the Yuan, Ming, and Qing dynasties led to economic development and population growth in China. This encouraged substantial development of irrigation with system planning being more scientific, technologies more mature and project styles more diversified, in addition to facilitating the expansion of irrigation throughout the country.

During this period, the polder concepts and technology in the Taihu Lake basin greatly improved. The polders in the Poyang Lake and Dongting Lake of the Yangtze River basin and Dike enclosure at the Pearl River delta all entered the large-scale development period, which led the Yangtze and Pearl River basins to become economic centers of China. In Northern China, ancient irrigation districts in the Yellow River basin, such as Ningxia and Hetao, further developed. Farmers in semi-arid areas such as Hebe, Shanxi and Shaanxi and other provinces, excavated wells and springs with the purpose of using groundwater. The Ming and Qing dynasties adopted the practice of stationing troops that would cultivate and guard frontier areas as an important state policy for developing border areas and consolidating frontier defense. This made the irrigation in remote areas advance with unprecedented development. Among these irrigation projects, karezs (qanats), mainly distributed along the Turpan Basin and Hami in Sinkiang, were well established irrigation systems. A karez mainly consisted of the shaft, the underground blind drain, the canal and the end pond. Shafts were laid out along the terrain from higher to lower elevations with the depth thereof changing with the ground elevation gradient. A blind drain was developed at the bottom of shafts so that these could be connected to divert water out of the drain. Water flowing from the drain was stored in ponds, and finally delivered to the farmland through canals.

The Sinkiang area was characterized by a dry climate with high evaporation. With water flowing through the ground blind drain, the high evaporation losses associated with general water diversion works could be avoided, so that precious water resources could be effectively spared and subsequently used. In addition, the development of Tien Lake water conservancy project in Yunnan Province made Kuming a beautiful and prosperous city in southwestern China.

8. Irrigation in Modern Times

8.1. The Ottoman Period (ca 16th Century–1923)

As indicated before, during the early Islamic period irrigation was further developed and up-scaled to large schemes fed by long water conveyance canals. This was done by the Abbasid dynasty, which was headquartered in Baghdad (762–1258 AD) [12]. A major expansion of the qanat technology to North Africa, Cyprus, and Sicily also took place during this period.

Afterwards, Turks founded the Ottoman Empire in 1299 near present-day Bursa. The Empire expanded its territory, eventually covering a large part of Europe, the Middle East and North Africa at the end of the 16th century. The Ottomans constructed important engineering works in that part of the world, and numerous bridges and irrigation systems, including dams and canals, can be found even today in Algeria, Syria, Anatolia, and the former Yugoslavia. The particular techniques Ottomans used to build those engineering structures are not known, unfortunately, as they did not report or publish their knowledge so that one could be aware of solutions used for solving different kinds of problems. The early Ottoman dams were used to store water for domestic purposes, and special devices were used for delivering water to users [132].

The Sultan Mehmed II (1451–1481) commanded that urgent repairs be made to the existing water systems. During his reign, a water department was established, underscoring the relevance of water supply to the Ottomans, as it was for earlier civilizations. During the reign of Mehmet II's son Sultan Bayezid II (1481–1512) the Bayezid waterway was built and during that of Bayezid II's son Selim I (1512–1520), various and diverse waterworks were constructed. Aqueducts in the form of arched bridges had been used since Roman times to convey water across valleys and streams, dividing two areas of high grounds so that water did not lose head. During the reign of Süleyman the Magnificent (1520–1566), the former Roman water system, which conveyed water from the Belgrade Forest to İstanbul, was rebuilt with additions and extensions by Mimar Sinan and became known as the Kırkçeşme water system. Preliminary solutions for the water demand problems of İstanbul were handled with the construction of 40 fountains in Fatih Sultan Mehmed (also known as Mehmed the Conqueror) era. This duty was given to Mimar Sinan in the era of Sultan Süleyman. In this period, new water aqueducts were constructed, while repairing works were performed for the existing ones [133,134]. In the period of Sultan Süleyman, not only numerous fountains for each district of Istanbul were constructed, but also many waterways were either constructed or repaired in the Medina and Kudus provinces. This indicates how the Ottomans effectively associated the importance of urbanization with water management [133,134]. Most of the water conveyance structures built in Ottoman times are still in use today [135].

Irrigation was actively performed in the Fertile Crescent (Mesopotamia, Egypt, Jordan, etc.) and other adjacent regions during the Ottoman period. Ottomans developed irrigated agriculture in river basins' areas such as Danube, Nile, Euphrates, Tigris, Sakarya, Red, Yeşilırmak, Çoruh, Seyhan and Ceyhan, which are next to the sea in three continents.

One of the important irrigation projects in Anatolia was constructed during the Ottoman period. There were no large water structures except for the Konia Plain Irrigation (Figure 18) in the Ottoman period [136], and the last Sultan, Abdul Hamid (1876–1909), instigated this important irrigation project. The land to be irrigated begins at Konya and extends southeasterly, east, and west of the railway for a distance of 50–60 km, covering an area of some 500 km^2. This plain lies from 1000 to 1200 m above sea level. The rivers Beysehir and Carsamba convey water from Lake Beysehir to the Konia plain (in the present-day Turkey), where it is delivered into a system of secondary canals through three main supply canals, and then into tertiary canals. By cutting the banks of the last canals, farmers delivered water to these parts of their land for irrigation, and afterwards water could flow off into drains. A considerable amount of water was lost by evaporation, and the rest reached the main drains, which discharged onto low-lying grounds to the northeast and east of the plain [137].

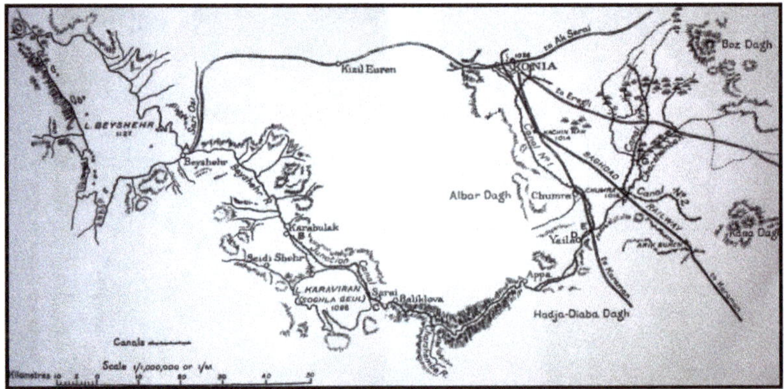

Figure 18. Sketch map illustrating the irrigation network of the Konia Plain (adapted from [137]).

During the Ottoman domination, irrigation projects developed in some areas of the present day Turkey and Egypt, such as Konya and El-Fayyum. The Fayyum lies in a large natural depression in Egypt's western desert, known as the Libyan Desert [138] (Figure 19), and irrigation provides a particularly good lens through which the history of Ottoman Fayyum becomes visible. Irrigation structures also allow understanding the region's relationship to the rest of the Ottoman Empire, because it was a local process, and differed according to each particular village environment, canal, sluice gate, and embankment. Water had to be managed and controlled by individuals on the ground with in-depth knowledge and experience of the local environments. At the same time, irrigation was a process of wide concern [139].

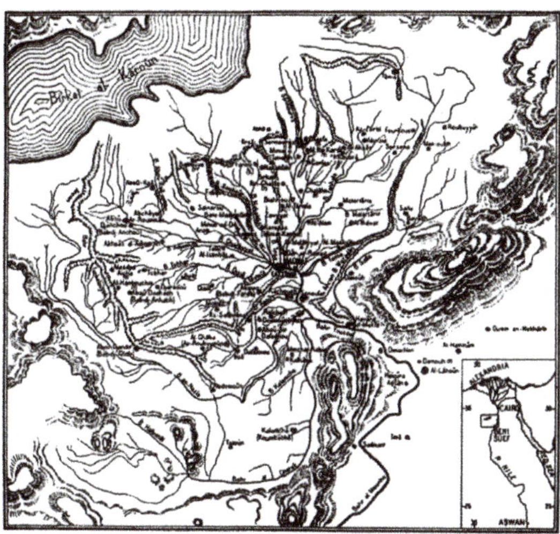

Figure 19. A map of El-Fayyum in Ottoman times (modified from [139]).

El-Fayyum is separated from the Nile Valley to the east by a high ridge of loose stones and soil. The area of El-Fayyum is roughly 1733 km^2, with nearly all of it being suitable for farming and crop production (Figure 20). This ridge is pierced at only one point by a natural opening, and through this inlet a canal conveys all of its water (except scant amounts from rain) to El-Fayyum for irrigation and other purposes. This extremely important waterway is known as Bahr Yusuf and branches off from the Nile at Beni Suef. The canal leaves the Nile Valley at an area known as al-Lahun and enters El-Fayyum later, at a point known as al-Hawara. The two most important irrigation features on Bahr Yusuf were the regulating dike of al-Lahun and the seawall-like dam of al-Gharaq [139,140]. The dike of al-Lahun was built at the narrow gap that allowed Bahr Yusuf enter into El-Fayyum, and thus was crucial for regulating the canal flow and the amount of water entering the region. The dam of al-Gharaq was located farther along the canal and was much larger in surface area than the dike of al-Lahun. Ottoman information sources from the period describe it as a huge dam of impressive stature that had been in existence since ancient times [139].

Thanks to the new/expanded/improved irrigation systems and use of modern agricultural tools, the variety of crops increased during the 19th century [141]. These systems also improved the socio-economic situation of the country.

Figure 20. El-Fayyum Lake (photo by M. Salgot).

8.2. Irrigation Evolution in Present Times (1800 onward)

Although farmland irrigation has been practiced for millennia to increase and secure food supply for an exponentially growing humanity, a vast expansion in irrigated land mainly took place during the 19th and 20th centuries, with irrigated agriculture becoming the principal water consumer in many countries. Some historians referred to the large development of irrigation in these two centuries as the true catalyst for the interaction of engineering, organizational, political and entrepreneurial skills and activities that contributed to food security and produced increasingly wealthier living conditions [142–144].

In terms of worldwide figures, it was estimated that around 1800 the extent of irrigated land was about 8 million hm^2, and it reached 47 million hm^2 around 1900. The four countries with the largest irrigated areas in 1900 were India, China, the USA and Pakistan, respectively [145].

During the 20th century the extent of areas equipped for irrigation doubled by 1945 and doubled again by 1980 [146]. The area equipped for irrigation between 1900 and 2000 at global level, in stacked order for the four major irrigation countries is shown in Figures 21 and 22; whereas the information about growth in irrigated areas by decades from 1950 to 1990 is presented in Table 1 as reported by [147].

However, in some regions such as sub-Saharan Africa, irrigation expansion and advances were limited with respect to the available land and water resources. In this region, many irrigation developments were attempted in the past, and several irrigation projects failed because of a combination of high investment costs, poor planning and lack of maintenance. As a result, until very recently, sub-Saharan Africa continues to have untapped water resources and a large potential for irrigation development [148], although there are various constraints to consider [149].

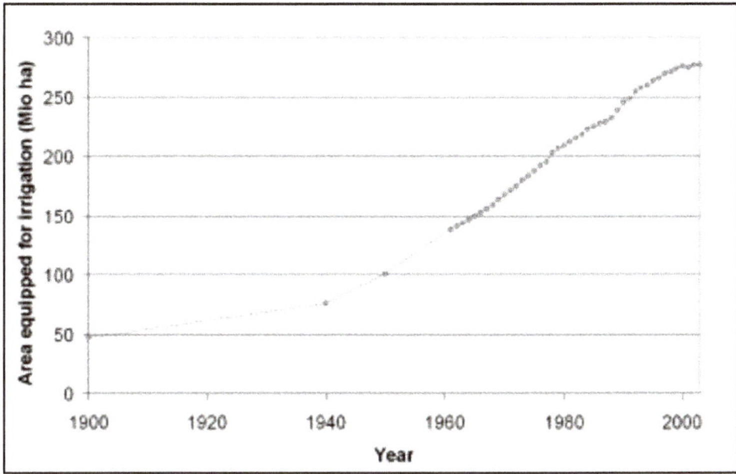

Figure 21. Global extent of the area equipped for irrigation during the period 1900-2000 (adapted from [146,150]).

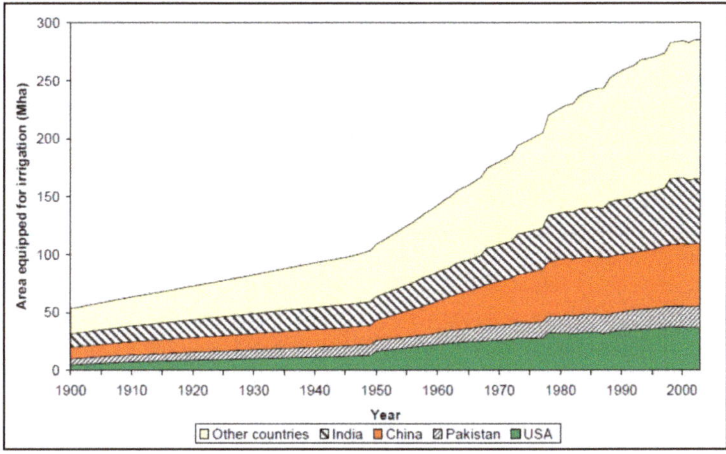

Figure 22. Area equipped for irrigation between 1900–2000 for the four major irrigated countries and the rest of the world (adapted from [145]).

Table 1. Growth in irrigated areas from 1950s to 1990s (adapted from [147]).

Decade	Annual Growth (M Ha)	Annual Rate of Growth (%)
1950s	16.5	4.30%
1960s	13.3	2.40%
1970s	16.8	2.50%
1980s	14.5	1.70%
1990s	12.6	1.30%

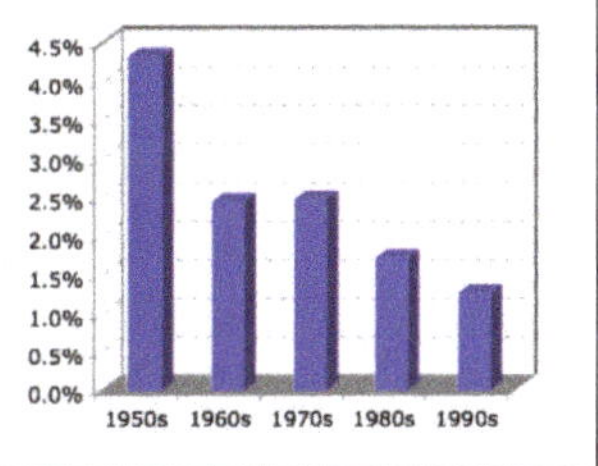

The UN Food and Agriculture Organization (FAO) [150,151] reported that the global area equipped for irrigation worldwide increased from 184 million hm^2 in 1970 to 258 million hm^2 in 1990 and finally reached 324 million hm^2 in 2012. In the 1980s, the rate of increase in irrigated areas slowed considerably [152,153]. Siebert and Doll [154] attributed this slower expansion of irrigated areas after 1980 to the fact that many large-scale irrigation schemes in Eastern Europe and the former Soviet Union (regions characterized at that time by the transition from central planning economies to market economies) went out of operation because water infrastructures were not sufficiently flexible to meet the requirements of new market-oriented private commercial farming models. In other irrigated areas, the lack of adequate drainage infrastructures caused water-logging and salinity problems that impaired the productivity of irrigation schemes and created urgent needs for costly rehabilitation. At the same time, limited water resources, increasing competition for water by other sectors, and environmental regulations strongly restrained irrigation expansion in many arid areas. For the period after 2000, the World Water Resources Report [146] indicates that in a number of developing and developed countries the extent of irrigated land has stabilized or even diminished due to the very high cost of irrigation networks, salinity-induced problems, depletion of water-supply sources, and rising public concerns on environmental protection.

The rapid irrigation expansion during the 19th and 20th centuries can be related to several factors. Sojka et al. [155] pointed out that, while the main physical relationships related to water flows, i.e., among mass, energy and turbulence, were well investigated and mastered at remarkably high levels of proficiency in the ancient cultures, the understanding of physical and chemical soil–water interactions was somehow inadequate until the beginning of 19th century. In ancient irrigation development, the combination of soils, climate, water quantity and quality knowledge were more established at some locations than others. At some schemes, in fact, irrigation has continued to the present day without skilled or sophisticated management, either because seasonal rains provided sufficient leaching, or because soils were sufficiently permeable and well drained to prevent water-logging and salinity accumulation. In other cases, irrigation water had favorable chemical composition that avoided the occurrence of land-impairing situations. In other areas, increased soil salinity and/or sodicity, as well as raised water tables, have either limited the functionality of irrigation schemes or adversely affected land productivity. The success or failure of irrigation schemes, as well as their productivity and sustainability, became tightly dependent on skilled design and management, as populations grew and the need for increased food supplies encouraged irrigation development on more marginal areas with less productive soils, poorer drainage, and greater natural or induced salinity and sodicity problems. In turn, these aspects required the knowledgeable application and adaptation of scientific principles that started being developed and consolidated from the mid-19th century onwards.

Then, a conjunction of progress and learning in several scientific disciplines, including soil science, hydrogeology, chemistry, physical-chemistry, physics, and plant physiology occurred and contributed to foster irrigation development. These disciplines were adapted, blended and applied in the sub-disciplines of soil chemistry, soil physics, soil biology, crop physiology and agronomy, whose fundamentals proved being essential knowledge for irrigation design, construction and operation, as well as for their economic, social, and environmentally sustainable management.

Similar patterns occurred both in the old and new continents. For instance, from the turn of 19th century onwards, Southern Europe experienced a vigorous upturn and expansion of irrigation, which was mainly fostered by an upcoming industrial economy, high demographic pressure and adapting agriculture, as well as by the availability of affordable energy. Leibundgut and Kohn [156] report that specific new laws for field irrigation and grassland (meadow) cultivation frequently encouraged the modernization of existing and the implementation of new irrigation systems, as well as the foundation of large irrigation cooperatives/associations. In comparison to more traditional irrigation schemes developed in historical periods (e.g., middle ages), the newly developed irrigated areas of 19th and 20th centuries were usually large-scale upstream-controlled systems laid out by engineers. Such systems were characterized by gravity-fed designs, whose implementations required labor- and capital-intensive works to reshape and modify valley-floor plains, as opposed to earlier projects largely adapted to natural landscapes and land contours. Official regional statistics and land register records begun to be available from the 19th century onwards, which provide quantitative information about the overall irrigated area [157].

Leibundgut and Kohn [156] indicate that, in terms of spatial extension and geographical distribution, irrigation in Europe probably reached its peak around the turn of the 20th century. Modern irrigation developments were launched in Spain soon after the end of the Civil War in 1939, based on a national water development plan designed by the government prior to the war. Such developments, originally established to settle small farmers, constituted the foundation of a vibrant irrigated agriculture today, which expanded to 3.7 million hm^2 and is the largest irrigated area in the European Union.

In the USA, modern irrigation developments are reported to have probably begun with the Mormon settlement of the Utah Great Salt Lake Basin in 1847, followed by the cultivation of nearly 2.5 million irrigated hm^2 across the inter-mountain western USA by 1900. America's Mormon pioneers chose to settle in a remote salt-impaired desert habitat, and thus were forced to use trial and error and application of all available new knowledge to reclaim lands from the desert and practice sustainable irrigated crop husbandry, as described by [155]. According to Reisner [158], the Mormon pioneers were very successful in their efforts to the point that the practices they followed in reclaiming, managing, and irrigating arid and salt-affected lands provided the main guiding principles for irrigation developments that occurred throughout the western USA before and under the Reclamation Act since 1902. The passage of the Desert Land Act of 1877 and the Carey Act of 1894 spurred irrigation developments further in the western USA, providing the legal framework to land acquisition for settlement and support of governmental infrastructure for development.

Bucks et al. [145] refer that the lessons learned in the settling of the American West from 1847 onward provided many practical modern principles and methodological approaches of irrigation systems design and operation, and irrigated soil management until the end of World War II, when the total U.S. irrigated area had grown to 7.5 million hm^2. Concurrently, the innovations and designs, developed in this context, had worldwide importance. Irrigated agriculture remained an engine of the western U.S. development until the 1970s [159].

After World War II, another wave of rapid irrigation expansion occurred worldwide, owing to a fast population growth and to the increasing demand for safe food supplies. For the largest part, this population expansion resulted from the progresses obtained in public health and the successful control of malaria and other insect-borne diseases in many regions, which in turn significantly increased life expectancies. In addition, the first and the second World Wars spurred advances in technology that were applied to many production areas, including agriculture. Among those, electrical, steam and internal combustion power sources became available to lift and pressurize water. New pump designs and the patenting of sprinkler delivery systems came together in a few decades between and immediately following the wars and revolutionized the ability to withdraw, convey and deliver water. Before it, with the invention of internal combustion engines and shallow well pumps, irrigation largely expanded beyond riparian and gravity flow service areas.

Available records show that irrigation gained even more prominence under the Green Revolution of the second half of the 20th century. The Green Revolution was initiated to address the issue of malnutrition in the developing world, and combined plant genetic improvements with agronomy (greater use of fertilizers, pesticides, and irrigation) to increase crop yields. In several parts of the world, the availability and use of irrigation water represented a key factor for the positive results accomplished by the Green Revolution, such as in the Indian sub-continent. The UN-FAO [160] estimated that in 2003 the total world-wide irrigated area was 277 million hm^2, corresponding to about 18% of the total cultivated lands (Table 2), with the largest proportion of this area (70%) being in Asia, as shown in Figure 23.

Table 2. Area equipped for irrigation and percentage of cultivated land for selected years (adapted from [160]).

Region	Irrigated Area (Million hm^2)			Irrigated Area as % of Cultivated Land		
Year	**1980**	**1990**	**2003**	**1980**	**1990**	**2003**
World	193	224.2	277.1	15.8	17.3	17.9
Africa	9.5	11.2	13.4	5.1	5.7	5.9
Asia	132.4	155	193.9	28.9	30.5	34
Latin America	12.7	15.5	17.3	9.4	10.9	11.1
Caribbean	1.1	1.3	1.3	16.4	17.9	18.2
North America	21.2	21.6	23.2	8.6	8.8	9.9
Oceania	1.7	2.1	2.8	3.4	4	5.4
Europe	14.5	17.4	25.2	10.3	12.6	8.4

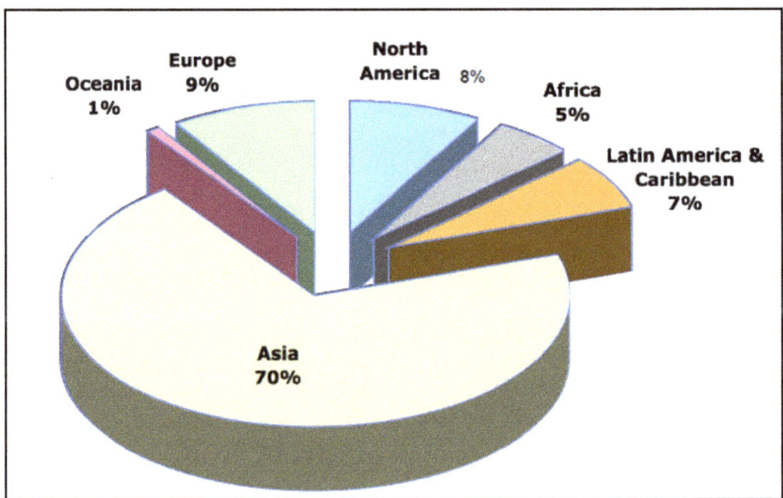

Figure 23. Distribution of irrigated surface in the World by major geographic regions (adapted from [160]).

The introduction of advanced western hydraulic engineering technologies in China and other Asian countries between the end of the 19th century and the beginning of the 21st century gradually encouraged the reconstruction and expansion of traditional irrigation districts and the construction of modern irrigation projects. A group of intellectuals appeared that had been exposed to both western education and oriental cultural influence, who became the promoters of hydraulic reforms during this period. Among all these practices, eight irrigation canals in the Central Shaanxi Plain, including Jinghui and Luohui canal and others, were built by Li Yizhi, who was the major representative of these promoters. Since the founding of the People's Republic of China in 1949, irrigation in China has been developing in full scale, improving significantly the ability to resist drought and flood disasters in agriculture.

From the 1970s onwards, increasing development costs, reducing governmental support and financing, rising demand for municipal and industrial water supplies, diminishing sources of fresh water supply, and a growing concern for the environment have forced water managers and planners to begin rethinking traditional approaches to agricultural water management [161]. Most of projects developed around the world during the 1980s were fueled by financial resources from national or international agencies and were largely expected to result in national self-sufficiency in producing staple foods. As a result, these irrigation schemes tended to be centrally organized and operated by state authorities (government-based irrigation systems), which often also directed input supply and cropping patterns. Several of these projects failed to meet expectations, as farmers had very little opportunities to get involved in management and decision-making, and felt no incentive and commitment to proper water use, system maintenance, forced crop selection, and high-productivity. A second generation of agency-funded projects encouraged farmers assuming a sense or responsible ownership of irrigation schemes as major key for success. This objective was pursued through the self-organization of water users, who could understand the economic potential of farmer-managed irrigation systems and thus assume full responsibility. Funding agencies therefore fueled irrigation expansion through development of farmers-managed irrigation schemes and through irrigation management transfer (IMT) programs, i.e., transferring responsibility, decision-making and financing of irrigation systems from public sector to water users' organizations. In this context, one of the main funding agencies' goal was to ensure that irrigation systems would provide adequate, flexible and more dynamic water delivery services to meet the needs of market-oriented agriculture.

9. Main Agricultural Water Management Challenges and Future Trends

9.1. Present Day Challenges in Irrigated Agriculture

Irrigation has been playing a crucial role in the economic development of many countries that extends far beyond the production of food and fiber. In several areas of the world, it made the settlement and establishment of active communities possible, while transforming lands with little or no apparent economic value into highly productive and market-oriented farming systems. An often-overlooked key benefit of irrigation, yet not easy to quantify, is food security, i.e., the security derived from stability of food production. Weather vagaries and seasonal rainfall fluctuations could make rain-fed farming a risky venture, whereas irrigation reduces some of the uncertainties and promotes increased and stable crop production [121]. According to the Second United Nations World Water Development Report [162], there is a strong positive link between investments in irrigation, poverty alleviation and food security. Irrigation allows minimizing abrupt and unpredictable yield reductions that many world's food production areas would face due to recurring or sporadic droughts.

Presently, more than 60% of the cereal production worldwide and 50% of the value of all crops harvested come from irrigated agriculture. Irrigated land produces on average two to two and half times the yield and three times the crop value per hectare relative to non-irrigated farmland, with the irrigation portion amounting to only one sixth of the world's total production area, including cropland, rangeland, and pasture [145,150,163–165]. However, the success of irrigation projects has been often achieved through strong governmental involvement and support (both in terms of engineering and financial assistance), and has been fueled by economics and market patterns. In the last two decades, government involvement and support for irrigation development have greatly diminished, and the role of agriculture has changed in the economy of many developed countries, with some of the nations' priorities shifted towards increased environmental concerns and the need to compete in the global economy.

The world community has become aware that irrigation causes serious land transformations, that are often permanent and irreversible and take place mainly in two ways: (a) by direct modifications of the land surface, occurring when water conveyance and distribution networks are constructed, and lands are cleared, shaped and leveled for farming and irrigation; and (b) by in-depth transformations taking place when the water and salt balances in the region are changed as a consequence of importing large quantities of water and salts into the area [121]. As such, it is now recognized that irrigation development is a dynamic process disturbing the natural water and salt balance of a region, and eventually causing irreversible damages to the local ecosystems, from the projects' onset to the medium and long term. Among the major adverse ecological effects, the diversion and storage of water for irrigation profoundly alter the natural hydrology of streams and the habitats of native plants and animals depending on them. Moreover, the application of large quantities of water to irrigated lands may result in soil erosion and sedimentation of streambeds. At the same time, salts, organic matter, solids of several sizes, fertilizers and pesticides may be leached out from the soil and transported into rivers, streams or aquifers, alongside with microorganisms and several pathogens.

Several surveys have indicated that of the existing irrigated lands, some 40–50 million hm^2 show measurable degradation from water logging, salinization, and sodification (e.g., [166,167]) leading often to desertification. All these adverse environmental and ecological consequences of irrigation, i.e., water diversion and consumption, impacts on water quality, effects on aquatic ecosystems, are increasingly being scrutinized and questioned [161], and there are calls for intensified research and wide adoption of conservation and resource-efficient water management practices. The challenge of soil losses due to combinations of these negative factors must also be addressed.

Irrigation developments have also several extended positive impacts, including social and economic benefits such as flood control, transportation and recreation, as well as hydropower, creation of employment opportunities and rural development. At the same time there are positive ecological effects resulting from transformation of drylands into green areas that produce large amounts of oxygen and fix carbon dioxide and accommodate through fields, ditches, canals and reservoirs, the habitats where a wide range of wildlife thrive, as opposed to deserts and undeveloped natural arid lands. Nevertheless, part of the mentioned impacts can be considered negative from the ecological point of view, since they alter or even destroy natural ecosystems (arid and semiarid).

Sojka et al. [155] highlighted that irrigated agriculture has greatly enhanced its ability to provide humanity's essential needs in closer harmony with environmental balances. There are opposite opinions indicating that the natural characteristics of deserts are being lost through irrigation, including specific flora and fauna. The case of the Segarra-Garrigues canal in Lleida province, Spain is paradigmatic, with the EU having forced to reduce the irrigable area in order to protect and conserve steppes' birds [168].

According to data from major international agencies, population growth is occurring mostly in developing countries, where there are added expectations of improved diet and standards of living that raise the need for increased food production per capita above a simple linear extrapolation based on population projections [155]. In fact, only intensive and high-yielding crop production from irrigated agriculture has the potential to achieve these projected performance targets. At the same time, the production of biofuels represents an additional demand on water resources, and also competes with food for limited water and land [169].

Besides meeting the increasing needs for food and energy (biofuels), the main challenge faced by irrigation planners is no longer the high agricultural productivity only, but rather the long-term economic and environmental sustainability of irrigated areas. The potential hazards intrinsic to irrigation developments need to be carefully evaluated at the planning stage, and minimized during the operation phase, if the stability and permanency of irrigation projects are to be ensured. Several authors [166,170,171] report that the adequate knowledge and technology exist to design, implement and operate irrigated cropping systems in an environmentally compatible way, and thus prevent progressive and sometimes irreversible land and water resources impairments. However, in most cases, when designing and developing a large-scale irrigation project there is a great political emphasis and pressure to develop as many irrigated hectares as possible at the project outset. As a result, often not sufficient financial and technical resources are allocated to implement known scientific principles and technologies in a timely manner, as part of irrigation systems design and management. Also, in many projects there is not sufficient provision of technical and social support to assist the farming community with the transition from rain-fed or no agriculture to irrigated agriculture. Often, this insufficient assistance is perceived at later stages as failures of irrigation, but in reality, those are failures in the human institutions and project governance, as pointed out by Sojka et al. [155]. In this regard, a real challenge for irrigation expansion in the near future is reaching consensus around social, political, economic and institutional considerations rather than only promoting technical advances on water availability and use.

Overall, the availability of suitable land and water is likely to remain the principal determinant of the extent of irrigation and its expansion at the global scale. Water-related costs are rising and demands on the available water resources are growing rapidly in many areas of the world, generating increasing competition for fresh water supplies, and both trends are very likely to continue in the future. Irrigated agriculture is currently, and will most likely in the future, be particularly subject and vulnerable to water supply fluctuations. Alongside, irrigation methods, systems and management practices will continue evolving towards advanced technologies to provide better water control, communication and record-keeping, improve water and nutrient management, enhance soil and water resource use efficiency, and minimize their degradation.

All the above-indicated factors call for inevitable changes in how water is managed, allocated and valued in the future. Among others, the major key change drivers are population growth, urban expansion and resulting loss of agricultural acreage, increasing competition over fresh water supplies, changing of policy objectives, and increasing environmental concerns. The possible response to those drivers may be developments in science and technology and changes in management practices, as well as governance, institution, and policy reforms.

In terms of science and technology, in the last decade efforts have shifted away from construction of large irrigation infrastructures, such as dams, reservoirs, and large-scale water supply and delivery systems (Figure 24), and more attention is being given to improve on-farm irrigation systems and management practices to reduce the total amount of water diverted from various sources. There is a clear trend, fostered by financial incentives and technical support programs, towards broader adoption of micro-irrigation methods that apply water at slow rate in the vicinity of plant roots, thus enabling improved water and nutrient crop management and efficiencies. Research institutions are also working to breed and test new plant varieties that are drought tolerant, drought resilient, or better adapted to water stress. However, substantial water savings from irrigation technologies and genetic engineering do not appear imminent, although large efforts are being made to develop durable techniques and strategies. In terms of irrigation technology, it is tempting, and has been so for decades, to address the need for better on-farm irrigation efficiency only with technological interventions rather than by application of better management techniques [172].

(**a**) (**b**)

Figure 24. Aswan dam in Egypt: (**a**) view of today; and (**b**) past view (Photos by M. Salgot).

Innovative irrigation technology is generally promoted as enhanced water-use efficiency along with multiple other benefits, but often these remain elusive in practice. In many cases, individual farmers and/or their organizations made significant investments in irrigation technologies, but their implementation has not been systematically evaluated for effectiveness through better knowledge regarding soil–water–plant relationships. As a result, frequent investments in technological improvements have been causing higher water prices without gaining the full potential benefits due to increased water efficiency, as indicated by Levidow et al. [173]. Under such circumstances, irrigated agriculture will likely maintain relatively low water-efficiency levels, and farmers will have no clear incentives to improve, and most likely will make no efforts towards implementing more efficient practices. Reversing these tendencies will require continuous knowledge-exchange and knowledge-delivery programs, and that all relevant stakeholders share greater responsibility and benefits for increasing the efficiency across the entire water-supply chain. From this perspective, it will be crucial that more water-efficient management practices could also demonstrate how to combine wider environmental benefits with economic gains for farmers, without forgetting the possible negative impacts of new practices, i.e., pursuing eco-efficient water management practices.

From an institutional perspective, development agencies and local governments have also shifted the focus from policies for developing new irrigation schemes to better management and modernization of existing schemes, and towards increased environmental protection, while strongly reducing subsidies. In some cases, water-management agencies are also developing policies to implement water transfer programs among different water-use sectors.

9.2. Future Directions Anticipated in Irrigated Agriculture

With this historical review and modern perspective of current pressures and drivers on irrigated agriculture, we offer a few likely directions anticipated for the future.

Irrigation will continue playing a crucial role in the production of safe food and of biofuel supplies, most likely through agriculture intensification. The total irrigated area will likely remain similar or even decline, but the production from irrigated agriculture will remain constant or slightly increase, thanks to conversion of significant acreage to higher-value crops. In this context, the attention will be mainly given to irrigation performance and the economic efficiency of water use, i.e., water productivity.

Increasing pressure on fresh water supplies from different sectors will put agricultural water use under greater scrutiny and lead to possible decline of water allocations to irrigated agriculture, with increasing fractions transferred to municipal, recreational and environmental uses, as well as to natural landscapes. This will be particularly so in areas under water scarcity and during drought periods. Higher water tariffs, tiered pricing schemes, and water markets will likely be implemented to encourage more efficient agricultural water usage.

The irrigation sector will most likely be expected to comply with higher standards of efficiency as a condition of use, to be monitored through third-party auditing and certification processes. Tiered water pricing, new regulations and some sort of incentives to growers will have to be implemented to achieve higher irrigation efficiency, and support voluntary water transfer programs. This will also encourage research on the use of drought-tolerant varieties and on reuse of water from return flows.

Competition in the global economy, in conjunction with reduced protection to farming activities, will lead growers to deal with more risks and uncertainties. Among those, market conjunctures, fluctuations of crop prices, weather-related changes, and the cost of water and energy supplies, will all make farmers more vulnerable to vagaries and uncertainties. Minimizing some of these risks will entail relocating the production of high-value crops to areas with more dependable water supplies. Successful farmers will adapt to increased uncertainties through innovations in technology, resource-efficient practices, and cropping systems that are more resilient to climate variability and weather extremes.

In several large-scale irrigated areas, there will be a shift towards large-sized well-financed, diversified, and technology-oriented farming operations, run mostly by corporations. Smaller under-financed farms will tend to decline, due to less-skilled personnel and higher vulnerability to risks. Medium-sized farms will instead face hard challenges to stay competitive in the increasingly global economic context.

The increasing pressure for greater efficiency of water use will require research efforts for development and transfer of new technologies and water management practices, from both the public and private sectors (irrigation industry). Irrigation districts and water users' organizations will also have to engage more in testing and demonstrating the cost-effectiveness of new technologies, as well as in supporting education of irrigators to the correct deployment and best use of such innovations.

In the future, the major issue in the water supply will be related to the long-term availability and sustainability of existing and new water supply sources. Factors that will impact public water supplies are population growth, population demographics, limited availability of new natural sources of water, distribution of existing fresh water sources, increased contamination of these water sources with trace organics and nanoparticles, ageing infrastructures, and climate change [174]. In the next 30 years it is estimated that the world's population will increase from 7.3 billion today to 9.7 billion by 2050. Also by 2030, roughly 60% of the world's population will be living in urban areas. At the same time, it is anticipated that by 2030, a fraction of that 60% of the world's population will live near coastal regions, creating even more urban sprawl than already exists [175]. Based on the facts that the available fresh water globally will remain the same, combined with unequal distribution of the world's water resources in many parts of the world, large cities are already water stressed making it necessary to look for non-conventional water resources, e.g., treated wastewater, brackish water and desalinated waters [176,177]. In this context, wastewater reclamation and reuse is becoming increasingly necessary to reach equilibrium between supply and demand [178].

10. Epilogue

Throughout this review, our contributions clearly demonstrate that water and its proper management has always been a key element for food security. In addition, because of this need, the development of technologies and methodologies to use water efficiently has always been a concern, probably more linked to "sustainable" goals in the past, when nature was considered as part of the environment, than in the current time that are more dominated by economic interests. Nevertheless, an increasing awareness to maintain the planet's natural features is counteracting the economic objectives and then influencing irrigation patterns. A brief timeline of historical developments of agricultural irrigation globally is presented in Table 3.

Table 3. A Brief timeline for historical development of agricultural irrigation.

Period (ca)	Achievements	Comments
5000–2500 BC	The first confirmed evidence of habitation and the first farmers. The first successful effort in developing agricultural irrigation.	In Mesopotamia
5000–2200 BC	Emergence of the primitive irrigation and drainage engineering, and emergence of wells for irrigation.	In ancient China
ca. 3100 BC	Egyptians practiced basin irrigation using the loading of the Nile to inundate land plots, which had been surrounded by dykes.	In ancient Egypt
3050–2050 BC	Early irrigation systems in Egypt and Mesopotamia	
3000–1600 BC	Ancient Nubians developed a form of irrigation by using waterwheel-like device called sakia.	In ancient Sudan
2500–1450 BC	Irrigated and drainage agricultural lands in the eastern Crete. Major irrigated crops were olives and grapes	Minoan settlements (e.g., Choiromandres, Zakros, Agia Triada, Messara valley)
2500–1900 BC	Irrigation of agricultural lands.	In Mohenjo-daro and other centers located in modern Pakistan.
2200–771 BC	Emergence of farmland canal systems.	In ancient China
2000–1000 BC	Cultures and civilizations in the Niger river region practiced irrigation based on wet season flooding and water harvesting in the first or second millennium BC.	In sub-Saharan Africa

Table 3. *Cont.*

Period (ca)	Achievements	Comments
1500–800 BC	The realization of the importance of irrigation is evident already from the myths of ancient cultures. Irrigation of Kopais valley in the central Greece.	In various Mycenaean cities (in south Greece)
800–500 BC	Development of irrigation and drainage technology for field cultivation and gardening.	Mainland Greece, Western Greece and islands in the Aegean Sea
ca. 800 BC	The Qanat technology was developed which is among the oldest known irrigation methods and still use today in several parts of the world.	In Persia and other places
ca. 300 BC	The irrigation works in the reign of King Pandukabhaya and which were under continuous development for the following thousand years, were one of the most complex irrigation systems of the ancient world.	In ancient Sri Lanka
256 BC	In the Szechwan region, belonging to the State of Qin, the Dujiangyan irrigation system was built to irrigate an enormous area of farmland that still supplies the water.	In ancient China
246 BC	In the central Shaanxi Plain belonging to the State of Qin, the Zhengguo Canal irrigation system was built to irrigate an enormous area of farmland that today still supplies the water.	In ancient China
167 BC–330 AD	Romans invented the Roman concrete (opus caementitium) which allowed the construction of long canals, very large bridges and long tunnels in soft rock.	In Roman period
581–1279 AD	The pond-canal polder system in the Taihu Lake basin and projects for resisting seawater intrusion and storing freshwater in the southeast coast became the most representative water conservancy of this period.	In ancient China
1200–1500 AD	Use of Chinampas which is a crop growing method used by the Aztec civilization consisting of small manmade islands placed on lakes in marshes with no need for irrigation and optimizing the use of agricultural land, water and waste to raise yields	In Mexico
1279–1911 AD	The polders in the Poyang Lake and Dongting Lake of the Yangtze river basin and dike enclosure at the Pearl river Delta both entered the large-scale development period, that made the Yangtze river basin and the Pearl river basin become economic centers of China.	In ancient China

Table 3. *Cont.*

Period (ca)	Achievements	Comments
1299–1923 AD	Most of the water transmission lines built in Ottoman times. Irrigation activity was densely performed in Fertile Crescent region (Mesopotamia, Egypt, Jordan etc.) during Ottoman period	In Anatolia and neighboring regions
1900–Today	It was estimated that around 1800 the extent of irrigated land was about 8 M ha, and it reached 47 M ha million hm2 around 1900 with water consumption for irrigation 500 billion m3 per year. The top four countries with the largest area equipped for irrigation in 1900 were India, China, Pakistan and USA. Total water demand is expected to increase from 4000 today to 5500 billion m3/year in 2050 with irrigation to be the major use.	In the entire World

Irrigation has been implemented in Egypt and Mesopotamia since at least 5000 BC, where water of the flooding from Nile, Tigris, and Euphrates rivers was diverted to the agricultural fields for a couple of months during summer and fall. The excess water was then drained back into the rivers before the start of the crop growing cycle.

Chinese irrigation has a long history and a sustainable development. The origin of Chinese irrigation is very early, e.g., well water for irrigation emerged *ca* 6000 years ago; well-field irrigation system built in the Xia, Shang and Zhou dynasties, and large irrigation projects such as Dujiangy weir and Zhengguo Canal appeared 2000 years ago. Hereafter, due to long lasting and relative "isolation" from other civilizations, Chinese irrigation has always been developing according to its own rules, with its technologies continuously improving.

Secondly, China is vast in territory with diverse natural conditions, (such as topography and precipitation) and requirements of social and economic development, which vary enormously in different regions. These differences resulted in the development of a large number of irrigation projects, which featured various styles.

Thirdly, there is a close relationship between the development of Chinese irrigation and the shift of Chinese economic centers inside the country. The first economic center of China emerged in the lower reaches of the Yellow River, where irrigation was more advanced. With the spread of irrigation technologies, the Yangtze River basin became the economic center, followed then by the Pearl River basin. Even more, the development of these economic centers effectively facilitated the development of local irrigation. Historically, China lead to unique irrigation systems, which made Chinese irrigation become an indispensable and irreplaceable part of world's irrigation projects heritage.

On the contrary to most ancient civilizations, which developed and flourished near rivers and lakes where water supply for drinking and agricultural use was readily available, the Greeks established communities in arid and semi-arid regions since the early times [35]. Thus, the Greeks developed sophisticate irrigation practices and general water management practices due to the water scarcity during the Minoan Era and onwards. Ancient Greek water management techniques displayed a variety of strategies, concerns and, in some instances, ingenuity at a level which inspires to re-contemplate its potential in other contexts. Most obviously, Greek gardening displayed a concern with the multifarious engagement of irrigation, manure, manpower and intensive cultivation of confined spaces. It is borne out of the evidence that the combination of manuring and labor-intensive cultivation with irrigation induced by control and storage of seasonal rainfall can provide a highly effective concept for intensive gardening.

Less relevant may be the large-scale irrigation projects of the Classical and Hellenistic periods (*ca* 500–100 BC), which both had the effect of re-molding the landscape, changing it from rugged, marginal lands into fertile soils of the plains, and providing abundant water to regular field and tree crops cultivation. The few extant examples of large-scale irrigation projects to survive in the material record suggest that irrigation strategies undoubtedly helped addressing highly complex local conditions and problems. Again, however, it may be relevant to contemplate scenarios where erratic seasonal changes of rainfall in semi-dry regions compelled farmers to both control and store rainfall in order to implement larger combined drainage and irrigation strategies.

Terrace irrigation is documented in pre-Columbian America, early Syria, India, and China, as well as in the Mediterranean Basin. In the Zana Valley of the Andes mountains in Perú, archaeologists found remains of three irrigation canals that were radiocarbon-dated from ca. the 4th millennium BC, the 3rd millennium BC and the 9th century AD, respectively [179]. These canals are the earliest records of irrigation in the New World; traces of a canal possibly dating from the fifth millennium BC were found under that fourth millennium canal [180]. Sophisticated irrigation and storage systems were developed by the Indus Valley civilization in present-day Pakistan and North India [181]. In those places, large-scale agriculture was practiced and an extensive network of canals was used for irrigation purposes.

The Aztec culture, although not as ancient as some of the other cultures mentioned, showed a full understanding of the benefits of managing water jointly with land and biodiversity. Certainly, this was because the Aztecs had to face the lack of arable land combined with the presence of water resources that were partly saline. Unfortunately, with the arrival of a new culture, i.e., the Spanish, this philosophy on the integrated management of water and other natural resources was almost lost. Nevertheless, it is now being "re-discovered" by some experts for the aim to deal with similar problems that are affecting many other regions of the world.

While the Aztec water management system was almost lost as indicated, other Central and South American cultures maintained traditional ways of farming the land, and indeed the modern Mayan and Andean people continue practices that reverberate the ancient techniques. Ancient Andean irrigation canals continue to be functional in the contemporary world. Indeed, indigenous Andean farmers still use a similar system to allot water to farms throughout the river valleys [117]. Thus, an ancient Andean system that was institutionalized by the Inca and Spanish imperial rule resonates today in the daily lives of modern Andean farmers.

Indigenous Central American farming techniques in particular have important implications for modern farmers living in tropical areas. Many farmers throughout Southern Mexico, Belize, and Guatemala practice traditional farming methods, planting rain-fed milpas, and produce enough food to feed their families and sell part of the crops on the market. These practices are more sustainable, as they seem better suited to the local environmental conditions than modern Western agricultural practices. Many farmers shun the hybrid corn varieties planted by growers in the United States and Canada, for example, preferring local corn varieties that are better adapted to the seasonal rain and the local traditions of implementing irrigation strategies.

Presently, irrigation is being considered as a way to improve standards of living for farmers, which rely more and more on highly technified ways of applying water to the soil-plant system. The water-energy-food nexus is now guiding brand new approaches to make irrigation a sustainable practice from various viewpoints. The most important is perhaps the capacity of ensuring sufficient and healthy food supply for an ever-growing world population, i.e., food security and safety. Water for irrigation is a basic commodity but it is not to forget that, without soils in good productive conditions, agriculture would not thrive even if there is enough water to grow plants for food production.

Finally, the lessons learned from the past and the existing concerns point to several future main water-related challenges and trends, as indicated hereafter:

- water supply will be limited and regulated for all users, with water rights and water pricing schemes to be re-visited and adapted to changing conditions;

- larger quotas of water will be transferred from agriculture to urban areas, recreational activities and the environment;
- increased climatic variability will exacerbate problems of water supply and expose farmers to increasing vagaries and uncertainties;
- degradation of land and water resources may become more substantial and as well draw increasing public attention;
- more resource-efficient management practices will be needed at farm level that could also combine wider environmental benefits with economic gains for farmers;
- cropping systems implemented in irrigated agriculture in the arid and semiarid areas will need to adapt to less water availability and to lower-quality supplies;
- new varieties and cultivars of plants more adapted to drier conditions are and will be essential for a better use of water in agriculture; and
- the knowledge of the relationships of water and nutrients with the different types of soils will be also essential for proper agricultural water management.

Lessons from the past must not be forgotten, as they will help avoiding a broad range of mistakes perpetrated nearly consistently along the millennia.

Author Contributions: A.N.A. conceived the idea of this review article, contributed to the project idea development, prepared the first draft of the manuscript (writing the original draft, reviewing and editing, collecting and preparing the figures), collected the information for developing the body of text and supervised the research. D.Z. wrote the last three sections (Irrigation in the modern times, Main agricultural water management challenges and future trends, and Epilogue). He also revised and edited the manuscript. J.K. mainly contributed to the Section 3 on Historical Times, but also revised and edited the manuscript. M.S. provided contribution to the Section 6 on Medieval Times and revised the manuscript. M.B. contributed to the Sections 1 and 2 on Prolegomena and Prehistoric Times, respectively. P.R. has contributed to the Section 3 on Historical Times. B.J. elaborated the Section 3 on Late Pre-Columbia Civilizations. A.K. provided contribution to Section 2 on Prehistoric Times. W.Y. has mainly contributed to the narrative about irrigation in China. A.B. contributed to the Section 2 on Prehistoric Times. J.A.H. provided contributions to Section 2.4 on the Maya Civilization. A.G.-J. has mainly contributed to the Section 3 on Late Pre-Columbia Civilizations. E.F. has revised and edited the manuscript, providing contributions to various sections. All authors have read and agreed to the published version of the manuscript.

Funding: This research received no external funding.

Acknowledgments: We greatly acknowledge Raffaella Sibilla, Renata Teixeira de Almeida Minhoni and Giuseppe Inchingolo, all collaborators in the Agricultural Water Management Lab of the LAWR Department of UC Davis, for their valuable contributions in editing and reviewing multiple sections of the manuscripts. The authors are also very grateful to Prof. Mark Grismer from the LAWR and BAE Departments of UC Davis for conducting a detailed review of the final version of the manuscript.

Conflicts of Interest: The authors declare no conflict of interest.

References

1. Breasted, J.H. *Ancient Times: A History of the Early World*; The Athenaeum Press: Boston, MA, USA, 1916.
2. Koutsoyiannis, D.; Angelakis, A.N. Agricultural Hydraulic Works in Ancient Hellas. In *The Encyclopaedia of Water Science*; Stewart, B.A., Howell, T., Eds.; Markel Dekker Inc.: New York, NY, USA, 2004; pp. 1–4.
3. Sallares, R. *The Ecology of the Ancient Greek World*; Cornell University Press: New York, NY, USA, 1991.
4. Lebesi, A. *The Miniature House from Archanes*; Archaiologiki Ephemeris: Athens, Greece, 1965; pp. 12–43.
5. Angelakis, A.N.; De Feo, G.; Laureano, P.; Zourou, A. Minoan and Etruscan hydro-technologies. *Water* **2013**, *5*, 972–987. [CrossRef]
6. Luthin, J.N. *Drainage of Agricultural Lands*; American Society of Agronomy: Madison, WI, USA, 1957.
7. Krasilnikoff, J.A. Irrigation as innovation in ancient Greek agriculture. *World Archaeol.* **2010**, *42*, 108–121. [CrossRef]
8. Bazza, M. *Experience on Wastewater Reuse in the Near East*; FAO Regional Office for the Near East: Cairo, Egypt, 2003.
9. Bazza, M. Overview of the history of water resources and irrigation management in the Near East region. *Water Sci. Technol.* **2007**, *7*, 201–209. [CrossRef]

10. Bazza, M. Overview of the history of water resources and irrigation management in the near east region. In Proceedings of the 1st IWA International Symposium on Water and Wastewater Technologies in Ancient Civilizations, Iraklion, Greece, 28–30 October 2006; pp. 593–604.
11. Goblot, H. Dans L'ancien Iran: Les Techniques de L'eau et la Grande Histoire. *Ann. Hist. Sci. Soc.* **1963**, *18*, 499–520. [CrossRef]
12. Water Encyclopedia. Available online: http://www.waterencyclopedia.com/Hy-La/Irrigation-Systems-Ancient.html (accessed on 30 December 2019).
13. Viollet, P.L. *L'hydraulique Dans les Civilisations Anciennes—5000 ans D'histoire*; Presses de l'Ecole Nationale des Ponts et Chaussées: Paris, France, 2005; p. 374.
14. Tamburrino, A. Water Technology in Ancient Mesopotamia. In *Ancient Water Technologies*; Mays, L.W., Ed.; Springer: New York, NY, USA, 2010; pp. 29–51.
15. Xiuling, Y. Several views about unearthed rice and bone plough in the fourth cultural layer of Hemudu archaeological site. *Archaeol. J.* **1976**, *8*, 23–26.
16. IA-CASS. *Institute of Archaeology—Chinese Academy of Sciences. Zhongguo Kaoguxue Zhong Tanshisi Shujuji 1965–1991 (Radiocarbon Dating Data Set in Chinese Archaeology 1965–1991)*; Cultural Relies Publishing House: Beijing, China, 1991.
17. Zhenglang, Z. Research on Technologies of Ancient Cultivation According to Oracles of Shang dynasty. *Archaeol. J.* **1973**, *1*, 97–124.
18. Moody, J.; Grove, A.T. Terraces and Enclosure Walls in the Cretan Landscape. In *Man's Role in the Shape of the Eastern Mediterranean Landscape*; Bottema, S., Balkema, A.A., Eds.; CRC Press: Boca Raton, FL, USA, 1990; pp. 183–191.
19. Lyrintzis, A.; Angelakis, A.N. Is the Labyrinth a Water Catchment Technique? People and water management in Minoan Crete. In Proceedings of the 1st IWA International Symposium on Water and Wastewater Technologies in Ancient Civilizations, Iraklion, Greece, 28–30 October 2006; pp. 163–174.
20. Angelakis, A.N.; Spyridakis, S.V. The status of water resources in Minoan times: A preliminary study. In *Diachronic Climatic Impacts on Water Resources with Emphasis on Mediterranean Region*; Angelakis, A.N., Issar, A.S., Eds.; Springer: Heidelberg, Germany, 1996; pp. 161–191.
21. Angelakis, A.N.; Dialynas, M.G.; Despotakis, V. Evolution of Water Supply Technologies in Crete, Greece through the Centuries. In *Evolution of Water Supply throughout Millennia*; IWA Publishing: London, UK, 2012; pp. 227–258.
22. Yannopoulos, S.; Lyberatos, G.; Theodosiou, N.; Li, W.; Valipour, M.; Tamburrino, A.; Angelakis, A.N. Evolution of Water lifting devices (pumps) through the centuries worldwide. *Water* **2015**, *7*, 5031–5060. [CrossRef]
23. Water Management in Prehistoric Crete: The case of Choiromandres, Zakros. HYDRIA Project Collection, Storage & Distribution of Water in Antiquity Linking Ancient Wisdom to Modern Needs. Available online: http://www.hydriaproject.info/en/greece-crete-water-management-in-zakros-area/importance3 (accessed on 5 December 2019).
24. Flood, J.M. Water Management in Neopalatial Crete and the Development of the Mediterranean Climate. Master's Thesis, The University of North Carolina, Chapel Hill, NC, USA, 2012.
25. Betancourt, P.P. *The Dams and Water Management Systems of Minoan Pseira*; INSTAP Academic Press: Philadelphia, PA, USA, 2012.
26. Celka, S.M.; Puglisi, D.; Bendali, F. Settlement pattern dynamics and natural resources in MM-lM I Crete: The case of Malia. In *Physis: 'Environnement Naturel Et La Relation Homme-Milieu Dans Le MONDE Égéen Protohistorique*; Touchais, G., Laffineur, R., Rougemont, F., Eds.; Peeters, Leuven: Liege, Belgium, 2014.
27. Betancourt, P.P.; Davaras, C.; Simpson, R.S. *The Archaeological Survey of Pseira Island*; INSTAP Academic Press: Philadelphia, PA, USA, 2004.
28. Angelakis, A.N.; Spyridakis, S.V. Wastewater Management in Minoan Times. In Proceedings of the Meeting on Protection and Restoration of Environment, Chania, Greece, 28–30 August 1996; pp. 549–558.
29. Wright, R.P.; Bryson, R.A.; Schuldenrein, J. Water supply and history: Harappa and the Beas regional survey. *Antiquity* **2008**, *82*, 37–48. [CrossRef]
30. Lawler, A. Boring no more, a trade-savvy Indus emerges. *Science* **2008**, *320*, 1276–1281. [CrossRef]
31. Violett, P.L. *Water Engineering in Ancient Civilizations: 5000 Years of History*; International Association of Hydraulic Engineering and Research (IAHR): Madrid, Spain, 2007.

32. Kenoyer, M.J. *Ancient Cities of the Indus Civilization*; Oxford University Press: Oxford, UK, 1998; p. 260.
33. Kenoyer, J.M. Urban Process in the Indus Tradition: A Preliminary Model from Harappa. In *Harappa Excavations 1986–1990*; Meadow, R.H., Ed.; Prehistory Press: Madison, WI, USA, 1991; pp. 29–60.
34. Irrigation Development in India: History & Impact, Water Technology Centre, WTC, IARI, New Delhi. Available online: http://indiairrigation.blogspot.com/2009/01/history-of-irrigation-development-in_01.html (accessed on 30 December 2019).
35. Krasilnikoff, J.; Angelakis, A.N. Water management and its judicial contexts in ancient Greece: A review from the earliest times to the Roman period. *Water Policy* **2019**, *21*, 245–258. [CrossRef]
36. Harper, R.F. *The Code of Hammurabi, King of Babylon, about 2250 BC (Translated)*; University of Chicago Press: Chicago, IL, USA, 1904.
37. Richardson, M.E.J. *Hammurabi's Laws: Text, Translation, and Glossary*; T&T Clark: London, UK, 2004.
38. Shaw, B.D. Water and society in the Ancient Maghrib: Technology, property and development. *Antiq. Afr.* **1984**, *20*, 121–173. [CrossRef]
39. El Amami, S.; Chaabouni, Z. *Traditional Hydraulic Reshaping: A Means of Controlling Water Erosion*; Research Center of Rural Engineering: Tunis, Tunisia, 1981.
40. Criteria and Options for Appropriate Irrigation Methods. Available online: http://www.fao.org/3/W3094E/w3094e05.htm (accessed on 27 December 2019).
41. Kees, H. *Ancient Egypt: A Cultural Topography*; University of Chicago Press: Chicago, IL, USA, 1961.
42. Noaman, M.N.; El Quosy, D. Hydrology of the Nile and ancient agriculture. In *Irrigated Agriculture in Egypt: Past, Present and Future*; Satoh, M., Aboulroos, S., Eds.; Springer: Cham, Switzerland, 2017; Volume 2, pp. 9–28.
43. Murray, G.W. A note on the Sadd el-Kafara: The ancient dam in wadi Garami. *Bull. I'Institut d'Egvpte* **1947**, *28*, 33–43.
44. Garbrecht, G. The Sadd el Kafara, the world's oldest dam. In Proceedings of the Special Session on History of Irrigation, 12th ICID Congress, Fort Collins, CO, USA, 28 May–2 June 1984.
45. Joffe, G. *Irrigation and Water Supply Systems in North Africa*; Moroccan Studies: Casablanca, Morocco, 1992.
46. Wikander, O. Sources of Energy and Exploitation of Power. In *The Oxford Handbook of Engineering and Technology in the Classical World*; Oleson, J.P., Ed.; Oxford University Press: Oxford, UK, 2008; pp. 136–157.
47. Cohn, G.S. *L'origine des Norias de Fès*; Hespéris: Fez, Morocco, 1933; pp. 156–157.
48. Angelfire. Available online: http://www.angelfire.com//journal/millbuilder/album5.html (accessed on 15 May 2018).
49. Worldwatch. Available online: hrrp://www.worldwatch.org (accessed on 30 December 2019).
50. Fedick, S. Land Evaluation and Ancient Maya Land Use in the Upper Belize. *Lat. Am. Antiq.* **1995**, *6*, 16–34. [CrossRef]
51. Witschey, W.R.T. *Encyclopedia of the Ancient Maya*; Rowman & Littlefield Publishers: Lanham, MD, USA, 2015; p. 574.
52. Scarborough, V.; Dunning, N.P.; Tankersley, K.B.; Carr, C.; Weaver, E.; Gracioso, L.; Sadurní, N. *Diccionari de l'any 1000 a Catalunya*; Edicions 62: Barcelona, Spain, 1999; p. 24.
53. Doyle, J.A. Early Maya geometric planning conventions at El Palmar, Guatemala. *J. Archaeol. Sci.* **2013**, *40*, 793–798. [CrossRef]
54. Lucero, L.I.; Joel, D.; Gunn, J.D.; Scarborough, V.L. Climate change and classic Maya management. *Water* **2011**, *3*, 479–494. [CrossRef]
55. Scarborough, V.; Gallopin, G.G. A Water Storage Adaptation in the Maya Lowlands. *Science* **1991**, *251*, 658–662. [CrossRef]
56. Largest Ancient Maya Dam Found in Guatemala. Available online: http://www.sci-news.com/archaeology/article00470.html (accessed on 30 December 2019).
57. Isendalh, C. The Weight of Water: A New Look at Pre-hispanic Puuc Maya. *Anc. Mesoam.* **2011**, *22*, 185–197. [CrossRef]
58. Pohl, M.D.; Pope, K.O.; Jones, J.G.; Jacob, J.S.; Piperno, D.R.; de France, S.D.; Lentz, D.S.; Gifford, J.A.; Danforth, M.E.; Josserand, J.K. Early Agriculutre in the Maya Lowlands. *Lat. Am. Antiq.* **1996**, *7*, 355–372. [CrossRef]
59. Hodell, D.A.; Brenner, M.; Curtis, J.H.; Guilderson, T. Solar Forcing of Drought Frequency in the Maya Lowlands. *Science* **2001**, *292*, 1367–1370. [CrossRef]

60. China Daily. Nanhu Lake. *China Daily*, 10 May 2010.
61. Chi, C.T. *Key Economic Areas in Chinese History*; George Allenand Unwin: London, UK, 1936.
62. Rackham, O. Ancient Landscapes. In *The Greek City from Homer to Alexander*; Murray, O., Price, S., Eds.; Oxford University Press: Oxford, UK, 1990; pp. 85–111.
63. Amanatidis, G.T.; Paliatsos, A.G.; Repapis, C.C.; Bartzis, J.G. Decreasing precipitation trend in the Marathon area, Greece. *Int. J. Climatol.* **1993**, *13*, 191–201. [CrossRef]
64. Krasilnikoff, J.A. Innovation in Ancient Greek Agriculture: Some Remarks on Climate and Irrigation. *Class. Mediaev.* **2014**, *64*, 95–116.
65. Garnse-y, P. *Famine and Food Supply in the Graeco-Roman World. Responses to Risk and Crisis*; Cambridge University Press: Cambridge, UK, 1988.
66. Glick, T.F. *Irrigation and Society in Medieval Valencia*; Cambridge Belknap Press of Harvard University Press: Cambridge, MA, USA, 1970.
67. Crouch, D.P. *Water Management in Ancient Greek Cities*; Oxford University Press: Oxford, UK, 1993.
68. Krasilnikoff, J.A. Water and farming in Classical Greece: Evidence, Method and Perspective. In *Ancient History Matters: Studies Presented to Jens Erik Skydsgaard on His Seventieth Birthday*; Ascani, V., Gabrielsen, K.K., Rasmussen, A.H., Eds.; L'Erma di Bretschneider: Rome, Italy, 2002; pp. 47–62.
69. Carroll-Spilleche, M. *Der Antike Griechische Garten: Wohnen in der Griechischen Polis III*; Deutscher Kunstverlag: Munich, Germany, 1989.
70. Osborne, R. Classical Greek Gardens: Between Farm and Paradise. In *Garden History. Issues, Approaches, Methods. Dumbarton Oaks Colloquium on the History of Landscape Architecture XIII*; Hunt, J.D., Ed.; Harvard University Press: Cambridge, MA, USA, 1992; pp. 373–391.
71. Foxhall, L. *Olive Cultivation in Ancient Greece: Seeking the Ancient Economy*; Oxford University Press: Oxford, UK, 2007.
72. Bowe, P. Civic and other public planting in ancient Greece. *Stud. Hist. Gard. Des. Landsc.* **2011**, *31*, 269–285. [CrossRef]
73. Raaflaub, K.A. The Transformation of Athens in the Fifth Century. In *Democracy, Empire, and the Arts in Fifth-Century Athens. Center for Hellenic Studies Colloquia 2*; Boedeker, D., Raaflaub, K.A., Eds.; Harvard University Press: Cambridge, MA, USA, 1998; pp. 15–41.
74. Austin, M.M. *The Hellenistic World from Alexander to the Roman Conquest. A Selection of Ancient Sources in Translation*; Cambridge University Press: Cambridge, UK, 1981.
75. Knauss, J. *Wasserbau und Siedlungsbedingungen im Altertum: Die Melioration des Kopaisbeckens durch die Minyer im 2. Jt. v. Chr. (Kopasis 2)*; Technische Universität München, Versuchsanstalt für Wasserbau und Wassermengenwirtschaft: Munich, Germany, 1987.
76. Knauss, J.; Heinrich, B.; Kalcyk, H. *Die Wasserbauten der Minyer in der Kopais–die älteste Flussregulerung Europas. Bericht Nr. 50*; Institut für Wasserbau und Wassermengenwirtschaft und Versuchsanstalt für Wasserbau, Oskar v. Miller-Institut, Obernach Technische Universität München: Munich, Germany, 1984.
77. Horden, P.; Purcell, N. *The Corrupting Sea: A Study of Mediterranean History*, 8th ed.; Wiley-Blackwell: Hoboken, NJ, USA, 2000; p. 776.
78. Lohmann, H. *Atene: Forschungen zu Siedlungs-und Wirtschaftsstruktur des klassischen Attika*; Bohlau Verlag: Koln, Germany, 1993.
79. Price, S.; Nixon, L. Ancient Greek Agricultural Terraces: Evidence from Texts and Archaeological Survey. *Am. J. Archaeol.* **2005**, *109*, 665–694. [CrossRef]
80. Moreno, A. *Feeding the Democracy. The Athenian Grain Supply in the Fifth and Fourth Centuries BC*; Oxford University Press: Oxford, UK, 2007.
81. Prieto, A. Landscape Organization in Magna Graecia. Ph.D. Thesis, University of Texas, Austin, TX, USA, 2005.
82. Wilson, A. Land drainage. In *Handbook of Ancient Water Technology. Technology and Change in History, Vol. 2*; Wikander, Ö., Ed.; Brill: Leiden, Belgium, 2000; pp. 303–317.
83. Margaritis, E.; Jones, M.K. Greek and Roman agriculture. In *Handbook of Engineering and Technology in the Classical World*; Oleson, G., Ed.; Oxford University Press: Oxford, UK, 2008; pp. 567–578.
84. Cato, M.P.; Varro, M.T. *On Agriculture*; Harvard University Press: Cambridge, MA, USA, 1934.
85. White, K.D. *Roman Farming*; Thames & Hudson: London, UK, 1970.
86. Mays, L.W. A very brief History of Hydraulic Technology during Antiquity. *Environ. Fluid Mech.* **2008**, *8*, 471–484. [CrossRef]

87. Butzer, K.W.; Mateu, J.F.; Butzer, E.; Kraus, P. Irrigation agrosystems in eastern Spain: Roman or Islamic origins? *Ann. Assoc. Am. Geogr.* **1985**, *75*, 479–509. [CrossRef]
88. Lechtman, H.; Hobbs, L. Roman concrete and Roman architecture revolution. In *High Technology Ceramics: Past, Present, and Future: The nature of Innovation and Change in Ceramic Technology*; Kingery, W.D., Ed.; American Ceramic Society: Westerville, OH, UAS, 1986.
89. Pounds, N.J.G. *An Historical Geography of Europe 450 BC-AD 1330*; Cambridge University Press: Cambridge, UK, 1973.
90. Von Stackelberg, K.T. *The Roman Garden: Space, Sense, and Society*; Routledge: London, UK, 2009.
91. Fentress, E.; Quilici Gigli, S. La Domesticazione delle Piante e L'agricoltura: Mondo Greco e Mondo Romano. Available online: http://www.treccani.it/enciclopedia/la-domesticazione-delle-piante-e-l-agricoltura-mondo-grecoe-mondo-romano_%28Il-Mondo-dell$\delimiter"026E30F$T1$\delimiter"026E30F$textquoterightArcheologia%29/ (accessed on 20 October 2015).
92. Barker, G.; Gilbertson, D.; Jones, B.; Mattingly, D. *Farming the Desert: The UNESCO Libyan Valleys Archaeological Survey*; Graeme, B., Ed.; UNESCO/Department of Antiquities: Tripoli, Libya, 1996; Volume 1.
93. Kamash, Z. Irrigation technology, society and environment in the Roman Near East. *J. Arid Environ.* **2012**, *86*, 65–74. [CrossRef]
94. Smith, N.A. *History of Dams*; P. Davies: London, UK, 1971.
95. Belli, O. *Urartian Irrigation Canals in Eastern Anatolia*; Archeology and Art Press: Istanbul, Turkey, 1997.
96. Orhan, A.H. *Urartian Water Constructions and Hydraulics*; University of Yüzüncü Yıl: Tuşba, Van, Turkey, 2004.
97. Çifçi, A.; Greaves, A.M. Urartian Irrigation Systems. *Anc. Near East. Stud.* **2013**, *50*, 191–214.
98. Garbrecht, G. *Historische Wasserbauten im Ost-Anatolien*; DWhG: Siesburg, Germany, 2004.
99. English, P.W. The origin and spread of qanats in the Old World. *Proc. Am. Philos. Soc.* **1968**, *112*, 170–181.
100. English, P.W. Qanats and Lifeworlds in Iranian Plateau Villages. In Proceedings of the Conference on Transformation of Middle Eastern Natural Environment: Legacies and Lessons, New Haven, CT, USA, 28 October–1 November 1997.
101. Miu, Q.Y. *Research on History of Polders in the Taihu Lake Basin*; Agricultural Press: Beijing, China, 1985.
102. Wang, Z. *Annotation of Agricultural Treatise*; Shandong Publishing Group: Jinan, China, 1313.
103. Plattner, G. *Intergovernmental Panel on Climate Change. Climate Change 2014: Synthesis Report. Contribution of Working Groups I, II and III to the Fifth Assessment Report of the Intergovernmental Panel on Climate Change*; IPCC: Geneva, Switzerland, 2014.
104. Morehar, C.T. Mapping ancient chinampa landscapes in the Basin of Mexico: A remote sensing and GIS approach. *J. Archeol. Sci.* **2012**, *39*, 2541–2551. [CrossRef]
105. Armillas, P. Gardens on swamps. *Science* **1971**, *175*, 653–661. [CrossRef]
106. Smith, M. The Aztec Silent Majority: William T. Sanders and the Study of the Aztec Peasantry. In *Arqueología Mesoamericana*; Alba Guadalupe Mastache, A., Parsons, J.R., Santley, R.S., Serra Puche, M.C., Eds.; Instituto Nacional de Antropología e Historia: Mexico City, Mexico, 1996; Volume 1, pp. 375–386.
107. McCaa, R. The Peopling of Mexico from Origins to Revolution. In *The Population History of North America*; Richard Steckel, R., Haines, M., Eds.; Cambridge University Press: Cambridge, UK, 2000; pp. 241–304.
108. DGCOH Dirección General de Construcción y Operación Hidráulica. *El Sistema del Drenaje Profundo de la Ciudad de México*, 3rd ed.; Dirección General de Construcción y Operación Hidráulica, Dirección Técnica, Subdirección de Programación: Distrito Federal Mexico, Mexico, 1996.
109. Palerm, A. Obras hidráulicas prehispánicas en el sistema lacustre del Valle de Mexico. *Archaeol. Am. Anthropol.* **1973**, *78*, 1245.
110. Rojas, R.A.; Strauss, K.; Lameiras, J. Technological aspects of the colonial hydraulic infrastructure. In *New News on the pre Hispanic and Colonial Hydraulic Infrastructure in the Valley of Mexico*; Minister of Public Education, National Institute of History and Anthropology: Distrito Federal Mexico, Mexico, 1974.
111. Luna, G.; Gregory, G. Late Aztec Chinampa Agriculture and Settlement on Lake Xochimilco: A GIS Analysis of Lakebed Chinampas and Settlement. Ph.D. Thesis, The Pennsylvania State University, State College, PA, USA, 2014.
112. Villalonga, A. Imperialismo Hidráulico de los Aztecas en la Cuenca de México. *Rev. Tecnol. Agua* **2007**, *288*, 78–91.
113. Chinampa: Raised-Bed Hydrological Agriculture. Available online: https://anthrome.wordpress.com/2011/04/24/chinampa-raised-bed-hydrological-agriculture/ (accessed on 26 June 2013).

114. Canabal-Cristiani, B.P.; Tortes-Lima; Burela-Rueda, G. *La Ciudad y sus Shinampas*; Universidad Autonoma Metropolitana-Xochimilco: Mexico City, Mexico, 1992.
115. Smith, M.E. *The Aztecs*; Wiley Blackwell: Hoboken, NJ, USA, 2013.
116. Coe, M.D. The chinampas of Mexico. *Sci. Am.* **1964**, *211*, 90–98. [CrossRef]
117. Netherly, P.J. The Management of Late Andean Irrigation Systems on the North Coast of Peru. *Am. Antiq.* **1984**, *49*, 227–254. [CrossRef]
118. Boelens, R.; Gelles, P.H. Cultural Politics, Communal Resistance and Identity in Andean Irrigation Development. *Bull. Lat. Am. Res.* **2005**, *24*, 311–327. [CrossRef]
119. Barnebeck, A.T.; Sandholt, J.P.; Volmar, S.C. The heavy plow and the agricultural revolution in Medieval Europe. *J. Dev. Econ.* **2016**, *118*, 133–149.
120. Peix, J. *L'aigua i el Medi: Gestió dels Regadius*; Quaderns Agraris: Barcelona, Spain, 2002.
121. Squatriti, P. *Water and Society in Early Medieval Italy, AD 400–1000*; Cambridge University Press: Cambridge, UK, 2002.
122. Shanan, L. The impact of irrigation. In *Land Transformation in Agriculture*; Wolman, M.G., Fourier, F.G.A., Eds.; John Wiley & Sons: New York, NY, USA, 1987; pp. 115–131.
123. Avdat-Negev Attractions. Available online: http://www.planetware.com/Negev/avdat-isr-st-av.htm (accessed on 20 December 2013).
124. Salgot, M.; Angelakis, A. The historical development of water supply technologies in Barcelona, Spain. In *Evolution of Water Supply through the Millennia*; IWA Publishing: London, UK, 2019.
125. Christensen, P. Middle Eastern Irrigation: Legacies and Lessons. *Bull. Ser. Yale Sch. For. Environ. Stud.* **1998**, *103*, 15–29.
126. Roldán, J.; Moreno, M.F. La ingeniería y la gestión del agua de riego en Al-Andalus. *Ing. Agua* **2007**, *14*, 223–236. [CrossRef]
127. Morell, A. «de la Vila». Fonts de Tramuntana. Available online: https://www.fontsdetramuntana.com/ (accessed on 18 November 2016).
128. Vivas, G.; Gómez-Landesa, E.; Mateos, L.; Giráldez, J.V. *Integrated Water Management in an Ancestral Water Scheme in a Mountainous Area of Southern Spain*; ASCE: Reston, VA, USA, 2009.
129. Mateos, L.; Vivas, G.; Giráldez, J.V.; González-dugo, M.P. Origin, tradition and decline of the ancestral irrigation systems in the high Alpujarra, Spain. In Proceedings of the Workshop History of Irrigation, Drainage and Flood Control, ICID 22nd European Regional Conference, Pavia, Italy, 2 September 2007.
130. De la Peña, J.L.; Ibarz, J.; Salgot, M. Water in the royal monastery of Santes Creus, Catalonia (Spain). In Proceedings of the 3rd Specialized Conference on Water and Wastewater Technologies in Ancient Civilizations, Istanbul, Turkey, 22–24 March 2012.
131. Das, S.K.; Gupta, R.K.; Varma, H.K. Flood and Drought Management through Water Resources Development in India. *Bull. World Meter. Organ.* **2007**, *56*, 3.
132. Postan, M.M. *The Agrarian Life of the Middle Ages*; Cambridge University Press: Cambridge, UK, 1966.
133. Şentürk, F. *Hydraulics of Dams and Reservoirs*; Water Resources Publications: Highlands Ranch, CO, USA, 1994.
134. Kuban, D. *Tarih Vakfı yurt Yayını, Istanbul, A History of City: Istanbul*; The Economic and Social History Foundation of Turkey: Istanbul, Turkey, 1996.
135. Inancık, H.; Arı, B. Urbanism in Turkish-Islamic-Ottoman Tradition and Studies of İnalcık. *H. J. Turk. Res. Lit.* **2005**, *3*, 25–56.
136. RTMCT. *Ottoman Period Waterworks and Water Admınıstratıon System*; Republic of Turkey Ministry of Culture and Tourism: Ankara, Turkey, 2005.
137. DSİ. *Water Resources in Turkey*; General Directorate State Hydraulic Works: Ankara, Turkey, 2012.
138. Money, R.I. The irrigation of the Konia Plain. *Geogr. J.* **1919**, *5*, 298–303. [CrossRef]
139. Keenan, G. Fayyum Agriculture at the End of the Ayyubid Era: Nabulsi's Survey. In *Agriculture in Egypt: From Pharaonic to Modern Times*; Bowman, A.K., Rogan, E., Eds.; Oxford University Press for the British Academy: Oxford, UK, 1999; pp. 287–299.
140. Alan, M. An Irrigated Empire: The Vıew from Ottoman Fayyum. *Int. J. Middle East Stud.* **2010**, *42*, 569–590.
141. Borsch, S.J. Environment and Population: The Collapse of Large Irrigation Systems Reconsidered. *Comp. Stud. Soc. Hist.* **2004**, *46*, 458–460. [CrossRef]
142. Quataert, D. *The Ottoman Empire–1700–1922*, 2nd ed.; Cambridge University Press: Cambridge, UK, 2005.
143. Mitchell, W. The Hydraulic Hypothesis. *Curr. Anthropol.* **1973**, *14*, 532–534. [CrossRef]

144. Worster, D. *Rivers of Empire: Water, Aridity & the Growth of the American West*; Pantheon Books: New York, NY, USA, 1985.
145. Reisner, M. *Cadillac Desert: The American West and Its Disappearing Water*; Penguin Books: New York, NY, USA, 1986.
146. Freydank, K.; Siebert, S. Towards mapping the extent of irrigation in the last century: A time series of irrigated area per country. In *Frankfurt Hydrology Paper 08*; Institute of Physical Geography, University of Frankfurt: Frankfurt am Main, Germany, 2008.
147. Shiklomanov, I.A. Appraisal and assessment of world water resources. *Water Int.* **2000**, *25*, 11–32. [CrossRef]
148. Brown, L.B. *Raising Water Productivity*; Earth Policy Institute, Ratgers University: Washington, DC, USA, 2008.
149. Kadigi, R.M.J.; Tesfay, G.; Bizoza, A.; Zinabou, G. *Irrigation and Water Use Efficiency in Sub-Saharan Africa*; Global Development Network: Washington, DC, USA, 2012.
150. Xie, Y.; Zhou, X. Income inequality in todays' China. *Proc. Natl. Acad. Sci. USA* **2014**, *111*, 6928–6933. [CrossRef]
151. Aquastat Database. Available online: http://faostat.org (accessed on 30 November 2019).
152. Faurès, J.M. *The FAO Irrigated Area Forecast for 2030*; Food and Agricultural Organization: Rome, Italy, 2002.
153. Postel, S. *Last Oasis: The Worldwatch Environment Alert Series*; W.W. Norton and Company: New York, NY, USA, 1992.
154. Shiklomanov, I.A. *Comprehensive Assessment of the Freshwater Resources of the World: Assessment of Water Resources and Water Availability in the World*; World Meteorological Organization and Stockholm Environment Institute: Stockholm, Sweden, 1997.
155. Siebert, S.; Doll, P. Irrigation water use. A global perspective. In *Global Change: Enough Water for All?* Lozán, J.L., Graßl, H., Hupfer, P., Menzel, L., Schönwiese, C.D., Eds.; Universität Hamburg: Hamburg, Germany, 2007; pp. 104–107.
156. Sojka, R.E.; Bjorneberg, D.L.; Entry, J.A. Irrigation: An historical perspective. In *Encyclopedia of Soil Science*, 1st ed.; Lal, R., Ed.; Marcel Dekker, Inc.: New York, NY, USA, 2002; pp. 745–749.
157. Leibundgut, C.; Kohn, I. European Traditional Irrigation in Transition Part I: Irrigation in Times Past—A historic land use practice across Europe. *Irrig. Drain.* **2014**, *63*, 273–293. [CrossRef]
158. Domokos, M.; Kálmán, M. Summary and conclusions, in Danube Valley. In *History of Irrigation, Drainage and Flood Control*; Csekő, G., Hayde, L., Eds.; ICID: New Delhi, India, 2004; pp. 679–715.
159. Bucks, D.A.; Sammis, T.W.; Dickey, G.L. Irrigation for Arid Areas. In *Management of Farm Irrigation Systems*; Hoffman, G.J., Howell, T.A., Solomon, K.H., Eds.; ASAE: Washington, DC, USA, 1990; pp. 499–548.
160. National Research Council (USA). *A New Era for Irrigation*; National Academy Press: Washington, DC, USA, 1996.
161. National Research Council (USA). *Water Transfers in the West: Efficiency, Equity, and the Environment*; National Academy Press: Washington, DC, USA, 1992.
162. UNESCO. *The 2nd UN World Water Development Report: "Water, a Shared Responsibility"*; UNESCO: Paris, France, 2006.
163. Gleick, P. An Introduction to Global Fresh Water Issues. In *Water in Crisis—A Guide to World's Fresh Water Resources*; Gleick, P., Ed.; Oxford University Press: New York, NY, USA, 1993.
164. Kendall, H.; Pimentel, D. Constraints on the expansion of the global food supply. *Ambio* **1994**, *23*, 198–205.
165. Howell, T.A. Irrigation's Role in Enhancing Water Use Efficiency. In *Proceedings of the 4th Decennial Symposium*; Evans, R.G., Benham, B.L., Trooien, T.P., Eds.; ASAE: Washington, DC, USA, 2000; pp. 66–80.
166. Rhoades, J.D. Sustainability of Irrigation: An Overview of Salinity Problems and Control Strategies. In Proceedings of the 1997 Annual Conference of Canadian Water Resources Association, Footprints of Humanity: Reflections on Fifty Years of Water Resources Development, Lithbridge, AB, Canada, 3–6 June 1997; pp. 1–40.
167. Ghassemi, F.; Jakeman, A.J.; Nix, H.A. *Salinisation of Land and Water Resources*; University of New South Wales Press Ltd.: Canberra, Australia, 1995.
168. Cerrillo, A. Los ocho naufragios del canal Segarra-Garrigues (The eigth failures of the Segarra-Garrigues canal). *La Vanguardia Newspaper*, 18 November 2005.
169. WWAP. *The United Nations World Water Development Report 4: Managing Water under Uncertainty and Risk, Vol. 1*; UNESCO: Paris, France, 2012.

170. Jensen, M.E.; Burman, R.D.; Allen, R.G. *Evapotranspiration and Irrigation Water Requirements*; ASCE Manuals and Reports on Engineering Practice, 70; ASCE: New York, NY, USA, 1990.
171. Sojka, R.E. Understanding and Managing Irrigation-Induced Erosion. In *Advances in Soil and Water Conservation*; Pierce, F.J., Frye, W.W., Eds.; Sleeping Bear Press: Ann Arbor, MI, USA, 1998; pp. 21–37.
172. Walker, W.R.; Zaccaria, D. Agricultural Water Management Challenges in the Western US. Discussion Paper. Presented at the NSF Water Workshop, Cairo, Egypt, 10–14 August 2014.
173. Levidow, L.; Zaccaria, D.; Maia, R.; Vivas, E.; Todorovic, M.; Scardigno, A. Improving water-efficiency irrigation: Prospects and difficulties of innovative practices. *Agric. Water Manag.* **2014**, *146*, 84–94. [CrossRef]
174. Angelakis, A.N.; Voudouris, K.S.; Tchobanoglous, G. Evolution of Water Supplies in the Hellenic World Focusing on the Water Treatment and Modern Parallels Water Supply. Available online: https://iwaponline.com/ws/article/doi/10.2166/ws.2020.032/72561/Evolution-of-water-supplies-in-the-Hellenic-world (accessed on 30 December 2019).
175. UN. *Population 2030: Demographic Challenges and Opportunities for Sustainable Development Planning*; UN: New York, NY, USA, 2015.
176. Paranychianakis, N.V.; Salgot, M.S.; Snyder, A.; Angelakis, A.N. Quality Criteria for Recycled Wastewater Effluent in EU-Countries: Need for a Uniform Approach. *Crit. Rev. Environ. Sci. Technol.* **2015**, *45*, 1409–1468. [CrossRef]
177. Tzanakakis, V.E.; Koo-Oshima, S.; Haddad, M.; Apostolidis, N.; Angelakis, A.N. The History of Land Application and Hydroponic Systems for Wastewater Treatment and Reuse. In *Evolution of Sanitation and Wastewater Management through the Centuries*; Angelakis, A., Rose, J., Eds.; IWA Publishing: London, UK, 2014; pp. 459–482.
178. Salgot, M.; Oron, G.; Cirelli, G.; Dalezios, N.R.; Díaz, A.; Angelakis, A.N. Criteria for Wastewater Treatment and Reuse under Water Scarcity. In *Handbook of Drought and Water Scarcity: Environmental Impacts and Analysis of Drought and Water*; Eslamian, S., Eslamian, F., Eds.; CRC Press: Boca Raton, FL, USA, 2016.
179. Hill, D.R. *A History of Engineering in Classical and Medieval Times*; Croom Helm: London, UK, 1984.
180. Dillehay, T.D.; Eling, H.H., Jr.; Rossen, J. Preceramic Irrigation Canals in the Peruvian Andes. *Proc. Natl. Acad. Sci. USA* **2005**, *102*, 17241–17244. [CrossRef] [PubMed]
181. Rodda, J.C.; Ubertini, L. *The Basis of Civilization—Water Science*; ISHS Proceedings & Reports: Wallingford, UK, 2004; Volume 286, p. 342.

Article

The First Drying Lake in Chile: Causes and Recovery Options

Rodrigo Valdés-Pineda [1,2,3,*], Pablo García-Chevesich [1,4], Juan B. Valdés [1] and Roberto Pizarro-Tapia [3]

1 Hydrology and Atmospheric Sciences Department, University of Arizona, J W Harshbarger Bldg., 1133 James E. Rogers Way, Tucson, AZ 85721, USA; pchevesich@mines.edu (P.G.-C.); jvaldes@email.arizona.edu (J.B.V.)
2 Piteau Associates, Water Management Group, 2500 North Tucson Boulevard, Suite 100, Tucson, AZ 85716, USA
3 Technological Center of Environmental Hydrology, University of Talca, Mailbox 747, Avenida Lircay s/n, 3460000 Talca, Chile; rpizarro@utalca.cl
4 Department of Civil and Environmental Engineering, Colorado School of Mines, Golden, CO 80401, USA
* Correspondence: rvaldes@email.arizona.edu; Tel.: +1-520-4816397

Received: 18 November 2019; Accepted: 13 January 2020; Published: 19 January 2020

Abstract: Located southwest of the city of Santiago (Chile), the Aculeo Lagoon used to be an important body of water, providing environmental, social, and economic services to both locals (mostly drinking water and small-scale agricultural irrigation) and tourists who visited the area for fishing, sailing, and other recreational activities. The lagoon dried completely in May of 2018. The phenomenon has been attributed to the current climatic drought. We implemented and calibrated a surface-groundwater model to evaluate the hydrogeologic causes of the lagoon's disappearance, and to develop feasible solutions. The lagoon's recovery requires a series of urgent actions, including environmental education and significant investment in infrastructure to import water. Ultimately, there are two goals: bringing back historic water levels and ensuring the sustainability of water resources at the catchment scale.

Keywords: Aculeo Lagoon; drought; Chile; water scarcity; water demands; water management

---

1. Introduction

The Aculeo Lagoon (Figure 1) is one of the largest natural bodies of water in central Chile, with an historical water surface of 11.7 km^2 [1] and a maximum depth of 8 m. It is located 50 km southwest of Santiago (33°50′ S–70°54′ W) at an altitude of ~350 m.a.s.l. [2]. The lake is located between the inner foothills of La Costa mountain range in the Paine commune of the Metropolitan Administrative Region. Its name comes from the Mapudungun (Mapuche) dialect Acud-Leu, which means "where the river ends". Only small direct tributaries naturally recharge the Aculeo lagoon every winter and sometimes during the summer. The climate of Aculeo is characterized by a well-defined annual cycle with peak of precipitation accumulation in the austral winter (April–September), and much lower values in the austral summer (October–March) [3].

1.1. *Paleoclimatological Records of the Lagoon*

According to paleoclimatological studies [2], the lagoon had dried between 9500 and 7500 cal yr B.P. Rates of pollen accumulation allowed scientists to infer that the lagoon was a "d[illegible]ent" associated with more arid and warmer weather conditions. No evidence has been found of human settlements in the Aculeo Basin during this period. In modern history, this is the first total drying of the lagoon—a socially and environmentally significant phenomenon, given that during the last 100 years the Aculeo Lagoon has been transformed primarily by the increase in large-scale agricultural activities and other human activities. Therefore, this is the first drying of the lagoon concurrent with the presence of human settlements, and this factor must be considered in the water balance calculations of the catchment, the local aquifer, and its interaction with the lagoon.

From a paleoclimatological perspective, it is also important to consider that the five years in which the lagoon was reduced from its historical levels to complete dryness—as shown in Figure 1—is a short period compared to the resolution of the paleoclimatological processes described by [2]. The authors suggested that the dry conditions persisted during the early- and mid-Holocene; therefore, the paleoclimatological fluctuation of the seasonal lagoon's water levels occurred over more protracted periods of time than the hundred years it took humans to transform the basin, or the ten years of climate drought that supposedly triggered the lagoon dry conditions. These conclusions support the hypothesis that the hydrological processes that triggered the hydrologic drought of the lagoon in the mid-Holocene probably occurred at a much slower rate than those associated with the current drought and drying.

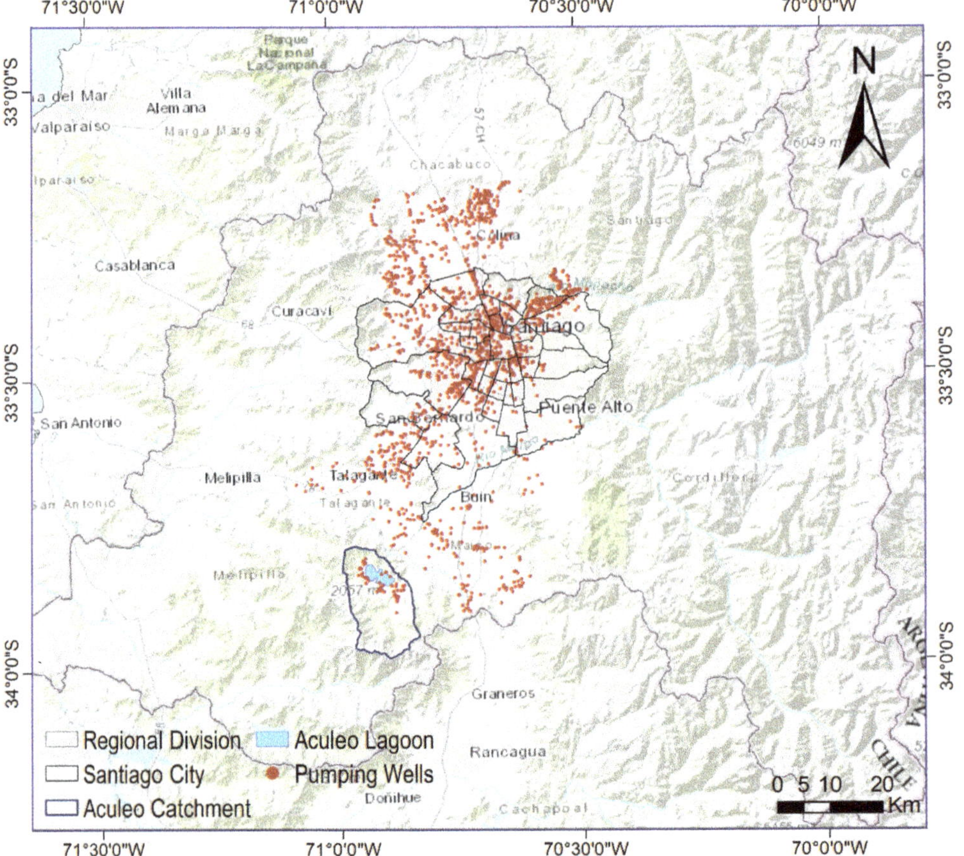

Figure 1. Location of the Aculeo Lagoon within the Metropolitan Region (Santiago), Chile. Approximately 2150 legally registered pumping wells are displayed in the map.

1.2. Identifying Climatic, Hydrologic, and Socioeconomic Droughts

Since 2012, the hydrogeological sector in which the Aculeo Lagoon is located has slowly progressed to a state of extreme dryness (Figure 2). Many researchers have linked the total drying of the lagoon, which occurred in May 2018, to the climatic mega-drought that has been affecting central Chile since

2010 [4]. It is important to recognize that all types of droughts may originate from a sustained reduction in accumulation, i.e., a climatic drought. From a hydrological perspective it is always important to investigate how the deficit could affect a hydrological or hydrogeological system and be a determining factor in a hydrologic drought. Both climatic and hydrologic droughts are labels that refer to the physical process itself [5]. However, the presence of human settlements adds another concept to the list, the socioeconomic drought; this type of drought occurs when the water demands are higher than the available resources (in Chile, this type of drought is commonly known as water scarcity), resulting in a breakdown of the hydrodynamic balance of a hydrologic system, whether superficial, underground, or the interaction between both. It is possible that the drying of the lagoon stems from a combination of these processes that have interacted over a short period of time.

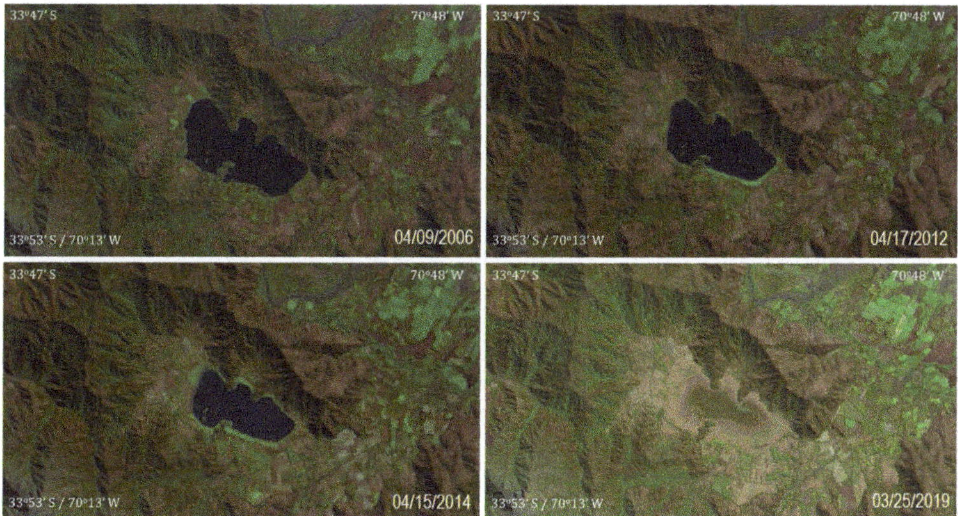

Figure 2. Evolution of the Aculeo Lagoon between 2006 and 2019. Its complete disappearance process started in 2012 and finally occurred on May 2018. The composite is based on Landsat ETM+ images.

2. Materials and Methods

2.1. Hydrogeological Setting of the Aculeo Lagoon

The initial groundwater levels measured in 57 legally registered pumping wells located in the lagoon's domain ranged from artesian to approximately 20 m of depth (see Supplementary Material for details about the input data). The main area of the lagoon is formed by deposits of lake sediments with a significant proportion of clays, which decreases toward the northeast portion of the lagoon. The geologic map of Chile [6] shows that the mountain range surrounding the Aculeo Lagoon is characterized by cretacic formations constituted of sedimentary and volcanic sequences located to the west and northeast of the lagoon. Intrusive rocks have been identified mostly in the southeast portion of the lagoon [7]. A lithographical conceptual model for the lagoon's basin was constructed from 57 boreholes and 12 Vertical Electrical Sounding locations (Figure 3). The borehole intervals were obtained from all legal wells drilled and registered at the National Directorate of Water Resources of Chile (DGA); this is public information. Unfortunately, there is not standard required by DGA to perform these tests and, therefore every drilling company provides their own sediment/rock intervals as part of the borehole information. This is the main reason why it was necessary to simplify and homogenize the information available at each location for the final construction of the lithographic model.

The raw lithography obtained from the boreholes and vertical sounding intervals was classified and simplified using the Soil Texture Calculator (USDA, 2019). The simplification also helped to minimize the convergence errors during the simulation of groundwater fluxes. The simplified conceptual lithographic model was used to construct a surface sedimentary model for the lagoon's basin using the Leapfrog model software [8]. The sedimentary surface chronology was assumed to have the youngest or finer sediments in the upper horizons, and the older or coarser sediments in the lower horizons (see Supplementary Material for details of the categories). All the sediments were also assumed to be seated in a rock basement at approximately 200–300 m of depth. The goodness of fit of the surface chronology model was significant and allowed for the construction of a hydrogeological model with low structural complexity for the simulation of groundwater fluxes in the lagoon and the basin.

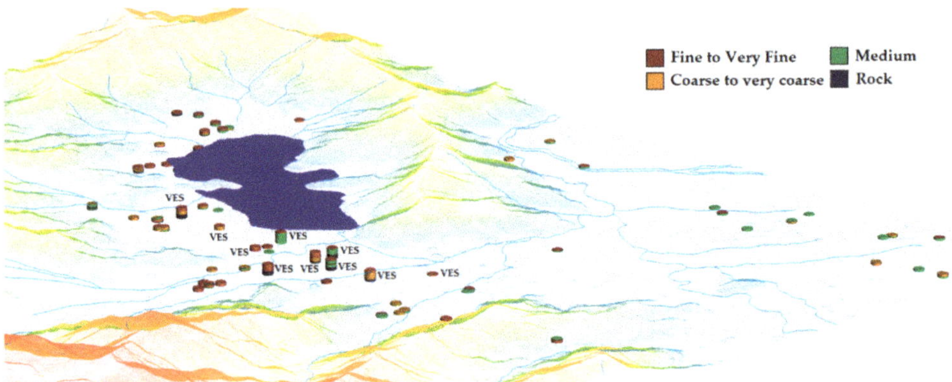

Figure 3. Boreholes and vertical electric sounding (VES) locations used to construct a geological model for the Aculeo domain. The legend represents a simplification of the textural classes found in the lithographic sampling intervals.

2.2. *Understanding the Drying of Aculeo Lagoon*

To understand what other factors, in addition to the climatic drought, could have triggered and accelerated the lagoon's disappearance, we performed the first groundwater modeling for the whole Aculeo basin using MODFLOW [9–12]. MODFLOW is a numerical model that solves the three-dimensional partial differential equation of groundwater flow for a porous medium by using a finite-difference method (see Equation (1)). The model is composed of different packages and programs (see Supplementary Material for details about the packages used in this study) that form the main structure of the input datafiles for the model. The packages used aimed to simulate 5 years of the regional steady and semi-transient daily flow of groundwater in the basin (at 250 m of spatial resolution), to understand the lagoon–aquifer interaction and the possible causes of the drying. The 5-year period for the simulations was selected because it approximately corresponds to the number of years that it took for the lagoon to dry out. The initial conditions of each 5-year period (2006–2011, 2012–2018, and 2018–2023) were used for the calibrations and simulations. The groundwater flow equation utilized in MODFLOW is defined as

$$\frac{\partial}{\partial x}\left(K_{xx}\frac{\partial h}{\partial x}\right)+\frac{\partial}{\partial y}\left(K_{yy}\frac{\partial h}{\partial y}\right)+\frac{\partial}{\partial z}\left(K_{zz}\frac{\partial h}{\partial z}\right)+W=S_S\frac{\partial h}{\partial t} \quad (1)$$

where,

- K_{xx}, K_{xx}, K_{zz} are the values of hydraulic conductivity along the x, y, and z coordinate axes, respectively, which are assumed to be parallel to the major axes of hydraulic conductivity (L/T).
- h is the hydraulic head (L).

- W is a volumetric flux per unit volume representing sources or sinks of water It is negative for groundwater outflow and positive for groundwater inflow of the hydrogeologic system (1/T).
- S_S is the specific storage of the porous material (1/L).
- t is time (T).

2.3. *Aquifer–Lagoon Interaction*

The MODFLOW package LAK3 (Lake package) was used to numerically simulate the aquifer–lagoon interaction. In this package, the lagoon is represented as a water body in contact with the surface grid of the groundwater simulation, which represents the adjacent aquifer. The simulation of the interaction between the Aculeo lagoon and its local shallow aquifer is assumed to occur vertically or laterally. This indicates that the lagoon stage can be significantly affected by the water volume that is transmitted though the lagoon's bed. The transmission of water between the aquifer and the lagoon also depends on the elastic properties of the aquifer, the size of the simulation grid, and the assumptions related to the flow of groundwater through the lagoon's bed. The mathematical formulation utilized to quantify the hydraulic relationship between the lagoon and the adjacent aquifer is based on the application of Darcy's Law (see Equation (2)). The hydraulic level (stage) at the lagoon and the groundwater level in the aquifer can be used to calculate the specific groundwater flux as

$$q = K\frac{h_l - h_a}{\Delta l} \quad (2)$$

where,

- q is the specific (recharge or discharge) flux of water between the lagoon and the aquifer (L/T).
- K is the hydraulic conductivity of the lagoon's bed that separates the lagoon and the aquifer (L/T).
- h_l is the water level in the lagoon (L).
- h_a is the groundwater level in the aquifer (L).
- Δl is the distance between the measurement's points of h_l and h_a.

Note that the specific flux is positive when the water is recharged from the lagoon to the aquifer ($h_l > h_a$) and negative when the water is discharged from the aquifer to the lagoon ($h_l < h_a$). However, in most natural waterbodies, a mixed flux of water is observed where some areas of the lagoon's bed discharge water from the aquifer to the lagoon and some areas recharge water from the lagoon to the aquifer. Furthermore, Equation (2) also can be expressed as volumetric flux (L^3/T) by integrating the specific flux over the transversal area of the plane perpendicular to it as follows.

$$Q = qA = \frac{KA}{\Delta l}(h_l - h_a) = c(h_l - h_a) \quad (3)$$

The parameter $c = \frac{KA}{\Delta l}$ is a term known as conductance (L^2/T). This parameter also can be dimensionless assuming a unit area. In the MODFLOW simulation, the area perpendicular to the groundwater flux can have a horizontal component (X–Y), or a vertical component (Y–Z or X–Z). In the implementation of LAK3, the lagoon occupies the whole volume of each cell with no superposition between cells. The conceptualization of LAK3 also assumes that the lagoon–aquifer interaction occurs in two different materials: (1) the lagoon's bed and (2) the aquifer materials, which depend on the geological model and the discretization applied. More details about the assumptions and limitations of LAK3 package can be found in [13,14].

2.4. Water Balance for the Lagoon

The aquifer–lagoon interaction is updated at every time interval used in the temporal discretization of MODFLOW. The water balance for the lagoon uses Equation (4) to resolve the water level at each time step. The explicit form of the equation is

$$h_l^n = h_l^{n-1} + \Delta t \frac{p - e - rnf - w - sp + Q_{si} - Q_{so}}{A_s} \quad (4)$$

where,

- h_l^n, h_l^{n-1} are the water levels of the lagoon for the current and the previous time interval (L).
- Δt is the time interval used in the MODFLOW simulation (T).
- p is the rainfall accumulation for the lagoon in the current time step (L).
- e is the evapotranspiration from the lagoon for the current time step (L).
- rnf is the surface runoff towards the lagoon for the current time step (L/T).
- w is the additional lagoon's inflow or outflow due to anthropogenic influence.
- Q_{si} is the lagoon's surface inflow from rivers or streams.
- Q_{so} is the lagoon's surface outflow to river or streams.
- A_s is the lagoon's area for the current time step.
- s_p is the lagoon's recharge or discharge flow due to its interaction with the adjacent aquifer.

2.5. Calibration of the Groundwater Model

The process of calibration was divided in two steps: (1) manual calibration performed by perturbing the parameters utilized in the input files of the simulation packages and (2) performing sensitivity analysis of the results. This calibration was based on the validation of known factors, for example, the water levels, the water volumes, and the total area of the Aculeo Lagoon. On the other hand, the Automatic Calibration process was implemented utilizing UCODE [15], a program developed by the USGS and the US Army Corps of Engineers that allows the implementation of inverse modeling through no lineal regression to estimate the elastic parameters of the aquifer, i.e., hydraulic conductivity, storage, and anisotropy [15]. UCODE utilizes the static levels of the wells to assimilate the response of MODFLOW and modify the parameter values. A comparison between observed and simulated groundwater levels in the lagoon was used to evaluate the calibration performance. The coefficient of determination was significant for each of the three 5-year simulation periods: the base or control year (2006–2011) under normal hydroclimatological conditions, the recorded beginning of the drought (2012–2017), and the total drying of the lagoon (2018–2023).

3. Results

The simulation results suggested that the water imbalance was caused by the combination of the following possible factors; (1) high groundwater demands, which were kept constant at about 450-500 L/s during the three simulation periods; (2) the drying of the lagoon's natural and/or man-made stream tributaries, which together contributed at least 300–400 L/s to the water body; and (3) the decrease in rainfall defined as climatic mega-drought (see Supplementary Material for details of the rainfall records). This decrease has affected the aquifer recharge which normally ranged between 200 and 250 L/s as an average for the basin. All the above factors affected the natural process of water capture, storage, and natural drainage, which resulted in a complete drying over a period of about five years. For example, it has been historically suggested that the pumping rate of 500 L/s was sustainable for the lagoon's water volume and levels; however, the simulation showed that this rate is sustainable only under normal hydrogeological conditions, with tributaries running towards the lagoon, and with normal rainfall accumulation. On the other hand, the model showed that this pumping rate was not sustainable for the lagoon's water volume and levels, suggesting that pumping rates should be adjusted taking into consideration the most recent hydroclimatological and hydrogeological conditions of the basin and the lagoon. A more detailed discussion of each process is conducted in the following sections of this paper.

3.1. Capture and Natural Drainage

The results revealed the presence of an important area of groundwater capture in the Aculeo Lagoon domain. The groundwater captured by the lagoon covers an area greater than 60% of it (considering an area of 11.5 km^2) and dominated the water storage in the first 30 m below ground surface. This capture by the Aculeo Lagoon is probably generated by the combination of several factors, as, for example, the groundwater pressure gradient generated from the mountainous areas surrounding it, or the hydraulic characteristics of the transport phenomena in the Lagoon, which can produce groundwater flow in a vertical (bottom) jet-like manner due to mixing of water with differing temperatures and densities [16]. Additionally, the presence of confining layers or areas of greater water transmissibility, such as those observed in the lagoon, could also lead to the development of artesian gradients toward the water body, which would explain how this natural capture process occurs. The shallow water and groundwater drainage (outflow) occur in a vertical and horizontal jet like-manner in an area close to the outlet (east) of the Aculeo Lagoon (Figure 4). The natural drainage in the lagoon occurs in a smaller area compared to the capture area. However, the rate of groundwater drainage flow from the Lagoon into the local aquifer can be up to four times faster than the captured flow (Figure 5).

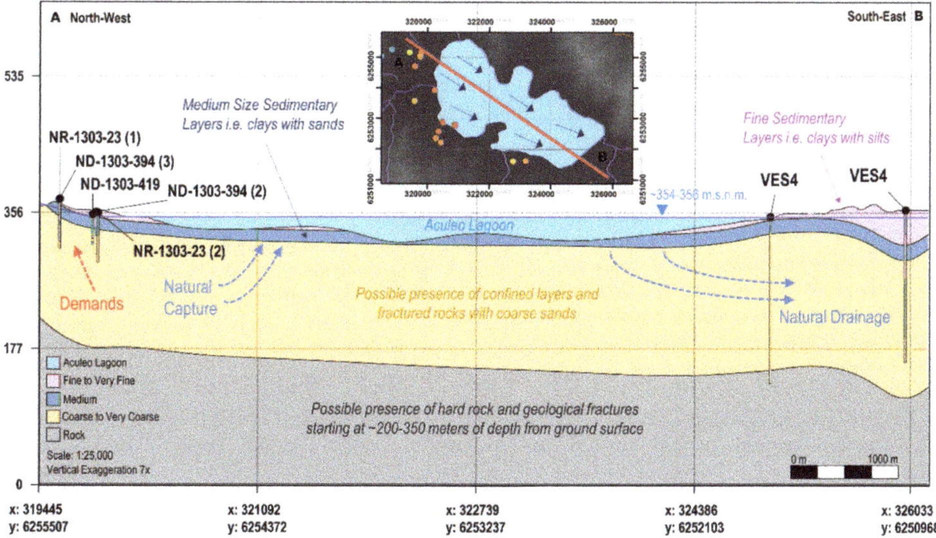

Figure 4. Cross-section along the Aculeo Lagoon (northwest to southeast). The main geological setting together with the hydrogeologic processes of the Lagoon are also conceptualized.

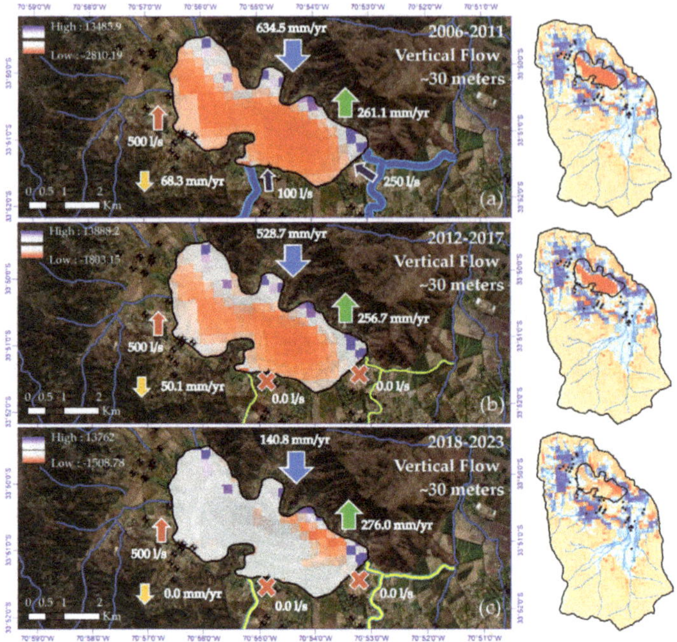

Figure 5. Vertical (recharge and discharge) instantaneous flow (m^3/s) at 30 m deep (boundary between layers 2 and 3) at the end of each MODFLOW simulation for the Aculeo basin/lagoon. The (daily time-step) simulation was carried out for a spatial resolution of 250 square meters and using the initial conditions of three 5-year periods (**a**) 2006–2011, (**b**) 2012–2017, and (**c**) 2018–2023. The lagoon's water levels calibrated for each period are 350.92, 349.66, and 345.0 m.a.s.l., which are associated with depths of 5.9, 4.6, and 0 m, respectively. The aquifer area was delineated in concordance with [17]. The fluxes: annual rainfall accumulation (blue arrow) in mm/year; annual evapotranspiration (green arrow) in mm/year; the recharge (yellow arrow) in mm/year; the pumping rate (red arrow) in L/s; and the lagoon's tributaries (dark-blue arrows) in L/s. The map on the right side shows the discharge and recharge rates for the whole Aculeo basin.

3.2. Long-Term Variability of Natural Capture and Drainage

The capture and drainage areas of the lagoon did not vary significantly between the three 5-year simulation periods (2006–2011, 2012–2017, and 2018–2023). For example, at 30 m deep, it was observed that areal capture decreased by only 4.8% between 2006 and 2012 (Figure 5a,b), and by approximately 3.9% between 2006 and 2018 (Figure 5a,c). However, the groundwater flow captured by the lagoon decreased by 21.5% between 2006 and 2018. On the other hand, the groundwater outflow (or recharge from the lagoon to the aquifer) increased by 25.4% between 2006 and 2018; it was mainly observed at around the first 30 m of depth, and it was located mainly in the northeastern and eastern portions of the lagoon (closer to the outlet of the basin).

3.3. Surface Water Balance

The Aculeo Lagoon and its local aquifer constantly interact through a process of natural capture (storage gain) and drainage (storage loss). The changes in the volume of the lagoon over the three simulation periods strongly correlate with the changes observed in the rainfall accumulation with the presence of surface tributaries to the lagoon and with the changes observed in the surface water and groundwater drainage of the Lagoon. Between 2006 and 2011 as well as 2012 and 2017, the decrease in rainfall (~16.7%) did not generate negative changes in natural recharge. However, this situation changed radically during the period 2018–2023, when the amount of rainfall captured by the basin dropped by approximately 77.8% compared to 2006, resulting in a total loss of natural recharge (Figure 6). The reduction of rainfall presumably made less water available for the evapotranspiration process; this hydrological partition flux, however, did not reveal significant changes. In fact, evapotranspiration volumes only decreased by 25.9% (36.6%) between 2006 and 2011 as well as 2012 and 2017 (2006–2011 and 2018–2023). The simulation results showed that the relationship between rainfall accumulation and evapotranspiration (P/EVT) increased from 20% to 70% (Figure 6). The hydrological partition generated accumulated dry conditions and affected the aquifer recharge rates, which dropped dramatically to nearly zero according to the simulation results (see Figure 6).

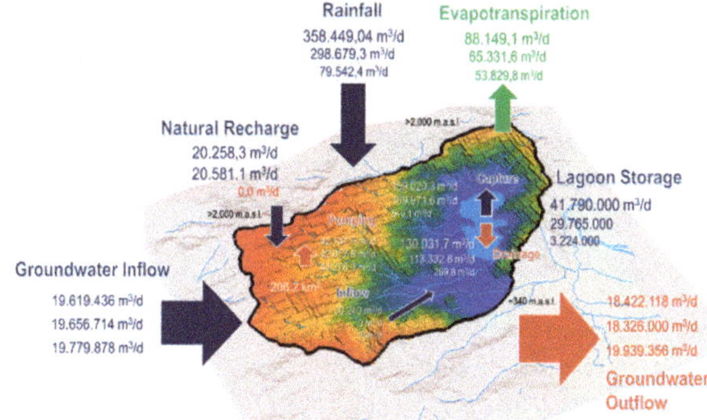

Figure 6. Simulation results of the surface and groundwater balance of the Aculeo basin and its lagoon, for each 5-year simulation period (2006–2011, 2012–2017, and 2018–2023). Each flux of the water balance has three values representing each simulation period, going from normal to dry conditions (top to bottom).

3.4. Water Balance for the Lagoon–Aquifer Interaction

The deviation of the surface tributaries to the lagoon, combined with both the sustained decrease in rainfall and the high demands for groundwater pumping (which remained constant over the study period at ~500 L/s), resulted in important changes in the lagoon–aquifer interaction and the lagoon's water balance. Historically, the total volume of groundwater captured (storage gain) by the lagoon was greater than the volume of water drained. The climatic drought reflected in the initial conditions used for the 2018–2023 simulation suggested a significant decrease in the natural recharge and significant reductions in the total capture–drainage rates of the lagoon. The 41.8 Mm^3 of water present in the lagoon during 2006 was balanced with a net gain or storage (capture minus drainage) of ~20,000 m^3/day of groundwater, plus 30,000 m^3/d of surface tributaries. However, during 2018–2023, the net storage was ~600 m^3/day, and there was zero surface water capture to contribute to this. This reduction in surface water capture, when combined with the constant groundwater demands and climatic drought, have put a significant strain on the groundwater resources (0–30 m deep) in the lagoon area, and the simulation results for the latest period suggest no positive future changes if the final water balance in the basin continues in the same manner. Under this watershed management scheme, the extreme dryness of the Lagoon will continue affecting the economy of the basin, the social well-being in the area, the natural ecosystems, and probably the capacity of this small basin to regulate precipitation recycling processes as described by [18].

4. Discussion Regarding the Future implementation of a Monitoring and Recovery Plan for the Aculeo Lagoon

Based on our modeling results, the development of a Monitoring and Recovery Plan (MRP) for the Aculeo Lagoon should consider one or a combination of three feasible options: (1) the recovery of natural tributaries, (2) the reduction and regulation of the pumping and storage of groundwater, and (3) an analysis of the feasibility of sustainable water importation alternatives either from groundwater or from nearby basins with available water rights. To be able of efficiently inform the next round of regional groundwater modeling and monitoring strategies, the MRP implemented for the Aculeo Lagoon must contain at least the following elements.

- Recommendations on the establishment of channels for the transportation of surface or underground water from the water source (rivers or pumping wells) directly into the lagoon.
- Identification of the specific locations and characteristics of all the monitoring wells located within the basin, and of the wells that are required to pump water into the lagoon—which will minimize the groundwater drainage flow from the lagoon to the basin's underground exit.
- Establishment of detailed objectives for groundwater levels that will be required to provide water security in the Aculeo basin while also ensuring acceptable surface water levels in the short, medium, and long term. Ideally, such levels should mimic historical annual changes (increases during winter and decreases in the summer), to restore the area's ecological system; in other words, the importation of water should occur annually during winter months.
- Suggestions about the implementation of coupled hydroclimatic–hydrogeologic prediction models that will guarantee advance planning of surface and groundwater resources used and available in the basin, and available for the lagoon.
- Suggestions about the primary legal regulations that should be implemented to better control groundwater extraction or replenishment in the basin and around the lagoon's area, i.e., the enforcement of groundwater permits with real-time monitoring to regulate water extraction or replenishment in the whole basin.
- An environmental education plan to minimize water consumption within the basin. This is just as important as the other elements of an MRP.

All the above will require significant investments that we have estimated to be approximately US $10 M for the first two years of implementation.

5. Future Updates and Improvements to the Aculeo Lagoon's Hydrogeological Model

The implementation and execution of an MRP of the Aculeo Lagoon needs to be informed and evaluated by an updated version of the conceptual and numerical hydrogeological model, which must be calibrated and validated for the entire Aculeo Basin and its lagoon, considering the new data generated during the MRP execution. In this context, it will be possible to establish a scheme of continuous improvement of the current Aculeo Hydrogeological Model (AHM) constructed in this study. Periodical refinements to the AHM will have to be implemented based on the most recent information about the drilling of new wells (monitoring or pumping), as well as in relation to the activities associated with the hydraulic tests and the recovery operations of the Lagoon. With this strategy, it will be feasible to plan future MRP activities, prioritizing the acquisition of the best possible results. The new updated and improved versions of the AHM could also include changes or improvements in the Aculeo Geological Model (AGM), and in turn may count on the use of new time series of groundwater levels obtained through continuous monitoring of the target areas proposed in the MRP. In particular, for the Aculeo Lagoon Basin, it is necessary to identify with a greater level of detail (a) the most feasible areas for capturing groundwater flow within the basin, (b) the domain and boundaries of geological areas that will define a precise extent of the lagoon's recovery program, (c) the areas that will be less prone to developing groundwater drainage through the implementation of the MRP, and (d) the groundwater saturation zones that could generate the conditions conducive to the development of transient flows in the basin or in the vicinity of the Aculeo Lagoon. To improve the level of detail of the future modeling schemes, the following tasks must be completed.

- Incorporation of new geological data, as well as the identification of the most important geological faults (structures) within the domain of the basin and the Aculeo Lagoon.

- Developing and updating a new hydrogeological database, obtained from the MRP's drilling program, which should include hydraulic tests, traces of new drilled wells, updated groundwater levels measured from historical wells, and data from new explorations.
- Compilation and interpretation of all historical information as well as that generated during the implementation of the MRP, to determine the most important hydrogeological controls on the local aquifer's hydraulic properties, and its interaction with the lagoon.
- Development, monitoring, and simulation of groundwater hydrographs using historical data and data available from new monitoring points. The available hydrographs should be evaluated and grouped according to their location and hydrogeological response.

6. Conclusions

Results from the analysis revealed that the lagoon–aquifer interaction responds to a natural capture–drainage or storage–drainage process, whose local surface and underground flow are dependent on the hydrological and hydrogeological conditions accumulated over time. From a hydrogeological point of view, and taking into considerations our simulation results, it is possible to suggest that the hydrodynamic balance of the lagoon has been significantly affected mainly by (1) the drying of surface tributaries and streams that provided a continuous flow of water and helped maintain the balance of the hydraulic water level of the lagoon; (2) a high existing demand on subsurface (shallow) water and groundwater in the first 30 m of depth, particularly demands for large-scale agricultural activities and human consumption; and (3) the climatic mega-drought, which has brought a sustained reduction of rainfall and, consequently, a gradual but significant loss of natural water recharge during the last 10 years. From the modeling results, we suggest that the current groundwater pumping rates must be immediately reduced in the whole basin, or at least in the critical areas closer to the lagoon, to minimize the impact of the ongoing and future hydro-climatic conditions on the local aquifer. The execution of an MRP to continuously monitor the lagoon will require investments estimated in about US $10M in the first two years of implementation. However, the implementation of this type of MRP will additionally require efforts and commitment of the Chilean government to not only recover the historical water levels of the lagoon and the local aquifer, but also to recover the economy of the basin, the social well-being in the area, and most importantly, all the natural ecosystems that existed in the Aculeo Basin.

Supplementary Materials: The following are available online at http://www.mdpi.com/2073-4441/12/1/290/s1, Figure S1: (**a**) Daily rainfall (mm/d) and the lagoon's water levels between 2000 and 2018. The storm events larger than 40 mm/d are less frequent after 2010; (**b**) Daily streamflow records (m^3/s) for Pintué streamgauge and lagoon's water levels. No records of streamflow at Pintué have been available since 2011; (**c**) Daily Maximum (blue), average (yellow), and minimum (red) temperatures for the Aculeo Basin; (**d**) Monthly rainfall (mm) accumulation (left-axis), and monthly streamflow (m^3/s) at Pintué streamgauge (right-axis); (**e**) Average Monthly Water Levels at Aculeo Lagoon (meters); (**f**) Monthly maximum and average temperature at Aculeo Basin. Table S1: Geodatabase used in this study. Table S2: Simplified categories of lithographic sedimentation and rock basement used in the development of the Aculeo Geological Model (AGM). Table S3: Summary of the main MODFLOW packages used in this study.

Author Contributions: Conceptualization, R.V.-P.; methodology, R.V.-P.; software, R.V.-P.; formal analysis, P.G.-C., J.B.V., and R.P.-T.; investigation, R.V.-P.; writing—original draft preparation, R.V.-P., P.G.-C., and J.B.V.; writing—review and editing, R.V.-P., P.G.-C., J.B.V., and R.P.-T.; funding acquisition, P.G.-C. All authors have read and agreed to the published version of the manuscript.

Funding: This research was funded by Gobierno Regional (GORE) Metropolitano de Santiago, División de Planificación y Desarrollo, Departamento de Planificación Regional, 2019.

Acknowledgments: The authors of this research sincerely acknowledge to GORE Metropolitano, and all Chilean students, and researchers for their support in the realization of this study. We also acknowledge the water users' organizations of the Aculeo Basin for their interest and support in the conclusions and suggestions of this study.

Conflicts of Interest: The funders of the project had no role in the design of the study; in the collection of field data, or in the analyses, or interpretation of data; in the writing of the manuscript; or in the decision to publish the results.

References

1. Valdés-Pineda, R.; Pizarro, R.; García-Chevesich, P.; Valdés, J.B.; Olivares, C.; Vera, M.; Balocchi, F.; Pérez, F.; Vallejos, C.; Fuentes, R.; et al. Water governance in Chile: Availability, management and climate change. *J. Hydrol.* **2014**, *519*, 2538–2567. [CrossRef]
2. Villa-Martínez, R.; Villagrán, C.; Jenny, B. The last 7500 cal yr BP of westerly rainfall in Central Chile inferred from a high-resolution pollen record from Laguna Aculeo (34° S). *Quat. Res.* **2003**, *60*, 284–293. [CrossRef]
3. Valdés-Pineda, R.; Valdés, J.B.; Diaz, H.F.; Pizarro-Tapia, R. Analysis of spatio-temporal changes in annual and seasonal precipitation variability in South America-Chile and related ocean–atmosphere circulation patterns. *Int. J. Climatol.* **2016**, *36*, 2979–3001. [CrossRef]
4. Garreaud, R.D.; Alvarez-Garreton, C.; Barichivich, J.; Boisier, J.P.; Christie, D.; Galleguillos, M.; Zambrano-Bigiarini, M. The 2010–2015 megadrought in central Chile: Impacts on regional hydroclimate and vegetation. *Hydrol. Earth Syst. Sci.* **2017**, *21*, 6307–6327. [CrossRef]
5. Wilhite, D.A.; Glantz, M.H. Understanding: The drought phenomenon: The role of definitions. *Water Int.* **1985**, *10*, 111–120. [CrossRef]
6. SERNAGEOMIN. Geologic Map of Chile, Scale 1:1.000.000. 2003. Available online: https://www.sernageomin.cl/ (accessed on 14 November 2019).
7. Eridandus. *Monitoreo Ambiental de Ecosistemas Acuáticos Estratégicos—Laguna de Aculeo. Informe Final—Revisión 1*; Ministerio del Medio Ambiente—SEREMI del Medio Ambiente: San Diego, Chile, 2016.
8. ARANZ, G. Leapfrog Geo. 2018. Available online: http://www.leapfrog3d.com/ (accessed on 28 November 2019).
9. Harbaugh, A.W.; Banta, E.R.; Hill, M.C.; McDonald, M.G. User Guide to Modularization Concepts and the Ground-Water Flow Process. In *MODFLOW-2000 the U.S. Geological Survey Modular Ground-Water Model: U.S. Geological Survey Open-File Report*; USGS: Reston, VA, USA, 2000; pp. 92–121.
10. Harbaugh, A.W. MODFLOW-2005, the U.S. Geological Survey Modular Ground-Water Model—The Ground-Water Flow Process. *USA Geol. Surv. Tech. Methods* **2005**, *6*, 16.
11. Niswonger, R.G.; Panday, S.; Ibaraki, M. MODFLOW-NWT, a Newton formulation for MODFLOW-2005. *USA Geol. Surv. Tech. Methods* **2011**, *6*, 44.
12. Panday, S.; Huyakorn, P.; Therrien, R.; Nichols, R. MODFLOW–USG version 1: An unstructured grid version of MODFLOW for simulating groundwater flow. *Adv. Water Resour.* **2013**, *25*, 497–511.
13. Merritt, M.L.; Konikow, L.F. *Documentation of a Computer Program to Simulate Lake-Aquifer Interaction Using the MODFLOW Ground-Water Flow Model and the MOC3D Solute-Transport Model (No. 4167)*; US Department of the Interior, US Geological Survey: Reston, VA, USA, 2007.
14. Hunt, R.J.; Haitjema, H.M.; Krohelski, J.T.; Feinstein, D.T. Simulating ground water-lake interactions: Approaches and insights. *Groundwater* **2003**, *41*, 227–237. [CrossRef] [PubMed]
15. Poeter, E.P.; Hill, M.C. UCODE, a computer code for universal inverse modeling. *Comput. Geosci.* **1999**, *25*, 457–462. [CrossRef]
16. Starosolszky, Ö. Lake Hydraulics. *Hydrol. Sci. J.* **1974**, *19*, 99–114. [CrossRef]
17. Dirección General de Aguas (DGA). *Reevaluación de la Delimitación y de la Disponibilidad de Recursos Hídricos Subterráneos en el Acuífero El Monte, en la Región Metropolitana*; Informe Técnico DARH No. 89; DGA: Santiago, Chile, 2018.
18. Pizarro, R.; García-Chevesich, P.; Valdés-Pineda, R.; Dominguez, F.; Hossain, F.; Ffolliott, P.; Olivares, C.; Morales, C.; Balocchi, F.; Bro, P. Inland water bodies in Chile can locally increase rainfall intensity. *J. Hydrol.* **2013**, *481*, 56–63. [CrossRef]

Communication

Acute Water-Scarcity Monitoring for Africa

Amy McNally [1,*], Kristine Verdin [1], Laura Harrison [2], Augusto Getirana [3], Jossy Jacob [4], Shraddhanand Shukla [2], Kristi Arsenault [1], Christa Peters-Lidard [5] and James P. Verdin [6]

1 SAIC, Reston, VA 20190, USA and NASA GSFC, Greenbelt, MD 20771, USA; kristine.l.verdin@nasa.gov (K.V.); kristi.r.arsenault@nasa.gov (K.A.)
2 UC Santa Barbara Climate Hazards Center, Santa Barbara, CA 93106, USA; harrison@geog.ucsb.edu (L.H.); sshukla@ucsb.edu (S.S.)
3 UMD ESSIC, College Park, MD 20740, USA and NASA GSFC, Greenbelt, MD 20771, USA; augusto.getirana@nasa.gov
4 SSAI, Lanham, MD 20706, USA and NASA GSFC, Greenbelt, MD 20771, USA; jossy.p.jacob@nasa.gov
5 NASA GSFC, Greenbelt, MD 20771, USA; christa.d.peters-lidard@nasa.gov
6 USAID Office of Food for Peace, Washington, DC 20004, USA; jverdin@usaid.gov
* Correspondence: amy.l.mcnally@nasa.gov

Received: 15 August 2019; Accepted: 18 September 2019; Published: 21 September 2019

Abstract: Acute and chronic water scarcity impacts four billion people, a number likely to climb with population growth and increasing demand for food and energy production. Chronic water insecurity and long-term trends are well studied at the global and regional level; however, there have not been adequate systems in place for routinely monitoring acute water scarcity. To address this gap, we developed a monthly monitoring system that computes annual water availability per capita based on hydrologic data from the Famine Early Warning System Network (FEWS NET) Land Data Assimilation System (FLDAS) and gridded population data from WorldPop. The monitoring system yields maps of acute water scarcity using monthly Falkenmark classifications and departures from the long-term mean classification. These maps are designed to serve FEWS NET monitoring objectives; however, the underlying data are publicly available and can support research on the roles of population and hydrologic change on water scarcity at sub-annual and sub-national scales.

Keywords: drought; early warning; water scarcity; water supply; routine monitoring; hydrologic modeling; remote sensing; GIS

---

1. Introduction

Reliable and up-to-date information about chronic and acute water scarcity is a much-needed resource for tackling regional and global water issues. An estimated four billion people currently face water scarcity [1]. By 2030 global population may reach 9 billion [2], and demands on domestic, industrial and agricultural water needs are expected to substantially increase during that time, potentially by 40% [3]. In chronic water-scarcity areas, where demand persistently or seasonally exceeds supply, higher population and increased sectoral demands will likely require new strategies to strengthen long-term resilience. For acute, shorter-term, water-scarcity issues, droughts will continue to pose challenges in many regions. For Africa, large population growth and climate change-driven projected decreases in precipitation and runoff [4] are especially concerning for future chronic and acute water scarcity. Chronic and acute water scarcity have implications for national security [5], food security [6], and for achieving sustainable development goals through the food–water–energy nexus [7]. This study describes a resource that can inform decision making on water scarcity issues. Good assessments and planning require up-to-date and accurate information [8], but despite the need, there has been a lack of near real time, acute water scarcity monitoring tools that meet these standards.

One of the first global scale indices of water scarcity was by Falkenmark [9] who computed annual water per capita (total annual runoff (m^3/year)) and classified values ranging from "no stress" (>17,000 m^3/person/year) to "absolute scarcity" (<500 m^3/person/year). These thresholds are still commonly used and have been widely adopted given their ease of computation and interpretation. Ratios of supply-to-demand and different critical thresholds for municipal, industrial, and agricultural activities are sometimes used to identify sector-specific scarcity (e.g., [10]). Alternatively, WaterGAP [11] uses water use models to compute demand at each pixel. Pfister et al. [12] relates water withdrawals to water availability. More recently, the concept of a water footprint [13] has been used to describe sub-annual variations in water availability. In this approach, the blue water footprint represents the amount of water supplied to and consumed by domestic and industrial sectors, while the green water footprint represents the amount of water supplied to and consumed by vegetation (natural or agricultural). An added benefit is that this approach allows for the consideration of return flows from these sectors (e.g., [14]). Finally, a number of studies have examined long-term changes in water availability (e.g., [15]) and future projections of population, water use, and water supply (e.g., [16]). These are just a few indicators of chronic water scarcity that have been exhaustively reviewed [17–20]. The disadvantage of these indicators is that they are not routinely updated despite ever-changing water supply and water demand conditions.

Information about chronic water scarcity, the focus of many prior studies, is valuable for high-level policy guidance. In addition, humanitarian relief efforts require information that can provide a basis for responding to acute events, defined as a short-term phenomenon, related to a natural or human-made shock. Early identification of emerging acute water scarcity situations requires that information be reliable and timely, and thus be regularly updated at seasonal, monthly, or sub-monthly time scales. There are several hydrologic modeling systems that are used for drought monitoring, e.g., the North American Land Data Assimilation System (NLDAS) [21] and the Princeton Climate Analytics which includes the Africa Flood and Drought Monitor [22]. Monitoring systems like Princeton's and NLDAS provide timely hydrologic information but they only address the supply side of water availability. Our objective is to produce a timely monitoring product for acute scarcity that also incorporates societal water demand.

We propose to combine insights from both the chronic water-scarcity community and the drought-monitoring community, to provide accurate and low latency (~1 month) information for acute (near-term) decision support. As noted earlier, prior studies have explored a variety of ways to estimate water demand. However, these more complicated approaches are difficult to deploy and as a result their use is often limited to the original study [19]. Specifically, some water demand data is available in a tabular format in the literature [23,24] or via The United Nation's Food and Agriculture Organization's (FAO) Aquastat, however to our knowledge, geospatial databases of water withdrawals and consumption are only available from the World Resources Institute's (WRI) Aqueduct program [25]. These difficulties bring us back to the pioneering work by Falkenmark, who used population as a simple proxy for demand. This approach allows for routine monitoring of acute water scarcity to provide effective decision support while jointly considering dynamic changes in both supply and demand. In the case of the Famine Early Warning System Network (FEWS NET), monthly updated water scarcity maps have potential to provide input to monthly food assistance outlooks that support humanitarian decision making.

This paper presents a system to provide early warning and identify emerging regional hot spots for water scarcity associated with climate extremes. The system described in this paper focuses on Africa, but could be extended globally. The system's routine hydrologic analysis is based on satellite driven inputs to hydrologic modeling simulations and monthly water availability per capita estimates. Specifically we use the FEWS NET Land Data Assimilation System (FLDAS [26]) driven with Climate Hazards Group InfraRed Precipitation with Stations (CHIRPS) [27] rainfall, gridded population data, and a novel formulation of the Falkenmark Index. Given the lack of reliable, high-resolution water-use data, we use high-resolution population estimates from 2015 to inform water management in near real time. As described in the Materials and Methods section, the system builds upon FLDAS drought monitoring capabilities to also support investigations into human water scarcity.

The Results section demonstrates applications of our water scarcity index beginning with the Kariba Dam on the border of Zambia and Zimbabwe which experienced extreme dry conditions in 2019. Next, we ask questions regarding the sensitivity of water scarcity to both supply and demand. Specifically, we ask: over continental Africa, (1) how has water scarcity changed over time as a function of changes in streamflow? (2) How has water scarcity changed over time as a function of changes in population? Then for the Lake Victoria basin, where human pressures are challenging water managers [28], we explore the combined impacts of both streamflow and population change and how that manifests in our water scarcity index. We conclude with a discussion of strengths and weaknesses of our approach and avenues for future work and improvements.

2. Materials and Methods

We base routine water scarcity mapping (Figure 1) on outputs from the FLDAS which is a custom instance of the National Aeronautics and Space Administration (NASA) Land Information System (LIS [29]). The FLDAS's Noah 3.6 land surface model [30] is driven by CHIRPS rainfall [27] and NASA's Modern-Era Retrospective analysis for Research and Applications (MERRA-2) meteorological forcing [31]. The model partitions rainfall inputs into evapotranspiration, soil moisture storage, surface and subsurface runoff (i.e., baseflow). Surface runoff results from precipitation in excess of saturation and infiltration capacity, while subsurface runoff is gravity drainage from the bottom soil moisture layer. The sum, or total runoff, is routed through the river network with the Hydrologic Modeling and Analysis Platform version 2 (HyMAP-2) river routing scheme [32,33] using the local inertia formulation [34] to simulate surface water dynamics, including streamflow (m^3/s), in rivers and floodplains.

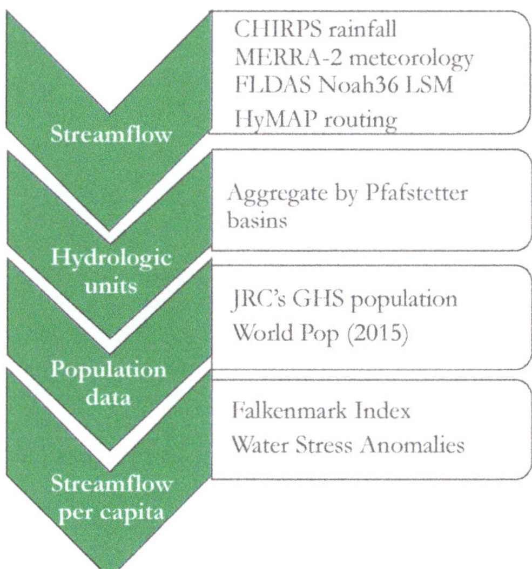

Figure 1. Our approach for water scarcity monitoring for Africa, with the Climate Hazards Group InfraRed Precipitation with Stations (CHIRPS) rainfall and Modern-Era Retrospective analysis for Research and Applications (MERRA-2) meteorological estimates used as input to the Famine Early Warning System Network Land Data Assimilation System (FLDAS) Noah 3.6 + Hydrologic Modeling and Analysis Platform (HyMap) routing scheme. Streamflow and population estimates are aggregated by Pfafstetter basins before computing Falkenmark water-scarcity categories, and water-scarcity anomalies. Routinely updated at https://lis.gsfc.nasa.gov/projects/fewsnet.

To define catchments we use boundaries from the U.S. Geological Survey (USGS) Hydrologic Derivatives for Modeling Applications (HDMA) database [35]. Catchments are attributed with Pfafstetter codes, based on a hierarchical numbering system, that carry important topological information. For instance, the Nile Basin is Pfafstetter level 1, and the Blue Nile basin is Pfafstetter level 2. For this application, we use Pfafstetter level 6 basins, to represent the relatively local nature of water supplies.

As a proxy for water demand, and consistent with the Falkenmark index, we use two population datasets. We use the WorldPop 2015 data [36] for routine mapping since these data provide better definition of the population distribution within the Pfafstetter level 6 basins than other data that we explored. For research questions regarding changes in population and water availability over time, we used the European Commission's Joint Research Center's (JRC) Global Human Settlement (GHS) data ([8]). These data are available for 1990, 2000 and 2015. Future work will explore how to use these data for time varying population in the routine scarcity maps, which will provide context to users on how both the changes in population and hydrology manifest as current water stress conditions.

As a proxy for demand, we found population to be adequate because, first, withdrawal intensity and population have the same spatial patterns. Second, population data are available for different points in time and, third, a goal of this effort was to produce an operational framework where alternative datasets can be tested in the future. We explored using the water withdrawals and consumptive use datasets generated by the WRI Aqueduct program [25]. However, we had difficulty reconciling the magnitude of streamflow simulations with WRI withdrawal and consumptive values. Our attempt at using sub-national data on surface water and groundwater use estimates (i.e., blue water footprint [13]) by sector [24] also was challenging because the data were not mapped. These and other alternatives should be explored for future efforts to better capture water demands from agriculture and other sectors.

The data and models used in our system, summarized in Table 1, are publicly available, with the exception of the HyMAP-2 river routing scheme, associated parameters, and streamflow outputs. HyMAP-2 and parameters are anticipated to be in a future LIS release, but is still in development at the time of this study. MERRA-2 and monthly FLDAS outputs are provided by the NASA Goddard Earth Sciences Data and Information Services Center (GES DISC). While the spatial domain for all of the inputs is global, for initial development of this routine water scarcity mapping workflow we focused on the Africa domain where many FEWS NET countries of interest are located.

Using these inputs, we developed a new, modified version of the Falkenmark Index for routine water scarcity monitoring. Falkenmark Index thresholds (Table 2) are specified annually, however, we required monthly updates of scarcity conditions. To accomplish this, we used a 12-month running total of the streamflow from the current and 11 previous months. This allowed us to use the standard Falkenmark thresholds while also reflecting the most up-to-date water scarcity conditions. We found this 12-month running-total approach to be superior to a simple monthly calculation (dividing the thresholds by 12), which resulted in the frequency of "no stress" or "absolute scarcity" conditions being too high. The reason for this is that deficits and surpluses were not carried over from month-to-month beyond what is implicit in the hydrological models.

The final products of our analysis that appear on the NASA FEWS NET project website (https://lis.gsfc.nasa.gov/projects/fewsnet) are streamflow anomalies, the water scarcity index, and water scarcity anomalies which are calculated as follows.

The average of the routed streamflow is calculated for each Pfafstetter basin level 6 from the HDMA database [35] and converted to a volume of water (m^3/month) (Figure 2a).

Table 1. Datasets used for routine water scarcity mapping.

Dataset or Model	Description	Data Citation	Data Availability
Climate Hazards Center InfraRed Precipitation with Station data (CHIRPS) rainfall	input to FEWS NET Land Data Assimilation System (FLDAS), available 1981–present	Funk et al. 2015 [27]	University of California, Santa Barbara (UCSB)
Modern-Era Retrospective analysis for Research and Applications Version 2 (MERRA-2) meteorology	input to FLDAS, availability 1979–present	Gelaro et al. 2017 [31]	NASA Goddard Earth Sciences Data and Information Services Center (GES DISC)
U.S. Geological Survey (USGS) Hydrologic Derivatives for Modeling and Applications (HDMA) basins	used for spatial aggregation	Verdin 2017 [35]	USGS
WorldPop	2015 population estimate	Linard et al. 2015 [36]	WorldPop
FLDAS-Noah.36	land surface model, monthly outputs available 1982–present	McNally et al. 2017 [26]	NASA GES DISC
Hydrologic Modeling and Analysis Platform (HyMAP-2) routing	river routing scheme, beta version	Getirana et al. 2012 [33]; 2017 [32]	NA

Table 2. Falkenmark categories.

Category	m^3/year/capita
no stress	>1700
stress	1000–1700
scarcity	500–1000
absolute scarcity	<500

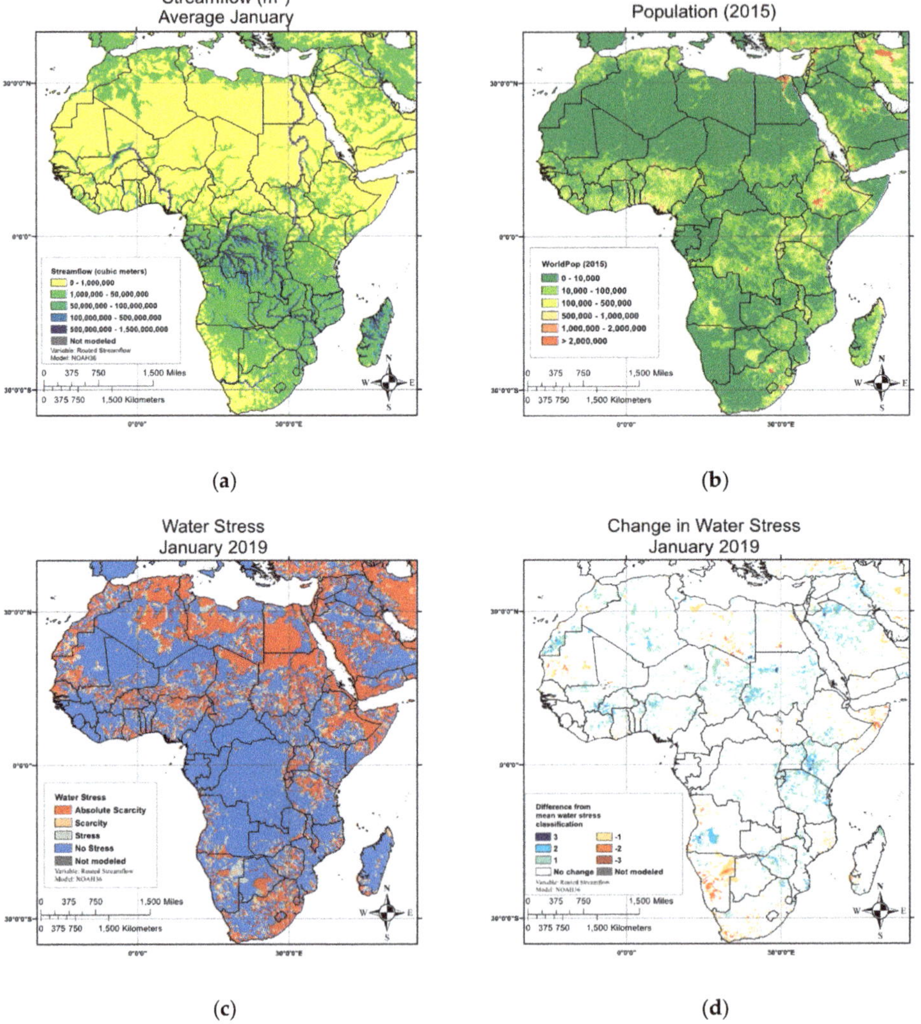

Figure 2. The components of the Africa water scarcity monitoring scheme using average January streamflow as an example: (**a**) Streamflow for January plus 11 previous months, averaged over Pfafstetter 6 basin units; (**b**) WorldPop 2015 population density averaged over Pfafstetter 6 basin units; (**c**) example map of the Falkenmark Index categories in January 2019; (**d**) Example map showing January 2019 departures from average January conditions, highlighting acute events.

Anomalies are calculated as a percent of the mean as:

$$\text{Anomaly (\%)} = [(\text{streamflow}_{m,y}\ \text{climatology}_m)/\text{climatology}_m] \times 100 \quad (1)$$

where m is a given month and y is a given year. Climatology_m is the mean streamflow for each 12-month period based on the 1982–2016 FLDAS historical record.

To compute water scarcity, first, we aggregate population estimates to the Pfafstetter basin level 6 (Figure 2b). Then we divide the given 12-month total spatially aggregated streamflow (m^3) by the population to produce an estimate of m^3/person. We then produce maps where each basin's water availability is classified by the values in Table 2 (Figure 2c).

Next, to highlight acute water scarcity conditions, we map the departure from average conditions for that 12-month period. Average water scarcity conditions are based on the ratio of the climatological mean 12-month streamflow total (1982–2016 FLDAS simulations) and the most recent 2015 WorldPop population. Water scarcity for both climatology and the current month is assigned a numeric value 1–4 (absolute scarcity to no stress, following categories listed in Table 2), which allowed us to compute the difference of the current water-scarcity classification from the climatology (Figure 2d). We show maps of the components of this system for January 2019 maps as examples of system inputs (Figure 2a,b) and outputs (Figure 2c,d). Streamflow and outputs vary by month and year, while population inputs are currently static in the system.

3. Results

The primary goal of this paper is to describe the routine water scarcity monitoring system (see Section 2. Materials and Methods), and show applications to acute water scarcity monitoring. We begin with an example for Lake Kariba, a dam along the Zambia and Zimbabwe border, where variations in measured lake levels generally correspond to time-varying information in water stress anomaly maps. We then explore combined impacts of both streamflow and population change and how that manifests in our water scarcity index, first for continental Africa, and then for the Lake Victoria basin, where human pressures are challenging water managers [28].

3.1. How Well Do the Maps Represent the Water Scarcity Events in Zambia and Zimbabwe That Affect the Kariba Dam?

Figure 3 shows monthly observed height levels for Lake Kariba from the US Department of Agriculture (USDA) Global Reservoir and Lakes (G-REALM) project, measured by satellite altimetry data [37], and water stress maps for three selected months for comparison. According to the January 1996 map (Figure 3, bottom left) numerous basins in the southern Zambia and northern Zimbabwe region were in a water stress category that was 1–3 classes drier than would be expected in a typical January, and this corresponds with measured lower than average Lake Kariba height. In January 2011 (Figure 3, bottom middle) numerous basins in that region show water stress conditions in a more positive state than a typical January, around 1–2 classes higher than average, which corresponds to measured higher than average reservoir height at that time. Most recently, in June 2019, Figure 3 data show impacts of a drier than average 2018–2019 rainfall season on basins in the region. Dry conditions are reflected in both reservoir heights and the water stress classification. News reports from 2019 confirmed low levels (29% full [38]) in the Kariba Dam, reducing hydropower supply and threatening complete shut-down [39,40]. Water shortages were also reported in Botswana [41] in 2019 and in Namibia, where severe drought left many people without access to clean water for crops and drinking water for people and livestock [42]. The June 2019 map indicates higher than average water stress categories in these areas

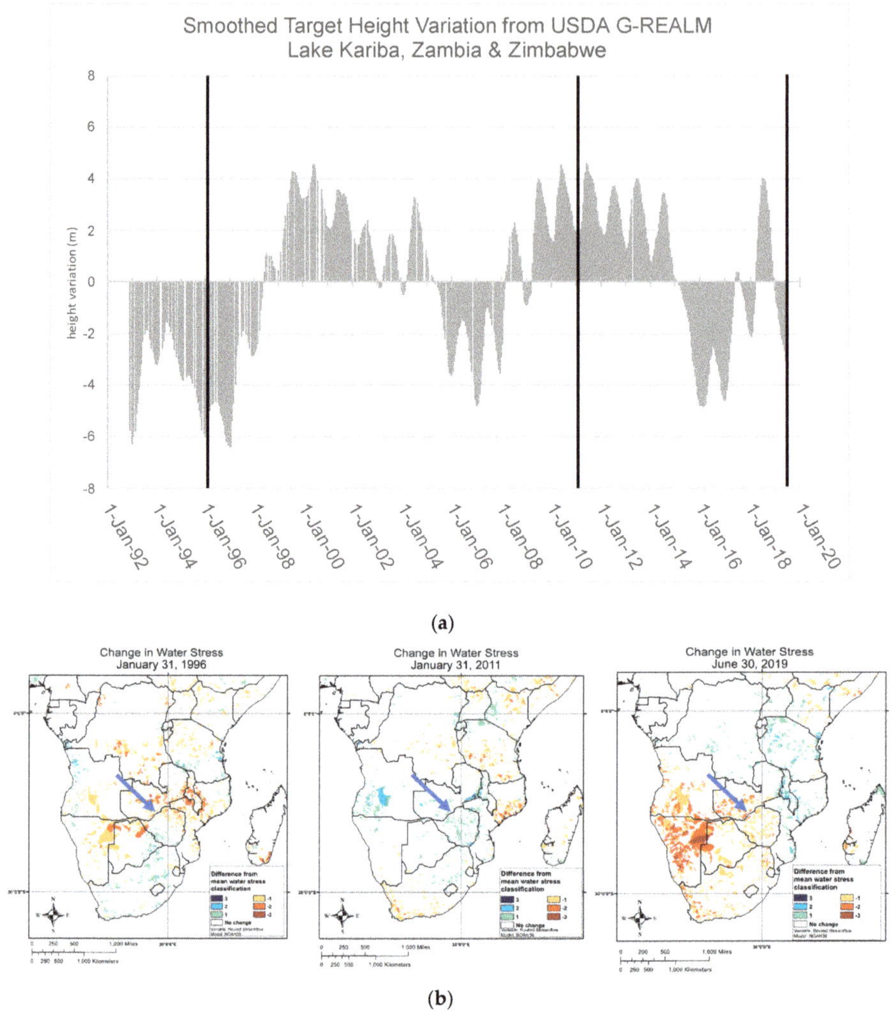

Figure 3. (**a**) US Department of Agriculture (USDA) Global Reservoir and Lakes (G-REALM) altimetry data for Lake Kariba, located on the border with Zambia and Zimbabwe, black lines denote dates shown in maps. (**b**) Maps of change in water stress class with respect to average. Blue arrow denotes location of Kariba Dam.

3.2. How Have Water Scarcity Classes Changed over Time as a Function of Changes in Streamflow?

In addition to monitoring conditions, we wanted to understand how water scarcity has changed over time exclusively as a function of streamflow, holding population constant. To answer this question, first we computed the difference in mean annual streamflow for an early period (1990–2002) and a late period (2003–2015). The difference between the periods is shown in Figure 4a. Notable patterns in this figure are the decrease in streamflow for central Africa, southern Tanzania, southern Kenya, Somalia, Guinea, southern Africa, and western Madagascar. Meanwhile other regions have shown a marked increase: Lake Victoria region, Malawi, Zambia, northern Mozambique, and Ghana and neighboring countries.

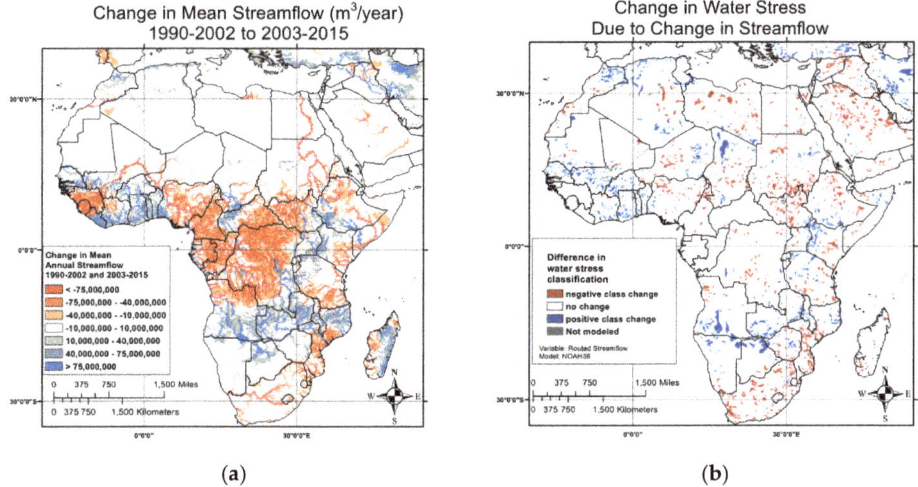

(a) (b)

Figure 4. (**a**) Difference in mean annual streamflow between 1990–2002 and 2003–2015; (**b**) change in Falkenmark water-scarcity class due to hydrologic change (using WorldPop 2015 population). Change is ±1 class following patterns in streamflow change, with the exception of a few basins in central Kenya, central Mozambique, and South Africa.

We explored using a linear trend for this analysis but favor period-differences for several reasons. First, linear trends are sensitive to extreme events, particularly at the end-point of the time period. Second, linear trends are sensitive to the time period that is used, making stated trends non-robust and easy to misinterpret. A third reason is that we are particularly interested in how, over the past 20–30 years, have people been experiencing changes in water availability on a decadal scale.

The patterns in Figure 4a are generally confirmed by literature that references water availability trends estimated from the Gravity Recovery and Climate Experiment (GRACE) satellite (2002–2014). GRACE measures total water storage changes over time [43,44]. Changes in FLDAS streamflow are largely driven by increases or decreases in rainfall, which is consistent with results in Rodell et al. [43]. For example, Rodell et al. [43] and FLDAS streamflow show a wetting trend in southwestern Africa, in the Okavango Basin that includes Angola, Namibia, Botswana and Zimbabwe (Figure 4a). Farther south and east, in South Africa, eSwatini, Lethsoto, and Southern Mozambique there are drying trends due to decreasing rainfall. Both Rodell et al. and FLDAS streamflow show increases in the West African country of Ghana (and neighboring countries) with increases in rainfall. Rodell et al. attribute increases in the Lake Victoria region to increasing lake levels and groundwater. These characteristics are not modeled with FLDAS's Noah 3.6, but are ultimately driven by increases in rainfall.

While hydrologic trends are interesting, they do not address how much these changes matter in the context of water scarcity per capita. Here we ask, what is the effect of these changes in the context of the Falkenmark Index? To answer this question, we compute the Falkenmark index for the early and the late periods and map the changes in class (Figure 4b). The positive and negative spatial patterns are generally similar, but show interesting results with respect to the scope of societal impacts. In areas with a change in water stress class, the change associated with a streamflow change is most often to a class ±1 level higher or lower. Exceptions, with more extreme class changes, are seen in a few basins in central Kenya, central Mozambique, and South Africa. Also interesting is that results do not show prominent increases in water stress associated with large streamflow declines across the Democratic Republic of the Congo (DRC), Republic of the Congo, Gabon, and Cameroon, but they do show more prominent increases in water stress in South Sudan. This is related to differences in regional hydro-climatology, with most of central-west Africa being consistently wet enough to remain in the

"No Stress" Falkenmark category while less wet areas have more opportunity to transition between stress categories.

3.3. How Have Water Scarcity Classes Changed over Time as a Function of Changes in Population?

In addition to understanding changes in supply, we wanted to understand how water scarcity has changed over time as a result of demand, represented by changes in population. To answer this question, we use the average annual streamflow from 1990–2015, and computed the change in population from 1990 to 2015 using the JRC GHS data. Figure 5a shows the change in population between the two time periods, 1990–2002 and 2003–2015. Brown colors indicate increasing population, which is the case over much of the continent. Particularly high population growth is shown in Ethiopia, Nigeria, and Lake Victoria basin.

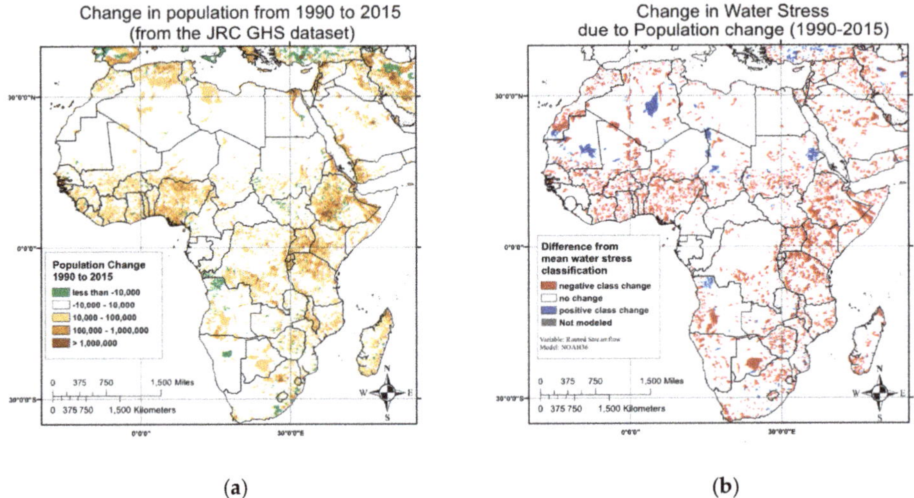

Figure 5. (**a**) Change in population between 1990–2002 and 2003–2015 from the JRC Global Human Settlement (GHS) dataset. Population is largely increasing, some places more than others. Ethiopia, Nigeria and Lake Victoria basins are notable; (**b**) using average streamflow conditions the effect of population has been to increase water scarcity. Change in classification due to population change is almost exclusively negative (i.e., scarcer).

Next, we ask: what is the effect of these changes in the context of the Falkenmark Index? Patterns in the change in water scarcity maps are similar to changes in population (Figure 5b). Unlike changes in hydrology that had a ±1 class impact on water availability, changes in population results in a −1 to −3 change in water scarcity class (not shown). While population is a simplistic proxy for demand, this result highlights the importance of including demand in the analysis of hydrologic changes if the results are to be applicable to water availability. Water-rich areas like the Congo and Liberia, did not change water-scarcity category despite increasing population. However, there are water-rich areas like Tanzania and Nigeria and other parts of the Gulf of Guinea coast where despite being water rich, population pressure is intense.

3.4. How Have Water Scarcity Classes Changed over Time as a Function of Changes in Population and Changes in Hydrology in the Lake Victoria Basin?

To demonstrate how changes in both population and hydrology impact water availability we use an example of the Lake Victoria Basin. In this basin, population has steadily increased since 1990, from about 400 million to 1 billion people, in 2017. Hydrologic conditions (water supply) over this time period have fluctuated with a downward trend from 1990–2009, and relatively wetter conditions from 2010–2017. A linear trend of the full time period, or differencing of two time periods (1990–2002 vs. 2003–2018) suggests no large change in water supply. When we plot population, streamflow, water scarcity, and water scarcity classes on the same figure (Figure 6) we can see that between 1990–2001 the typical classification was "stress". From 2001–2009 the typical classification was "scarcity" with three instances of "absolute scarcity", predominantly related to increases in population. From 2010–2017, the relatively wet period the classification improved to "scarcity" but only reached the less severe "stress" category in 2012, the wettest year in the period. This example highlights how population increases can shift scarcity categories, even in basins where hydrologic conditions are increasing water supply.

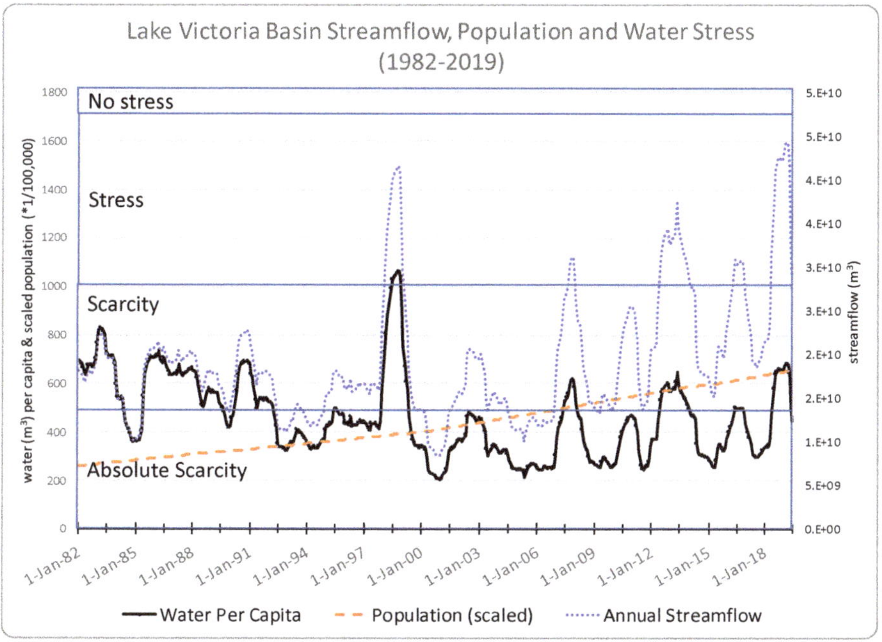

Figure 6. Time series of the Lake Victoria Basin showing changes in streamflow, population and water scarcity classification. The blue line shows annual streamflow at the mouth of Lake Victoria basin, with a positive trend since 2000. The orange line shows the total basin population, values interpolated from 1975, 1990, 2000, and 2015 GHS estimates. The black line is water availability per capita. Color blocks denote Falkenmark thresholds (Table 2). Despite the positive trend in streamflow since 2000, water available per capita has not increased.

4. Discussion and Conclusions

Our modified, operational Falkenmark Index, representing the current month's water availability based on population data and FLDAS streamflow, fills an important gap between acute drought-monitoring and chronic water-scarcity analysis. Drought monitoring efforts like NLDAS and Princeton Analytics provide routine updates regarding water supply but do not produce indices that account for water demand. We borrow from the chronic water-scarcity community to address water demand, and operationalize the Falkenmark index, traditionally based on annual water supply, by using a 12-month running total of streamflow. Our use of the Pfafstetter 6 hydrologic units further operationalizes previous water scarcity approaches by moving away from the national, or large river basin scale to compute water scarcity over smaller catchments that better represent the local nature of water availability.

This system provides monthly-updated maps of water scarcity (no stress, stress, scarcity and absolute scarcity) as well as maps that show how this classification deviates from the long-term average. These maps highlight acute water scarcity events and provide up-to-date information for decision makers who need to prioritize assistance at a regional scale and plan at a local scale. Compared to previously existing publicly available hydrologic and population datasets, this system provides decision makers with more timely information that is also interpretable, in that water stress changes can be traced back to FLDAS streamflow estimates based on CHIRPS and MERRA-2 climate inputs.

While it is difficult to validate specific locations' water scarcity classification, our maps show similar spatial patterns of scarcity in the Horn of Africa, southern Africa, and north Africa as studies using an inverse function of annual withdrawals to availability (Water Stress Index (WSI)) [12], the ratio of withdrawals to mean annual runoff (WSI; IWMI; [45], or the withdrawal to availability ratio (WRR [46]). However, our index, and the cumulative demand to withdrawal ratio (CWD) from Hanaski et al. [46], indicates more scarcity in densely populated regions of West Africa, and scarcity in Madagascar, both of which are confirmed in news reports (e.g., Ivory Coast [47], Madagascar [48]).

Our approach does not highlight water scarcity in sparsely populated areas (e.g., the Sahara Desert, Eastern Kenya, and the Kalahari Desert), due to low population. This has been highlighted as a drawback of the Falkenmark Index, and may under-represent the severity of water insecurity for poor populations, living in marginal areas with low population density and water supply. In many studies of chronic water stress, however, regions with low population density and, therefore, a small volume of absolute water demand are often masked out e.g., [45,49]. Alternatively, scarcity mapped with blue water footprints [1] do capture these low population density areas. To operationalize this or other more complicated approaches development of a high quality, moderate spatial resolution (e.g., 10 km^2) publicly available water use (consumption or withdrawals) maps, with water volumes expressed relative to a hydrologic model's naturalized streamflow, would be useful to bridge the gap between advances in research and more applications-oriented approaches.

In our exploration of changes of streamflow and population over time we found that annual water scarcity was more sensitive to population changes (Figures 4b and 5b). This is commensurate with Vorosmarty et al. [10] who highlight the expected changes to water scarcity due to climate change is far less than expected changes from population change. However, our routinely updated maps with static population show a range of ±3 classes (Figure 2). Meanwhile, the Lake Victoria Basin example (Figure 6) shows that year-to-year variation in a basin (Figure 5) changed the water stress by two classes from "absolute scarcity" to "stress", highlighting the importance of hydrologic variability on acute water scarcity. Additional modeling experiments and studies regarding hydrologic change and its relation to landuse/landcover change could be beneficial for planning for improved water use efficiency and management.

This work did not address the limitations related to uncertainty in the hydrologic or land surface modeling highlighted by Schere and Pfister [23]. The FLDAS model is uncalibrated and relies on global soil and vegetation parameters and parameters that are a source of considerable uncertainty [50]. However, select comparisons with data from the Global Runoff Data Center show that FLDAS performs well ($R > 0.70$) in naturalized flow regimes and larger basins (e.g., $R > 0.7$ in the Orange Basin [51] and $R > 0.8$ in the Upper Blue Nile Basin [52]). These data are also publicly available for independent verification by interested parties, which we encourage before applying FLDAS data to local scale studies and other applications. Future work should also consider the role of additional water sources, like groundwater, in our estimate of water scarcity.

Another limitation is that our routine water-scarcity index anomaly estimates are based on static population estimates. As highlighted in the Lake Victoria Basin example even when current hydrologic conditions are similar to past years population increase will result in greater water scarcity over time. Our routine mapping does not capture this, which likely underestimates the severity of current conditions in a historical context (assuming there have been no improvements in efficiency and management). How to implement a changing population over time in an operational context and further exploration of these assumptions, could be an avenue for future research.

Despite limitations, the system we describe here provides routinely updated situational awareness of water scarcity conditions in a transparent framework. Our straightforward approach provides easy to interpret information that could be applied to other sectors such as health, food security, natural hazards, and governance.

Author Contributions: Conceptualization, A.M., J.P.V., C.P.-L., K.V. and L.H.; methodology, A.M., K.V., J.J.; software, A.G., K.A.; formal analysis, A.M., K.V.; data curation and routine operations, J.J.; writing, A.M.; writing—review and editing, L.H., S.S., K.V., A.G., C.P.-L., K.A.; visualization, K.V.; supervision, C.P.-L.; project administration, J.P.V.; funding acquisition, J.P.V., C.P.-L., A.M.

Funding: NASA—USAID FEWS NET interagency agreement, USGS-UCSB cooperative agreement for FEWS NET. NASA SERVIR Applied Sciences Team.

Acknowledgments: We acknowledge the NASA GES DISC for data curation, NASA LIS team for software development. NCCS Computing for computational resources. Narcissa Princope and Greg Husak for early discussions on water availability. Lake products are courtesy of the USDA/NASA G-REALM program at https://ipad.fas.usda.gov/cropexplorer/global_reservoir/.

Conflicts of Interest: The authors declare no conflict of interest.

References

1. Mekonnen, M.M.; Hoekstra, A.Y. Four billion people facing severe water scarcity. *Sci. Adv.* **2016**, *2*, e1500323. [CrossRef] [PubMed]
2. UN DESA. *World Population Projected to Reach 9.7 Billion by 2050*; UN DESA United Nations Department of Economic and Social Affairs: New York, NY, USA, 2015.
3. US National Intelligence Council. *Global Trends 2030: Alternative Worlds*; Government Printing Office: Washington, DC, USA, 2013; ISBN 978-0-16-091543-7.
4. Gan, T.Y.; Ito, M.; Hülsmann, S.; Qin, X.; Lu, X.X.; Liong, S.Y.; Rutschman, P.; Disse, M.; Koivusalo, H. Possible climate change/variability and human impacts, vulnerability of drought-prone regions, water resources and capacity building for Africa. *Hydrol. Sci. J.* **2016**, *61*, 1209–1226. [CrossRef]
5. USAID. *U.S. Government Global Water Strategy 2017*; USAID: Washington, DC, USA, 2017.
6. USAID. *U.S. Government Global Food Security Strategy 2016*; USAID: Washington, DC, USA, 2017.
7. McNally, A.; McCartney, S.; Ruane, A.C.; Mladenova, I.E.; Whitcraft, A.K.; Becker-Reshef, I.; Bolten, J.D.; Peters-Lidard, C.D.; Rosenzweig, C.; Uz, S.S. Hydrologic and Agricultural Earth Observations and Modeling for the Water-Food Nexus. *Front. Environ. Sci.* **2019**, *7*, 23. [CrossRef]
8. Clark, W.C.; van Kerkhoff, L.; Lebel, L.; Gallopin, G.C. Crafting usable knowledge for sustainable development. *Proc. Natl. Acad. Sci. USA* **2016**, *113*, 4570–4578. [CrossRef] [PubMed]
9. Falkenmark, M. The Massive Water Scarcity Now Threatening Africa: Why Isn't It Being Addressed? *Ambio* **1989**, *18*, 112–118.
10. Vörösmarty, C.J.; Green, P.; Salisbury, J.; Lammers, R.B. Global water resources: Vulnerability from climate change and population growth. *Science* **2000**, *289*, 284–288. [CrossRef]
11. Döll, P.; Kaspar, F.; Lehner, B. A global hydrological model for deriving water availability indicators: Model tuning and validation. *J. Hydrol.* **2003**, *270*, 105–134. [CrossRef]
12. Pfister, S.; Koehler, A.; Hellweg, S. Assessing the Environmental Impacts of Freshwater Consumption in LCA. *Environ. Sci. Technol.* **2009**, *43*, 4098–4104. [CrossRef]
13. Hoekstra, A.Y.; Mekonnen, M.M. The water footprint of humanity. *Proc. Natl. Acad. Sci. USA* **2012**, *109*, 3232–3237. [CrossRef]
14. Wada, Y.; Wisser, D.; Bierkens, M.F.P. Global modeling of withdrawal, allocation and consumptive use of surface water and groundwater resources. *Earth Syst. Dyn.* **2014**, *5*, 15–40. [CrossRef]
15. Donchyts, G.; Baart, F.; Winsemius, H.; Gorelick, N.; Kwadijk, J.; van de Giesen, N. Earth's surface water change over the past 30 years. *Nat. Clim. Chang.* **2016**, *6*, 810–813. [CrossRef]

16. Mankin, J.S.; Viviroli, D.; Mekonnen, M.M.; Hoekstra, A.Y.; Horton, R.M.; Smerdon, J.E.; Diffenbaugh, N.S. Influence of internal variability on population exposure to hydroclimatic changes. *Environ. Res. Lett.* **2017**, *12*, 044007. [CrossRef]
17. Rijsberman, F.R. Water scarcity: Fact or fiction? *Agric. Water Manag.* **2006**, *80*, 5–22. [CrossRef]
18. Brown, A.; Matlock, M.D. A review of water scarcity indices and methodologies. *Food Beverage Agric.* **2017**, *106*, 1–19.
19. Liu, J.; Yang, H.; Gosling, S.N.; Kummu, M.; Flörke, M.; Pfister, S.; Hanasaki, N.; Wada, Y.; Zhang, X.; Zheng, C.; et al. Water scarcity assessments in the past, present and future. *Earths Future* **2017**, *5*, 545–559. [CrossRef] [PubMed]
20. Damkjaer, S.; Taylor, R. The measurement of water scarcity: Defining a meaningful indicator. *Ambio* **2017**, *46*, 513–531. [CrossRef] [PubMed]
21. Xia, Y.; Mitchell, K.; Ek, M.; Sheffield, J.; Cosgrove, B.; Wood, E.; Luo, L.; Alonge, C.; Wei, H.; Meng, J.; et al. Continental-scale water and energy flux analysis and validation for the North American Land Data Assimilation System project phase 2 (NLDAS-2): 1. Intercomparison and application of model products. *J. Geophys. Res. Atmos.* **2012**, 117. [CrossRef]
22. Sheffield, J.; Wood, E.F.; Chaney, N.; Guan, K.; Sadri, S.; Yuan, X.; Olang, L.; Amani, A.; Ali, A.; Demuth, S.; et al. A Drought Monitoring and Forecasting System for Sub-Sahara African Water Resources and Food Security. *Bull. Amer. Meteor. Soc.* **2013**, *95*, 861–882. [CrossRef]
23. Scherer, L.; Pfister, S. Dealing with uncertainty in water scarcity footprints. *Environ. Res. Lett.* **2016**, *11*, 054008. [CrossRef]
24. Mekonnen, M.M.; Hoekstra, A.Y. The green, blue and grey water footprint of crops and derived crop products. *Hydrol. Earth Syst. Sci.* **2011**, *15*, 1577–1600. [CrossRef]
25. Reig, P.; Shiao, T.; Gassert, F. *Aqueduct Water Risk Framework*; WRI Working Paper; World Resources Institute: Washington, DC, USA, 2013.
26. McNally, A.; Arsenault, K.; Kumar, S.; Shukla, S.; Peterson, P.; Wang, S.; Funk, C.; Peters-lidard, C.D.; Verdin, J.P. A land data assimilation system for sub-Saharan Africa food and water security applications. *Sci. Data Lond.* **2017**, *4*, 170012. [CrossRef] [PubMed]
27. Funk, C.; Peterson, P.; Landsfeld, M.; Pedreros, D.; Verdin, J.; Shukla, S.; Husak, G.; Rowland, J.; Harrison, L.; Hoell, A.; et al. The climate hazards infrared precipitation with stations–A new environmental record for monitoring extremes. *Sci. Data* **2015**, *2*, 150066. [CrossRef] [PubMed]
28. Juma, D.W.; Wang, H.; Li, F. Impacts of population growth and economic development on water quality of a lake: Case study of Lake Victoria Kenya water. *Environ. Sci. Pollut. Res.* **2014**, *21*, 5737–5746. [CrossRef] [PubMed]
29. Peters-Lidard, C.D.; Houser, P.R.; Tian, Y.; Kumar, S.V.; Geiger, J.; Olden, S.; Lighty, L.; Doty, B.; Dirmeyer, P.; Adams, J.; et al. High-performance Earth system modeling with NASA/GSFC's Land Information System. *Innov. Syst. Softw. Eng.* **2007**, *3*, 157–165. [CrossRef]
30. Ek, M.B.; Mitchell, K.E.; Lin, Y.; Rogers, E.; Grunmann, P.; Koren, V.; Gayno, G.; Tarpley, J.D. Implementation of Noah land surface model advances in the National Centers for Environmental Prediction operational mesoscale Eta model. *J. Geophys. Res. Atmos.* **2003**, 108. [CrossRef]
31. Gelaro, R.; McCarty, W.; Suárez, M.J.; Todling, R.; Molod, A.; Takacs, L.; Randles, C.A.; Darmenov, A.; Bosilovich, M.G.; Reichle, R.; et al. The Modern-Era Retrospective Analysis for Research and Applications, Version 2 (MERRA-2). *J. Clim.* **2017**, *30*, 5419–5454. [CrossRef]
32. Getirana, A.; Peters-Lidard, C.; Rodell, M.; Bates, P.D. Trade-off between cost and accuracy in large-scale surface water dynamic modeling. *Water Resour. Res.* **2017**, *53*, 4942–4955. [CrossRef]
33. Getirana, A.C.V.; Boone, A.; Yamazaki, D.; Decharme, B.; Papa, F.; Mognard, N. The Hydrological Modeling and Analysis Platform (HyMAP): Evaluation in the Amazon Basin. *J. Hydrometeor.* **2012**, *13*, 1641–1665. [CrossRef]
34. Bates, P.D.; Horritt, M.S.; Fewtrell, T.J. A simple inertial formulation of the shallow water equations for efficient two-dimensional flood inundation modelling. *J. Hydrol.* **2010**, *387*, 33–45. [CrossRef]
35. Verdin, K.L. *Hydrologic Derivatives for Modeling and Analysis—A New Global High-Resolution Database*; Data Series; U.S. Geological Survey: Reston, VA, USA, 2017; p. 24.
36. Linard, C.; Gilbert, M.; Snow, R.W.; Noor, A.M.; Tatem, A.J. Population Distribution, Settlement Patterns and Accessibility across Africa in 2010. *PLoS ONE* **2012**, *7*, e31743. [CrossRef]

37. Birkett, C.; Reynolds, C.; Beckley, B.; Doorn, B. From Research to Operations: The USDA Global Reservoir and Lake Monitor. In *Coastal Altimetry*; Vignudelli, S., Kostianoy, A.G., Cipollini, P., Benveniste, J., Eds.; Springer: Berlin/Heidelberg, Germany, 2011; pp. 19–50. ISBN 978-3-642-12796-0.
38. Latham, B.; Nhamire, B. Zimbabwe, Mozambique in power deal that could close Kariba hydro plant. *Business Day*, 11 July 2019.
39. Ndlovu, M. Kariba Dam left with 4 metres before switching off. *Bulawayo 24 News*, 30 June 2019.
40. Crecey, K. Power production on the verge of shutdown at Zim's Kariba Dam. *Fin24*, 31 May 2019.
41. Baaitse, F. Water finally flows in Shashe ward. *The Okavanogo Voice*, 18 June 2019.
42. Episcopal Relief and Development Responding to the Drought in Namibia. *Reliefweb*, 26 June 2019.
43. Rodell, M.; Famiglietti, J.S.; Wiese, D.N.; Reager, J.T.; Beaudoing, H.K.; Landerer, F.W.; Lo, M.-H. Emerging trends in global freshwater availability. *Nature* **2018**, *557*, 651. [CrossRef]
44. Scanlon, B.R.; Zhang, Z.; Save, H.; Sun, A.Y.; Schmied, H.M.; van Beek, L.P.H.; Wiese, D.N.; Wada, Y.; Long, D.; Reedy, R.C.; et al. Global models underestimate large decadal declining and rising water storage trends relative to GRACE satellite data. *Proc. Natl. Acad. Sci. USA* **2018**, *115*, E1080–E1089. [CrossRef]
45. Smakhtin, V.U.; Revenga, C.; Doll, P. *Taking into Account Environmental Water Requirements in Global-Scale Water Resources Assessments*; International Water Management Institute (IWMI), Comprehensive Assessment Secretariat: Colombo, Sri Lanka, 2004.
46. Hanasaki, N.; Kanae, S.; Oki, T.; Masuda, K.; Motoya, K.; Shirakawa, N.; Shen, Y.; Tanaka, K. An integrated model for the assessment of global water resources—Part 2: Applications and assessments. *Hydrol. Earth Syst. Sci.* **2008**, *12*, 1027–1037. [CrossRef]
47. Dontoh, E.; Cohen, M. Africa's Booming Cities Are Running Out of Water. *Bloomberg*, 18 March 2019.
48. Mosbergen, D. After 3 Years of Drought, A Starving Madagascar Teeters on The Brink Of "Catastrophe". Available online: https://www.huffpost.com/entry/madagascar-drought-hunger_n_580dbb32e4b0a03911ed7540 (accessed on 15 August 2019).
49. Vörösmarty, C.J.; Douglas, E.M.; Green, P.A.; Revenga, C. Geospatial indicators of emerging water stress: An application to Africa. *Ambio* **2005**, *34*, 230–236. [CrossRef]
50. Nearing, G.S.; Mocko, D.M.; Peters-Lidard, C.D.; Kumar, S.V.; Xia, Y. Benchmarking NLDAS-2 Soil Moisture and Evapotranspiration to Separate Uncertainty Contributions. *J. Hydrometeor.* **2016**, *17*, 745–759. [CrossRef]
51. McNally, A.; Verdin, K.; Jung, H.C.; Harrison, L.; Shukla, S.; Peters-Lidard, C. *Hydrologic Modeling for Monitoring Water Availability in Sub-Saharan Africa*; American Geophysical Union: Washington, DC, USA, 2017.
52. Jung, H.C.; Getirana, A.; Policelli, F.; McNally, A.; Arsenault, K.R.; Kumar, S.; Tadesse, T.; Peters-Lidard, C.D. Upper Blue Nile basin water budget from a multi-model perspective. *J. Hydrol.* **2017**, *555*, 535–546. [CrossRef]

Article

Consequences of the Integration of a Hyperbolic Funnel into a Showerhead for Droplets, Jet Break-Up Lengths, and Physical-Chemical Parameters

Maarten V. van de Griend [1,2,*], Luewton L. F. Agostinho [1,3], Elmar C. Fuchs [1], Nigel Dyer [1] and Willibald Loiskandl [2]

1 Wetsus, European Centre of Excellence for Sustainable Water Technology, 8911 MA Leeuwarden, The Netherlands; luewton.lemos@wetsus.nl (L.L.F.A.); elmar.fuchs@wetsus.nl (E.C.F.); nigel.dyer@wetsus.nl (N.D.)
2 Institute of Hydraulics and Rural Water Management, University of Natural Resources and Life Sciences, Muthgasse 18, 1190 Vienna, Austria; Willibald.loiskandl@boku.ac.at
3 Water Technology Group, NHL Stenden University of Applied Sciences, 8917 DD Leeuwarden, The Netherlands
* Correspondence: maarten.vandegriend@wetsus.nl

Received: 14 October 2019; Accepted: 18 November 2019; Published: 21 November 2019

Abstract: Introducing a hyperbolic vortex into a showerhead is a possibility to achieve higher spray velocities for a given discharge without reducing the nozzle diameter. Due to the introduction of air bubbles into the water by the vortex, the spray is pushed from a transition (dripping faucet) regime into a jetting regime, which results in higher droplet and jet velocities using the same nozzle diameter and throughput. The same droplet and jet diameters were realized compared to a showerhead without a vortex. Assuming that the satisfaction of a shower experience is largely dependent on the droplet size and velocity, the implementation of a vortex in the showerhead could provide the same shower experience with ~14% less water consumption compared to the normal showerhead. A full optical and physical analysis was presented, and the important chemical parameters were investigated.

Keywords: hydraulics; dividing flow manifold; showerheads; sprays; dissolved oxygen

1. Introduction

1.1. Motivation

The shortage of water resources due to population concentration in urban areas is a serious issue, which calls for water-saving measures on a global scale. With an expected world population from 7.7 billion people today to 9.7 billion people by 2050 and an urban population increase of 4.3 billion today to 6.7 billion in 2050 [1], this puts an additional strain on the availability of potable water. This is further increased by decreasing mountain snowpacks due to global warming, which feed reservoirs and streams in summer with meltwater. When this source provides less feed flow than usual, problems will ensue [2]. In addition to water scarcity, energy is required to transport and purify water before it reaches the consumer, which will partly increase CO_2 emissions [2]. Water reduction devices are commonly used to decrease domestic water consumption [3,4] and therefore contribute to decreasing in CO_2 emissions. Among such devices, those dedicated to shower water consumption are especially challenging as it is known that total flow influences consumer satisfaction [5]. More analytically, consumer (shower) satisfaction depends on total pressure exerted by water on consumer's skin, within a certain limit, i.e., both high- and low-pressure cause discomforts. Consequently, just reducing the water flow in order to save water and energy will also decrease customer satisfaction. For this purpose, smaller nozzles with higher pressures are often used, which are also more prone to clogging issues involving, e.g., particle deposition and lime precipitation. In this work, we presented an alternative approach, which produces higher spray velocities at the same flow rate compared to a normal showerhead. Since such a device builds on purely geometric modifications of the showerhead and there is no change in energy input, it provides a low cost and easy to implement a solution for more sustainable, equally comfortable showers. The consequences of such modifications on the overall water characteristics and shower performance would be presented in sequence.

1.2. Hyperbolic Vortices

Vortices are present in a number of natural phenomena. The most commonly known natural phenomenon, which can be described using vortex flow fields, is probably the tornado [6]. Another common example is the flow pattern observed when water, accumulated in a sink, flows down through the drain after opening its cap. Such flow also shows a resemblance to a hyperbolic velocity field.

Naturally occurring hyperbolic vortices were well described by the Austrian forester and bionics pioneer Viktor Schauberger [7] in the last century. Later on, his son Walter Schauberger [8] derived the mathematical formulation to describe the hyperbolic cone as a basic shape in which water vortices would appear.

For mathematicians, a hyperbola is a set of points, such that for any point P of the set, the absolute difference of the distances $|PF_1|$, $|PF_2|$ to two fixed points, F_1 and F_2 (the foci), is constant, usually denoted as $2a$ with $a > 0$. Such geometric space can be represented as

$$H = \{P \mid |PF_2| - |PF_1| = 2a\} \tag{1}$$

Looking more into the physical aspects of the phenomenon, hyperbolic flows can build very particular velocity and force (vector) fields. Such fields are directly related to the hydrodynamics of the process. Mostly, in literature, the physical analyses of such flows are done to model tornados and predict their formation and trajectory [9,10] using incompressible Navier–Stokes equations with specific boundary conditions. Here the pressure gradient is considered the most important one as it justifies the rotating and uplifting flow movements. Some examples are the Trap [11] model based on satellite obtained information with defined pressure gradient boundary conditions. Rotunno [12] assumed the existence of what they called "stagnation walls", which had basically no flow in or out in the vertical plane, which would force the flow to go up or down, generating thus the uplifting movement.

Below is a list of the most commonly covered aspects of this phenomenon (from [12]):

1. the velocity vector field of such structures is quite particular, meaning particles immersed in vortex structures, depending on their size, will be subjected to different tangential, axial, and radial velocities. This varies (considerably) with position and time,
2. when considering liquid-based hyperbolic flow structures, there is always a well-defined air-liquid internal interface, which could be eventually used to enhance gas-diffusion in the liquid,
3. for liquid structures, there is also, and necessarily, a solid-liquid interface, which would contribute to enhancing shear stresses and would be partially responsible (together with viscous stresses) for the axial velocity gradient and energy losses of the tangential component of the liquid velocity.

In this work, a setup was built using a modified showerhead, which would allow the formation of a hyperbolic vortex inside the head itself. Because it was expected that the presence of the vortex could allow extra aeration of the showered water, some characteristics of the water before and after the break-up of the liquid jets were analyzed and compared to a conventional showerhead. The authors also used the work of [5] to verify the possible influences on what they classified as "shower's user comfort", which, as explained by them, depends on flow rate, water temperature, and droplet impact pressure. Results showed that the modified tangential velocity field, created by the hyperbolically shaped geometry, accelerated the flow tangentially. After entering a spray plate, which acted as a dividing flow manifold, this tangential velocity directly impacted the pressure profile over the nozzles, which changed the total flow profile, causing different jet-break regimes. It would be shown that the modified showerhead produces break-up in the "jetting" regime and in the "transition" regime, where the water does not quite behave like a jet-producing small droplets (1.8 times nozzle diameter) but also not quite like a "dripping" regime, producing very large droplets formed when capillary forces are not enough to withstand gravity [13]. This regime change was possible because the modified geometry allowed air bubbles to enter the showerhead and mix with the water inside it, which brought both consequences to the break-up mechanism and, in a minor fashion, to the diffusion of gases in the liquid. These observations would be supported by the results of chemical and optical experiments.

2. Materials and Methods

The following sections describe the showerheads, the imaging system, and the setup used for the investigation of possible influences on chemical-physical parameters of the water. The particularities of each method have been discussed.

2.1. Showerheads

In this work, two showerheads were tested with identical external dimensions but different internal structures. Both were coupled with a tangential inlet with an inner radius of 8 mm. One showerhead (henceforth called 'regular') consisted of a short vertical cylindrical element of diameter 8 mm in the middle of the headspace, after which the flow was distributed over all nozzles in a narrow spacing (Figure 1a). The other showerhead (henceforth called 'vortex') had an internal hyperbolic-like funnel that compressed water through a narrow circular region of diameter 22 mm, generating large azimuthal velocities in the spray plate region. Both showerhead spray plates consisted of 90 nozzles positioned in concentric circles with 5, 9, 13, 17, 21, and 25 nozzles per circle, which had radial distances of 16, 36, 50, 64, 78, and 94 mm, respectively. All the nozzles' diameters were 1.2 mm. Schematics of both regular and vortex showerheads are shown in Figure 1a–c.

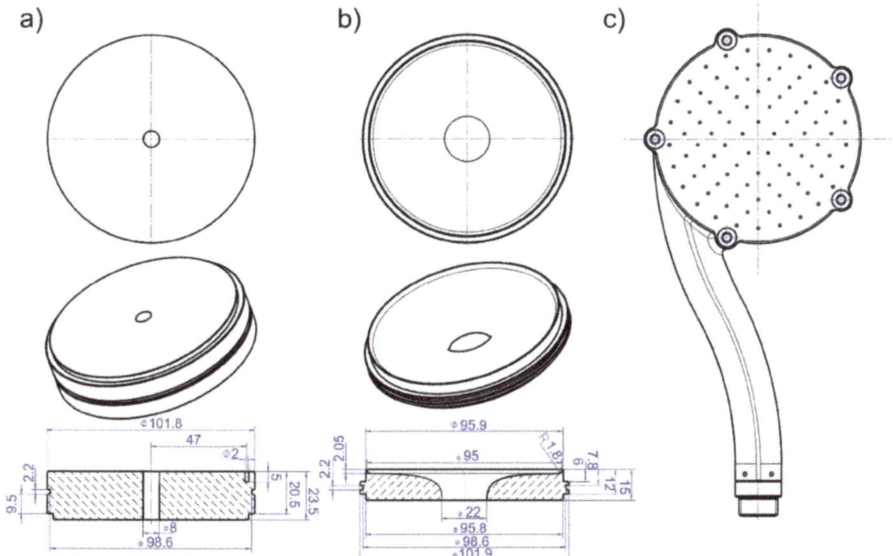

Figure 1. Schematics of the regular (**a**) and the vortex (**b**) showerhead. Both are encapsulated in the same handle and nozzle plate (**c**). The numbers in blue in subfigures (**a**) and (**b**) are the dimensions in mm.

2.2. Jet Break-Up Length, Jet Velocity, and Droplet Characteristics

In all experiments, tap water was used. Using high-speed imaging, the jet break-up and droplet characteristics could be visualized, and videos were recorded, which allowed droplet diameters, break-up length, etc. to be measured. The system consisted of a Photron SA-1.1 high-speed camera coupled with a macroscopic lens (35–70 mm F/3.3–4.5 Zoom-Nikkor) at a frame rate of 3000 fps and a shutter time of 1/6000 s for the movies video S1 and video S2, and at a frame rate of 2000 fps with a shutter time of 1/5000 s for the movies video S3 and video S4; all played back with 30 fps. To allow imaging of a specific row of jets and exclude the interference of out of focus jets and droplets, a tray with a slit was constructed to catch the flow in front or behind a specific row. The tray was positioned at approximately 6 mm below the nozzles' outlet, leaving only a single row of jets to pass undisturbed through the slit. The flow produced by the remaining jets was transported outside of the field of view of the camera. The field of view was selected in order to allow full visualization of the jet break-up lengths, i.e., from 6 mm below the nozzle tip (restriction from the tray) to the break-up point, as well as the first movements of the formed droplets. The recorded images were analyzed using ImageJ®. The program can differentiate between jets and droplets from one single image by using their circularity, i.e., the ratio between the smallest and the largest distance between two internal points of the projected object. Jets normally have circularity below 0.4 and droplets above this value, with perfect spheres having a circularity equal to 1. After differentiating the objects, ImageJ can also calculate the jet break-up lengths and droplet diameters by using their maximum Feret diameter [14], i.e., the maximum distance between two parallel lines enclosing the projected body. The jet length was calculated as the distance from the nozzle tip until the jet tip just after a break-up event. The liquid velocity was calculated using the movement of the jet between two consecutive break-up events. No retraction at the jet tip just after filament break-up was observed or considered.

2.3. Physical and Chemical Parameters

In order to verify possible changes in the physical-chemical parameters of the water running through a normal and a vortex showerhead, a recirculating shower setup was constructed (Figure 2). A recirculating system was tested because possible changes to the water properties would be amplified after multiple passes through the system. Also, in the future, showers that use a recirculation system will become more important because they are more sustainable, saving water and energy.

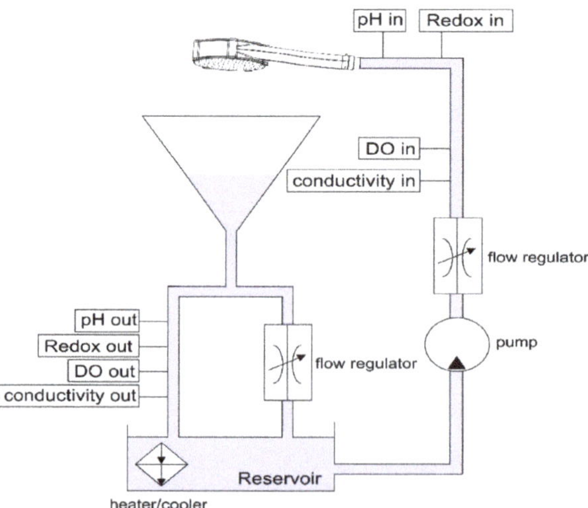

Figure 2. Recirculating setup for measuring chemical parameters for either the normal or the vortex showerhead.

In this system, water was pumped using a submerged pump from a ~20 L open reservoir to the showerhead positioned on top of this same reservoir. A flow and pressure regulator, an electrical conductivity probe (Endress Hauser), a pH probe (Endress Hauser), a redox potential probe (Endress Hauser), and a dissolved oxygen (DO) sensor (Presens) were mounted in the inlet (tubing) line, connecting the submerged pump to the showerhead. After passing through the showerhead, the sprayed water was collected by a plastic funnel, positioned between the showerhead and the reservoir, to which an outlet (tubing) line was connected. In the outlet line, the same set of probes and sensors installed in the inlet were connected. A T-junction split the shower flow between a line passing through a set of sensors, and a second line connected directly to the reservoir. This second outlet line was built for keeping a fixed water column height in the funnel, thus avoiding overflow, as well as ensuring that the outlet sensors were always submerged. The reservoir was temperature regulated through a secondary flow to a heater/cooler. Before each experiment, the tap water was purged with nitrogen to achieve a constant (low) dissolved oxygen concentration at the beginning of all the tests in order to more easily compare the diffusion of oxygen over time.

3. Results and Discussion

The following sections would discuss the properties of the measured sprays and chemical evolution of the recirculating system.

3.1. Optical Spray Analysis

A comparison of the spray from the regular and the vortex showerhead is given in Figure 3. The differences in spray characteristics between the two showerheads could be appreciated from the enclosed high-speed multimedia files (video S1 and video S2).

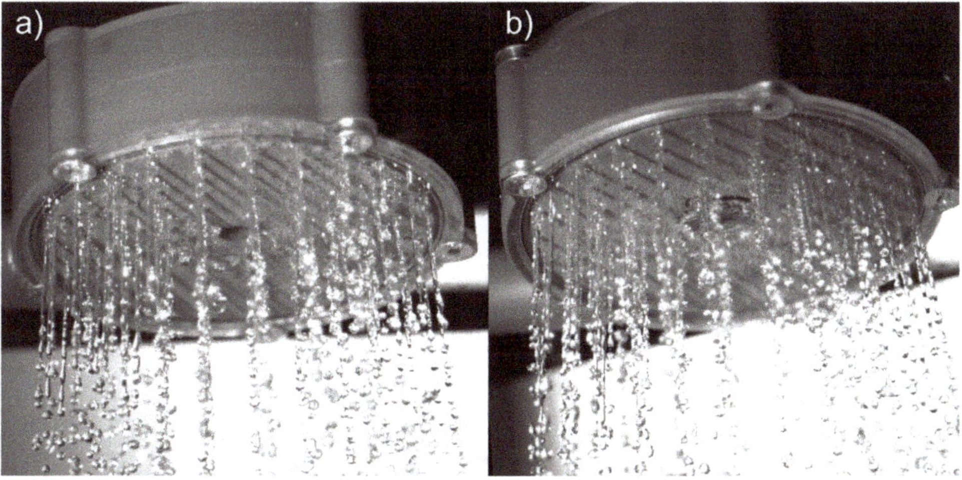

Figure 3. High-speed photography of the regular (**a**) and vortex (**b**) showerhead.

Figure 4a,b shows the spray of individual jets after placement of the tray filter. The jets were numbered 1 to 5 from the outermost to the innermost nozzle on both sides. The sixth jet was hidden behind the ruler. As can be seen in Figure 4, there were air bubbles inside the jets in both showerheads, with seemingly more air bubbles in the jets created by the vortex showerhead. Again, these differences could be better seen in the enclosed multimedia files (video S3 and video S4). For nozzles nr. 5b and 6b on the right side, the jet retracted back into the showerhead. From nozzle nr. 5b, only a droplet from a previous break-up could be seen directly right of the scale.

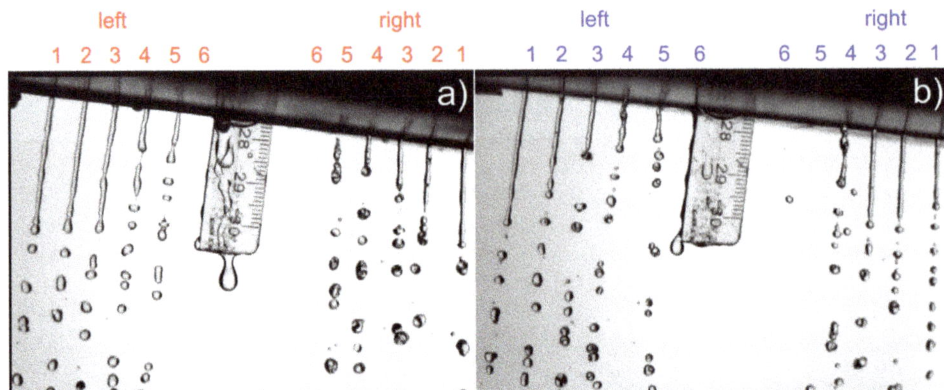

Figure 4. Individual jet comparison of the spray of regular (**a**) and vortex (**b**) showerhead at 8 L h^{-1} flow rate, 22.7 °C temperature using a tray filter, allowing only one line of nozzles to produce an undisturbed jet path to be imaged.

Both the regular and vortex showerheads produced a spray with a positive radial velocity gradient. This gradient was caused by centripetal forces due to the (peripheral) tangential inlet of both devices, increasing the pressure at the outermost nozzles (see Figure 1). This could also be seen in Figure 4a,b where the break-up length of the inner jets was clearly smaller than the outer jets, indicating lower velocities in the inner jets, which were a direct consequence of lower pressure flows. The average liquid velocities for the different showerheads and the different nozzles are shown in Figure 5 for a flow rate of 8 L·min^{-1}.

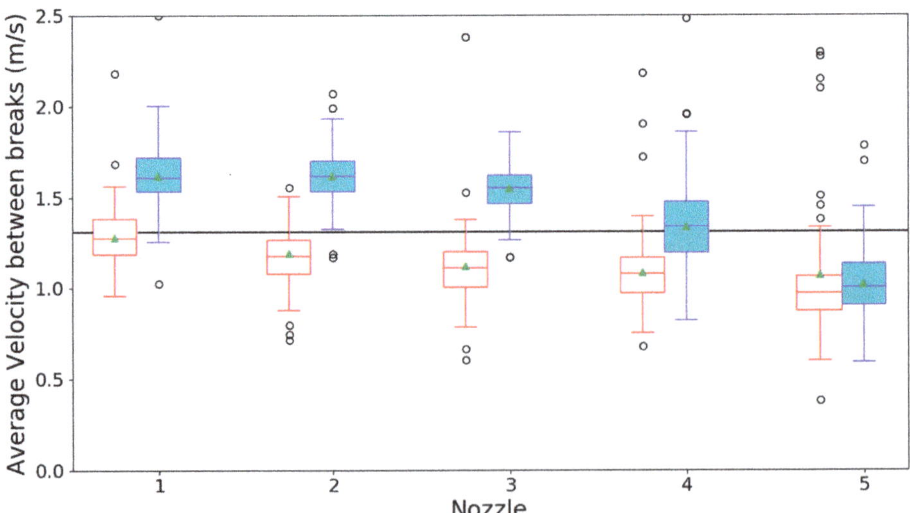

Figure 5. Boxplot graphs of the average liquid velocities in the experiments done with 8 L·min^{-1} for the regular (red boxes) and vortex (blue boxes) showerhead and nozzles 1 (outermost) to 5 (innermost). The green triangles are the calculated population mean. The horizontal black line is the expected liquid velocity calculated using continuity law [15] by taking the flow rate divided by the total nozzle surface area.

As could be seen from the boxplots, apart from the innermost nozzle, the vortex showerhead showed consistently higher velocities in comparison to the regular version. This effect was seen for all tested flow rates, i.e., 6, 7, 8, and 9 L·min^{-1} (additional data for 6, 7 and 9 L·min^{-1} in the Appendix A in Figures A1–A3, respectively). In an aqueous vortex, perturbations in the air/water surface create a significant air boundary layer that stays associated with the surface. As these perturbations move inwards, they pull the boundary layer with them, creating a force that draws the air inwards into the vortex [16,17]. Due to the resultant pressure gradient, a certain volume of air is drawn into the water and thereby increases the flow rate through the most peripheral nozzles. This would result in an increased liquid kinetic energy for these nozzles and a consequently bigger break-up length (also observable in Figure 4). Since the sub-pressure is highest in the center of the vortex, the air intake takes place primarily at the innermost nozzles (as, for example, nozzle 5, right side, in Figure 4b). In order to understand the consequences of this result for the "shower experience" [5], the obtained median velocities for the five nozzles were used to fit a proportionality constant between flow rate and velocity for each nozzle and each showerhead, using a simple least-squares linear regression model. Taking into consideration the number of nozzles present for each radius, a weighted average was taken of the ratio between these constants in order to make our results comparable to those of Okamoto et al. [5]. The results of this calculation showed that the vortex showerhead could provide the same jet velocity as a normal showerhead when using 14.4 ± 5.6% less water ($p < 0.01$). Or alternatively, when using the same amount of water, the velocities (see Figure 5) and jet lengths produced by a hyperbolic showerhead were higher than those produced by a normal showerhead. It has been shown that jet velocity is related to the "shower experience" [5]. Thus, according to the results of Okamoto et al. [5], a vortex showerhead could provide the same comfort level with less water. It should be pointed out, though, that the central nozzles are used as air inlets resulting in a different spatial spray distribution, which may or may not give a desirable effect.

Figure 6 shows a boxplot representation (per nozzle) of 219 to 285 independent jet length measurements per box right after the break-up of a droplet [5].

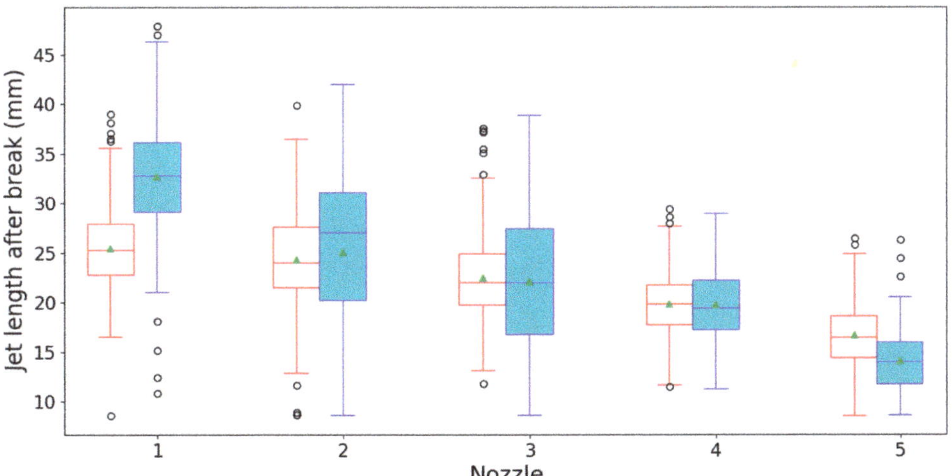

Figure 6. Jet lengths for the various showerheads and nozzles at 8 L min^{-1} and 35.8 °C taken from frames right after a filament had broken off.

When the liquid flow rate is increased to the point that inertial forces on the liquid overcome capillary upwards forces, the conventional image of a hanging droplet (quasi-static droplet formation) disappears, and a jet is formed at the nozzle tip from which droplets break-up [18,19]. The ratio between these two forces is the Weber number (Equation (2))

$$W_e = \frac{\rho \cdot Q^2}{D_0{}^3 \cdot \gamma} \tag{2}$$

with D_0 being the nozzle diameter, ρ the liquid density, Q the liquid flow rate, and γ the liquid surface tension [20]. Weber numbers smaller than one indicate the liquid break-up is happening in the so-called dripping regime. For Weber numbers between one and four, it is known as "transition regime", as the jet in that window is not yet completely formed. Rather, a small ligament forms at the nozzle tip from which droplets break-up [21]. For Weber numbers higher than four, a jet is clearly formed at the orifice. Normally, at this level, the break-up length is around 10 times bigger than the nozzle inner diameter [13]. This regime is known as the "jetting regime". Rayleigh [22] has thoroughly studied the physics behind these break-up mechanisms and defined that the diameter of the droplets formed from such break-up is around 1.8 times the jet diameter. The Weber numbers of the jets analyzed in this work were calculated from the flow rate according to equation 2. Their values were W_e = 11.3, 15.4, 20.1, 25.5 for flowrates of 6,7,8,9 L min^{-1}, respectively. Density and surface tension values of water at 35.8 °C were taken from the literature (ρ = 993.79 kg m^{-1} [23] and γ = 70.27 mN m^{-1} [24]). These calculations showed that the water was well within the jetting regime for all flowrates investigated.

It is known that, for inviscid liquids with break-up inside the jetting regime, higher liquid velocities lead to the formation of longer jets, since the characteristic timescale for the break-up is independent of jet velocity [21]. This effect is also visible in Figure 6 as a negative correlation was found (one-way ANOVA, $p < 0.001$ for both regular and vortex showerhead) between nozzle position and jet break-up length: the closer to the center the nozzle is, the shorter the jet break-up. Additionally, the variation of the data from the vortex showerhead was larger, indicating the presence of the transition regime. Moreover, when comparing Figures 5 and 6, it was possible to see that even though the outer nozzles of the vortex showerhead presented higher liquid velocities than the regular showerhead, the jet break-up lengths in both situations were rather similar. Only the first jet showed longer jets below the vortex showerhead when compared to the regular. This could be explained by the fact that the vortex showerhead creates a suction effect [17], which purges air bubbles from the central nozzles in the system. This effect decreases the total flow in the center nozzles, which can be also seen by the reduction of the jet break-up at this point. However, the additional intake of air increases the velocity in all nozzles, which compensates for this reduction (as can be seen in Figures 5 and 6) and consequently enhances the velocity at the peripheral ones. Whether this would cause higher comfort to the shower user is questionable, as the increase in peripheral jet velocity causes a consequent decrease in the velocity of the center jets.

When looking at the droplet size distribution, the minimal Feret diameter (minimum distance between two parallel lines in any orientation touching the particle) at different flow rates did not significantly shift but remained around 2.1 mm (Figure 7). Since all jets analyzed were inside the jetting regime, these droplet diameters were consistent with Rayleigh [22]. At higher flow rates, the distribution was wider for both cases.

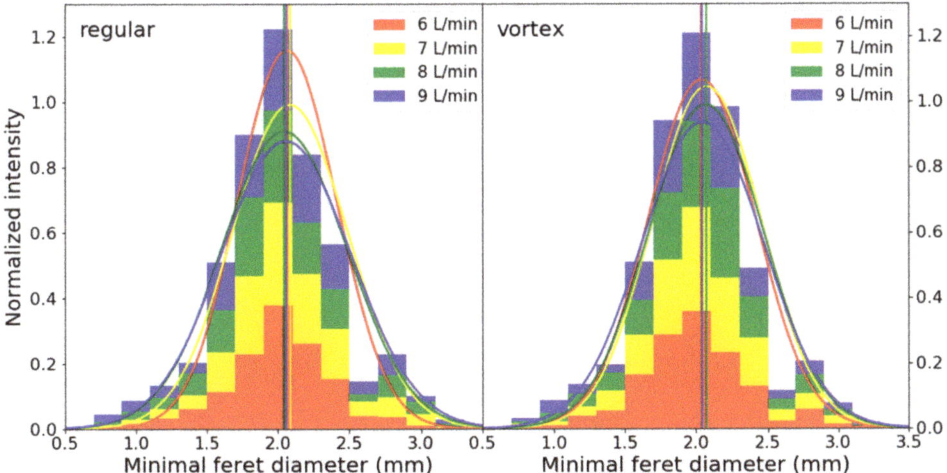

Figure 7. Minimal Feret diameter of droplets for different flow rates for regular showerhead (**left**) and vortex showerhead (**right**) at 35.8 °C. Normalized intensities (total surface is 1) and Gaussian distributions were drawn for the various flow rates, and the vertical lines indicate the means of their respective distributions.

3.2. Chemical Parameters

As shown in the experimental setup (Figure 2), the water flow for both showerheads was circulated with two oxygen sensors directly before and after the showerheads. When running for longer periods, the DO concentration increased, and exponentially approached its temperature-dependent saturation value because of the large interfacial area in the heads and sprays. An example of this effect is demonstrated in Figure 8 for the vortex showerhead. The difference between the two graphs represented the immediate effect of the shower (head + spray), while the overall increase was due to the cumulative effect resulting from water recirculation through the reservoir.

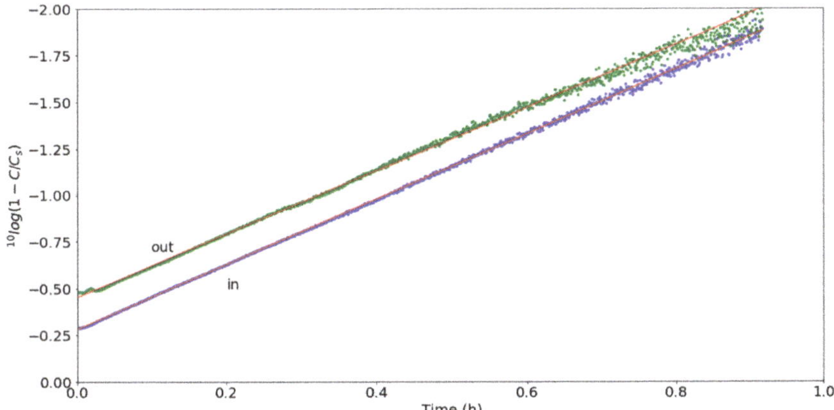

Figure 8. DO (dissolved oxygen) content before (blue) and after (green) the vortex showerhead expressed as the logarithm of $1-C/C_s$, where C is the concentration in ppm, and C_s is the saturation concentration determined from fitting the data to formula 3. The fits are shown in red. In this notation, 0 represents zero DO, and −2 represents a DO value of 0.99 C_s.

These data can be fit to the exponential function (assuming the driving force is proportional to the difference between the saturation point and the actual concentration):

$$C(t_2) = C_s - (C_s - C(t_1)) \times e^{-(t_2 - t_1)/\tau} \tag{3}$$

where $C(t_{1,2})$ is the dissolved oxygen concentration at times t_1 and $t_2 > t_1$, C_s is the saturation concentration, and $\tau = \frac{1}{ka}$ is a time constant typical for the system. K is the gas transfer coefficient, and a is the diffusion area divided by the total liquid volume. This equation can be rewritten to define a relative saturation coefficient F after [25]:

$$F \equiv \frac{C(t_2) - C(t_1)}{C_s - C(t_1)} = 1 - e^{-Ka\Delta t} \tag{4}$$

In Equation (4), t_1 and t_2 can also be replaced by the values at the inlet and outlet, respectively, to identify the instantaneous effect of the shower spray. F should be constant throughout the experiment, allowing us to calculate the K_a coefficient, which determines the efficiency of the system. Examples of various experiments are given in Table 1. In some experiments, the aeration was faster using the vortex showerhead (experiment 1), whereas, in others, there was no measurable difference (experiment 2). Moreover, the variation of the parameter τ was of the same order between showerheads and experiments. Naturally, the observed additional mixing of air to the water by the vortex is expected to increase the amount of dissolved oxygen. On the other hand, the time for this diffusion to happen is rather small. Hence, with the measurement precision available, no statistically significant difference in aeration could be found. However, some experiments indicated better aeration of the vortex showerhead compared to the normal one.

Table 1. Fitting parameters of equation 1 to data of the DO (dissolved oxygen) content for two identical experiments of regular and vortex showerheads before and after the shower using a least-squares method. t_1 indicates the start of the shower spray. C_s is the saturation concentration for oxygen found by the fit and $C(t_1)$ is the concentration at time t_1. The goodness of fit is indicated with R^2.

Experiment	Showerhead	Sensor	$C_s - C(t_1)$/ppm	τ/h	C_s/ppm	R^2
1	Vortex	In	3.13	0.21	6.13	0.9992
		Out	2.13	0.20	6.11	0.9972
	Regular	In	3.14	0.22	6.17	0.9998
		Out	2.09	0.24	6.17	0.9987
2	Vortex	In	3.23	0.25	6.17	0.9999
		Out	2.17	0.26	6.16	0.9989
	Regular	In	3.21	0.26	6.14	0.9998
		Out	2.20	0.25	6.12	0.9994

Apart from an increase in DO concentration, an increase of pH in the vortex showerhead compared to the regular one can be expected, since the additional mixing in the vortex will influence the carbonate/CO_2 equilibrium reaction so that CO_2 is expelled, comparable to stirring a glass of carbonated water,

$$H^+ + HCO_3^- \leftrightarrow H_2CO_3 \overset{stirring}{\rightarrow} H_2O + CO_2 \uparrow \tag{5}$$

Naturally, the process of spraying does this as well, as shown in Figure 9 for both showerheads. Since tap water was used, the initial pH values of the two experiments were slightly different, as could be seen in the initial difference between the two measurements at the water inlet of about 0.03 pH on the left of the Figure. The difference between the sensors was plotted in the bottom part. Although being close to the sensor resolution (ΔpH = 0.01), the vortex showerhead consistently showed a slightly smaller pH decrease than the regular showerhead.

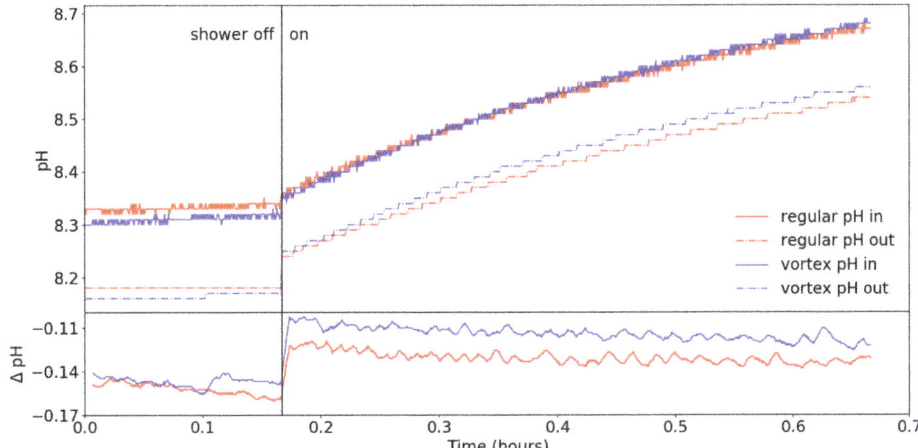

Figure 9. pH against time for both showerheads before (full) and after (dot-dashed) the spray in a circulating water setup at 35.8 °C. The lower part is the difference between the two pH sensors for both showerheads, shown as a moving average over 3 minutes.

Whereas the change of pH can be easily explained by the degassing of CO_2, the changes in the redox potential plotted in Figure 10 require some more in-depth discussion. The redox potential is an electrical characteristic of a solution that shows its tendency to transfer electrons to or from a reference electrode, describing a system's overall reducing or oxidizing capacity. In well-oxidized open waters, the redox potential is normally positive (above +300 to +500 mV), whereas, in reduced environments, it can be negative. Measuring redox potential in natural (potable) waters can yield different results depending on the method [26].

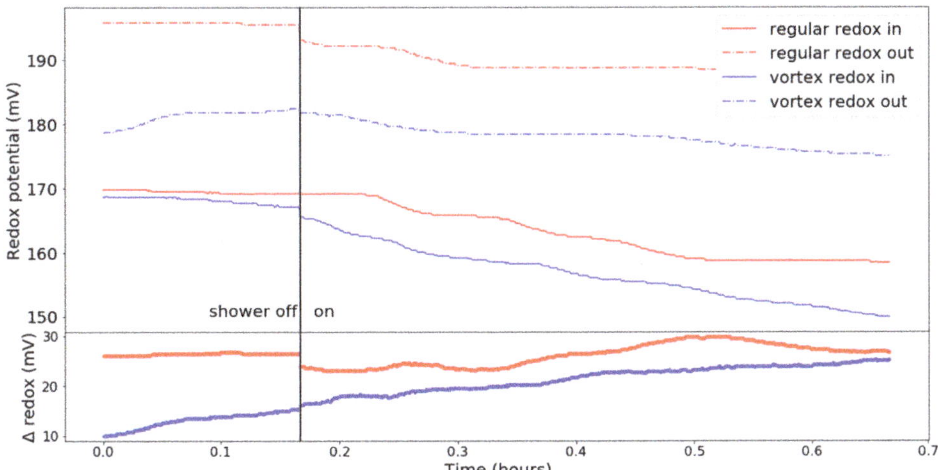

Figure 10. Redox potential against time for both showerheads before (full) and after (dot-dashed) the spray in a circulating water setup. The lower part is the difference between the two redox potential sensors for both showerheads.

Although the redox potential of water at equilibrium is relatively insensitive to a change in oxygen concentration and extent of saturation, it is, however, significantly changed by pH alterations [27]. In addition, it is highly dependent on the chemical composition of tap water. Therefore, it was to be expected that the initial values of the redox potential would differ for different measurements, as is shown in Figure 10, and made the absolute values difficult to compare. However, what could be compared was the evolution of the redox potential over time in both scenarios (thick red and blue lines in the lower part of Figure 10). The blue line showed a steeper inclination than the red line, indicating a faster rise of the redox potential difference for the vortex showerhead compared to the regular one. This could straightforwardly be explained with two other results:

- the change of pH,
- the (missing) increase in DO.

Solving the well-known Nernst equation [28],

$$E = E_0 + \left(\frac{RT}{zF}\right) ln\left\{\frac{\prod[A_{oxidised}]}{\prod[A_{reduced}]}\right\} \tag{6}$$

where E is the redox potential, E_0 is the standard potential at 25 °C, R is the general gas constant, T the absolute temperature in K, z the number of electrons transferred, F the Faraday constant, and A are the activities of the species involved. This allows us to derive a direct proportionality of the redox potential E and the pH, namely

$$E \sim -0.059V \cdot \text{pH}. \tag{7}$$

Therefore, an increase of one pH unit was accompanied by a decrease in the redox potential of 59 mV at 25 °C. The pH differences measured (see Figure 7) would thus account for 0.10 (−59 mV) = −5.9 mV for the regular and 0.15 (−59 mV) = −8.85 mV for the vortex showerhead, respectively. The realized measured reductions of ~20 and ~25 mV for normal and vortex showerhead were about three times larger. So, the pH change could only explain a part of the change of the redox potential. In order to explain the additional decrease of redox potential, let us assume that a part of the dissolved oxygen enters into a chemical reaction, with some components dissolved in the water, thus oxidation takes place and the concentration of dissolved oxygen decreases. The Nernst equation shows straightforwardly that such a process would also lead to a reduction of the redox potential. If we associate the remaining reduction in redox potential—14 mV and 16mV—with such reactions, it would require 2 and 2.3 ppm or 31 and 36 μmol of DO, respectively, to be consumed by chemical reactions, which are plausible amounts for the given circumstances.

4. Conclusions

Physical and chemical parameters of an aqueous spray through a regular and a vortex showerhead were investigated and compared. The inclusion of a hyperbolic vortex in a showerhead increased the flow rate through some individual nozzles compared to a showerhead without a vortex, while droplet and jet diameter was maintained. This was achieved because, in the vortex showerhead, air bubbles are introduced from the central part of the nozzle matrix in the sprayed liquid, which, in turn, causes higher liquid velocities and break-up length in the peripheral nozzles. Since droplet size and liquid velocity make up a significant part of the "shower experience" [5], the addition of a vortex allowed the same shower experience with lower flow-rates. By mixing air into the exterior jets, a vortex showerhead could save up to 14% of the water, when compared to conventional showerheads. In addition, an increased pH and a reduced redox potential were found when comparing the vortex showerhead to the regular showerhead, indicating an increased degassing of CO_2 and an increased intake of oxygen, part of which was immediately used for oxidation processes.

Supplementary Materials: The following are available online at http://www.mdpi.com/2073-4441/11/12/2446/s1, Video S1: Regular showerhead with bubbles. Video S2: Vortex showerhead with bubbles. Video S3: Spray of one row of jets from the regular showerhead. Video S4: Spray of one row of jets from the vortex showerhead.

Author Contributions: Conceptualization, methodology, M.V.v.d.G., E.C.F., L.L.F.A.; software, validation, formal analysis, investigation, resources, data curation, M.V.v.d.G.; writing—original draft preparation, M.V.v.d.G.; writing—review and editing, M.V.v.d.G, N.D., L.L.F.A., E.C.F.; supervision, L.L.F.A., E.C.F., W.L.; project administration, E.C.F., L.L.F.A.

Funding: This research received no external funding.

Acknowledgments: This work was performed at Wetsus, European Center of Excellence for Sustainable Water Technology (www.wetsus.eu). Wetsus is co-funded by the Dutch Ministry of Economic Affairs and Ministry of Infrastructure and Environment, the Province of Fryslân, and the Northern Netherlands Provinces. The authors would like to thank Jakob Woisetschläger (TU Graz, Austria) and the other participants of the research theme "Applied Water Physics" for the fruitful discussions and their financial support. A special thanks also go to Wiard Kuipers, who did a lot of experimental work and helped to build the setup.

Conflicts of Interest: The authors declare no conflict of interest.

Appendix A

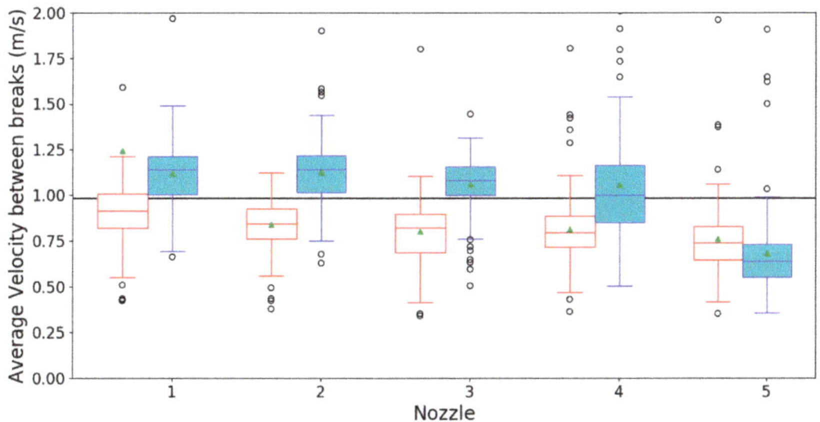

Figure A1. Boxplot graphs of the average liquid velocities in the experiments done with 6 L·min^{-1} for the regular (red boxes) and vortex (blue boxes) showerhead and nozzles 1 (outermost) to 5 (innermost). The green triangles are the calculated population mean. The horizontal black line is the expected liquid velocity calculated using continuity law [15] by taking the flow rate divided by the total nozzle surface area.

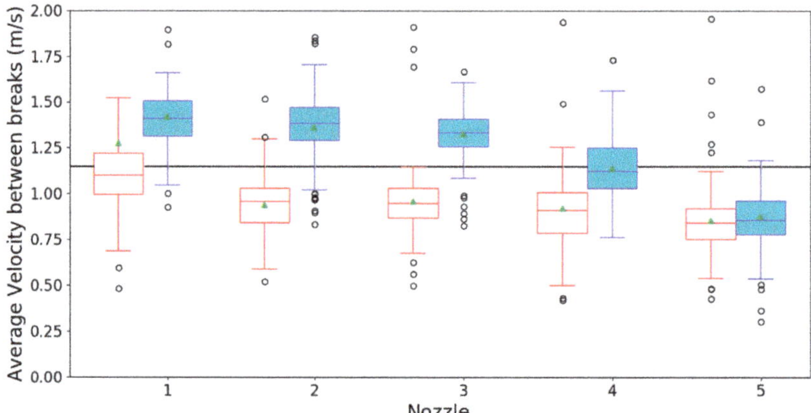

Figure A2. Boxplot graphs of the average liquid velocities in the experiments done with 7 L·min^{-1} for the regular (red boxes) and vortex (blue boxes) showerhead and nozzles 1 (outermost) to 5 (innermost). The green triangles are the calculated population mean. The horizontal black line is the expected liquid velocity calculated using continuity law [15] by taking the flow rate divided by the total nozzle surface area.

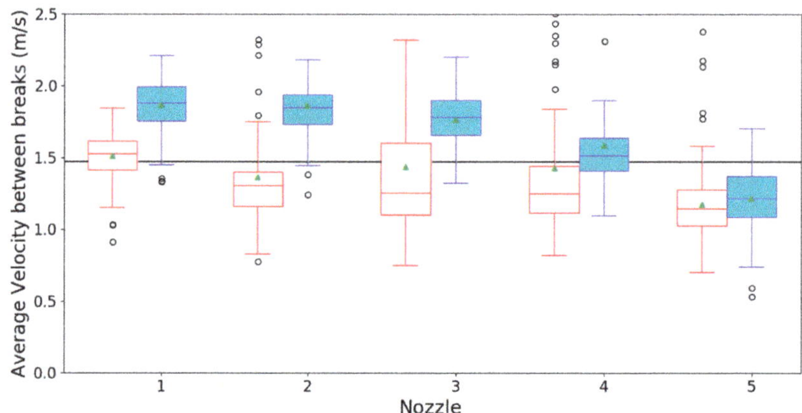

Figure A3. Boxplot graphs of the average liquid velocities in the experiments done with 9 L·min^{-1} for the regular (red boxes) and vortex (blue boxes) showerhead and nozzles 1 (outermost) to 5 (innermost). The green triangles are the calculated population mean. The horizontal black line is the expected liquid velocity calculated using continuity law [15] by taking the flow rate divided by the total nozzle surface area.

References

1. United Nations. *Department of Economic and Social Affairs, Population Division*; World Population Prospects; United Nations: New York, NY, USA, 2019.
2. Sokolow, S.; Godwin, H.; Cole, B.L. Impacts of urban water conservation strategies on energy, greenhouse gas emissions, and health: Southern California as a case study. *Am. J. Public Health* **2016**, *106*, 941–948. [CrossRef] [PubMed]
3. Shimizu, Y.; Dejima, S.; Toyosada, K. The CO2 emission factor of water in Japan. *Water* **2012**, *4*, 759–769. [CrossRef]
4. Hakket, M.J.; Gray, N.F. Carbon dioxide emission savings potential of household water user reduction in the UK. *J. Sustain. Dev.* **2009**, *2*, 36–43.

5. Okamoto, M.; Sato, M.; Shodai, Y.; Kamijo, M. Identifying the physical properties of showers that influence user satisfaction to aid in developing water-saving showers. *Water* **2015**, *7*, 4054–4062. [CrossRef]
6. Wood, V.T.; Brown, R.A. Simulated Tornadic Vortex Signatures of Tornado-Like Vortices Having One- and Two-Celled Structures. *J. Appl. Meteorol. Climatol.* **2011**, *50*, 2338–2342. [CrossRef]
7. Schauberger, V. Die Natur als Lehrmeisterin. *Implosion* **1963**, *7*, 21–27.
8. Schauberger, W. *Klaus Radlberger, Der Hyperbolische Kegel*; PKS Eigenverlag: Bad Ischl, Austria, 2002; ISBN 3950068619.
9. Drullion, F. Numerical Simulation of tornado-like vortices around complex geometries. *Int. J. Comp. Math.* **2009**, *86*, 1947–1955. [CrossRef]
10. Wan, J.W.L.; Ding, X. Physically-Based Simulation of Tornados. In *Workshop on Virtual Reality Interaction and Physical Simulation, Proceedings of the Second International Conference on Virtual Reality Interaction and Physical Simulation, Pisa, Italy, 8–11 November 2005*; Ganovelli, F., Mendoza, C., Eds.; ISTI-CNR: Pisa, Italy, 2005.
11. Trapp, R.; Fiedler, B. Numerical simulation of tornado-like vortices in asymmetric flow. In The Tornado: Its structure, dynamics, prediction, and hazards. *Geophys. Monogr.* **1993**, *79*, 49–54.
12. Rotunno, R. Numerical simulation of a laboratory vortex. *J. Atmospheric Sci.* **1977**, *34*, 1942–1956. [CrossRef]
13. Agostinho, L.L.F. Electrohydrodynamic Atomization in the Simple-Jet Mode: Out-scaling and Application. Ph.D. Thesis, TU Delft, Delft, The Netherlands, 2013.
14. Merkus, H.G. Particle Size Measurements: Fundamentals, Practice, Quality. Springer: Pijnacker, The Netherlands, 2009; ISBN 978-1-4020-9016-5.
15. Pedlosky, J. *Geophysical Fluid Dynamics*, 2nd ed.; Springer: New York, NY, USA, 1987; pp. 10–13, ISBN-13 978-0-387-96387-7.
16. Qian, Z.; Wu, P.; Guo, Z.; Huai, W. Numerical simulation of air entrainment and suppression in pump sump. *Sci. China Technol. Sci.* **2016**, *59*, 1847–1855. [CrossRef]
17. Blaszczyk, A.; Papierski, A.; Kunicki, R.; Susik, M. Surface Vortices and Pressures in Suction Intakes of Vertical Axial-Flow Pumps. *Mech. Mech. Eng.* **2012**, *16*, 51–71.
18. Lin, S.P.; Reitz, R.D. Drop and spray formation from a liquid jet. *Annu. Rev. Fluid Mech.* **1998**, *30*, 85. [CrossRef]
19. Van Hoeve, W.; Gekle, S.; Snoeijer, J.H.; Versluis, M.; Brenner, M.P.; Lohse, D. Breakup of diminutive Rayleigh Jets. *Phys. Fluids* **2010**, *22*, 122003. [CrossRef]
20. Van Hoeve, W. Fluid Dynamics at a Pinch: Droplet and Bubble Formation in Microfluidic Devices. Ph.D. Thesis, University of Twente, Enschede, The Netherlands, 2011.
21. Eggers, J.; Villermaux, E. Physics of liquid jets. *Rep. Prog. Phys.* **2008**, *71*, 036601. [CrossRef]
22. Rayleigh, L. On the instability of jets. *Proc. R. Soc. Lond.* **1879**, *10*, 4. [CrossRef]
23. Engineering ToolBox. Water—Density, Specific Weight and Thermal Expansion Coefficient. 2003. Available online: https://www.engineeringtoolbox.com/water-density-specific-weight-d_595.html (accessed on 31 October 2019).
24. Vargaftik, N.B.; Volkov, B.N.; Voljak, L.D. International Tables of the Surface Tension of Water. *J. Phys. Chem. Ref. Data* **1983**, *12*, 817. Available online: http://twt.mpei.ru/MCS/Worksheets/iapws/Surf-H2O.xmcd (accessed on 31 October 2019). [CrossRef]
25. Yin, Z.G.; Cheng, D.S.; Liang, B.C. Oxygen transfer by air injection in horizontal pipe flow. *J. Environ. Eng. ASCE* **2012**, *139*, 908–912. [CrossRef]
26. Matia, L.; Rauret, G.; Rubio, R. Fresenius J. *Anal. Chem.* **1991**, *339*, 455–462. [CrossRef]
27. Wetzel, R.G. Limnology. In *Chapter 14—Iron, Sulfur and Silica Cycles*, 3rd ed.; Academic Press: Cambridge, MA, USA, 2001; pp. 289–330. ISBN 9780127447605.
28. Orna, M.V.; Stock, J. *Electrochemistry, Past and Present*; American Chemical Society: Columbus, OH, USA, 1989; ISBN 978-0-8412-1572-6.

Review

Challenges and Opportunities for Sustainable Management of Water Resources in the Island of Crete, Greece

V. A. Tzanakakis [1,2,*], A. N. Angelakis [3,4], N. V. Paranychianakis [5], Y. G. Dialynas [6] and G. Tchobanoglous [7]

1 Hellenic Agricultural Organization Demeter (HAO-Demeter), Soil and Water Resources Institute, 57001 Thessaloniki, Greece
2 Department of Agriculture, School of Agricultural Science, Hellenic Mediterranean University, Iraklion, 71410 Crete, Greece
3 HAO-Demeter, Agricultural Research Institution of Crete, 71300 Iraklion, Greece
4 Union of Water Supply and Sewerage Enterprises, 41222 Larissa, Greece; angelak@edeya.gr
5 School of Environmental Engineering, Technical University of Crete, 73100 Chania, Greece; niko.paranychianakis@enveng.tuc.gr
6 Department of Civil and Environmental Engineering, University of Cyprus, Nicosia 1678, Cyprus; ydialy01@ucy.ac.cy
7 Department of Civil and Environmental Engineering, University of Davis, Davis, CA 95616, USA; gtchobanoglous@ucdavis.edu
* Correspondence: vtzanakakis@hmu.gr

Received: 12 April 2020; Accepted: 16 May 2020; Published: 28 May 2020

Abstract: Crete, located in the South Mediterranean Sea, is characterized by long coastal areas, varied terrain relief and geology, and great spatial and inter-annual variations in precipitation. Under average meteorological conditions, the island is water-sufficient (969 mm precipitation; theoretical water potential 3284 hm^3; and total water use 610 hm^3). Agriculture is by far the greatest user of water (78% of total water use), followed by domestic use (21%). Despite the high average water availability, water scarcity events commonly occur, particularly in the eastern-south part of the island, driven by local climatic conditions and seasonal or geographical mismatches between water availability and demand. Other critical issues in water management include the over-exploitation of groundwater, accounting for 93% of the water used in agriculture; low water use efficiencies in the farms; limited use of non-conventional water sources (effluent reuse); lack of modern frameworks of control and monitoring; and inadequate cooperation among stakeholders. These deficiencies impact adversely water use efficiency, deteriorate quality of water resources, increase competition for water and water pricing, and impair agriculture and environment. Moreover, the water-limited areas may display low adaptation potential to climate variability and face increased risks for the human-managed and natural ecosystems. The development of appropriate water governance frameworks that promote the development of integrated water management plans and allow concurrently flexibility to account for local differentiations in social-economic favors is urgently needed to achieve efficient water management and to improve the adaptation to the changing climatic conditions. Specific corrective actions may include use of alternative water sources (e.g., treated effluent and brackish water), implementation of efficient water use practices, re-formation of pricing policy, efficient control and monitoring, and investment in research and innovation to support the above actions. It is necessary to strengthen the links across stakeholders (e.g., farmers, enterprises, corporations, institutes, universities, agencies, and public authorities), along with an effective and updated governance framework to address the critical issues in water management, facilitate knowledge transfer, and promote the efficient use of non-conventional water resources.

Keywords: water resources management; water scarcity; water reuse; climate variability; circular economy; sustainability

1. Prolegomena

The vital role of water in sustaining life on earth was recognized early; Aristotle (384–322 BC) commenting Thales of Miletus (ca. 524–545 BC) declared: "*Besides this* (water is the primary principle), *another reason for the supposition would be that the semina of all things have a moist nature* ... "[*Metaph.* 983 b26-27]. Globally, population expansion and economic growth have increased the demand for adequate, clean, and safe water, challenging existing water resource policies and management, administrative schemes, and infrastructure development [1–3]. This situation is expected to deteriorate in the future, particularly in areas with pre-existing water shortages rendering them particularly vulnerable to climate change [4]; e.g. the Mediterranean basin [5–7]. Agriculture, domestic, and industry sectors are globally the ones with the biggest dependency on water availability and will likely be the most affected by the impact of climate change. New approaches are, thus, needed to ensure sustainable water use, while preserving water resources, environmental quality, and economic development [8–10].

The island of Crete is subjected to significant challenges regarding the management of water resources arising mainly from its location, climate, history, and culture. Crete, located in the South Mediterranean Sea, is characterized by long coastal areas, varied terrain relief, and great spatial and inter-yearly rainfall variations [11]. As a result, the island of Crete is subjected to uneven spatial availability of water resources [12] and variable water imbalances across the island. For instance, the western Crete tends to have a positive water balance, as contrasted to the eastern part which is more vulnerable to water shortages and drought events [13]. Such a water scarcity phenomena are expected to become more frequent in the Mediterranean in the near future and must be seriously considered by decision makers and water management frameworks.

Water management in Crete is primarily governed by the Water Frame Directive (WFD) 2000/60/EC and other relevant EU regulatory provisions supervised by a complex stakeholder network of national, regional, and local agencies [14,15]. Experience so far points to issues in the implementation of the existing policies, while several chronic problems remain to be addressed, further exacerbating the situation. These problems can be summarized as: (a) the labyrinthine legislation, (b) the confusing competences of the public services, (c) the delayed planning at national and regional level, and (d) the lack of modern perception on water management (particularly in the agricultural sector). The lack of a strategy on the use of non-conventional water sources, particularly in agriculture [16], or for aquifer recharge is a characteristic example of the latter.

The European Union (EU) has set a goal to increase water reuse from 1.7 to 6.6 billion m^3/year equivalent to 50% of the effluent produced by wastewater treatment plants (WWTPs) that may reduce fresh water use up to 5%. For Greece, it has been estimated that effluent from existing WWTPs can save up to 3.2% of the water used in agriculture [12]. Such a prospect is very promising for the local and regional economy with strong benefits on water resources availability, agricultural production, and development of rural communities [17–20]. However, to date in Crete, despite its high potential for water reuse [15], water recycling is not a common practice, and most of treated wastewater is discharged to the sea. Given the need to respond to intensive water shortages and the challenges imposed by climate variability, it is imperative to identify the causes of the low effluent reuse in order to provide effective solutions within an expanded and improved water management framework.

The main objectives of the present study are to (a) describe the available water resources of Crete; (b) identify barriers, constraints, opportunities, and future challenges in water resources management; and (c) propose solutions and directions for a more sustainable use of water resources, emphasizing the use of non-conventional sources (e.g., desalinated, recycled, and brackish water). For the purposes of this study, information has been compiled on the historical evolution of water management, spatial-temporal variation of climate, existing water governance framework, water availability considering all the potential sources, and infrastructure (i.e., dams), derived from governmental–private sources and literature, to provide insights in possible deficiencies, gaps, and challenges. Suggestions and possible solutions are presented to help improve water resources management, particularly in areas with water scarcity and to highlight the benefits derived by the proper use of non-conventional water in Crete, Greece, and in other EU countries.

2. Historical Evolution of Water Resources Management in Crete

Archeological evidence has revealed the development of an advanced civilization in Crete and in the islands of south Aegean Sea (Santorini) during the Bronze Age (ca. 3200–1100 BC). During that period, a number of technologies focused on water resources management were developed, including hydrogeology, design and construction of aqueducts and water impounds, water and wastewater mains, runoff management, agricultural irrigation, as well as use of water for recreation purposes [21]. The documented evidence provides support that the Minoan people were aware of the basic principles of water management well before the scientific approach of our times [22]. Their technologies were expanded to the mainland Greece during the dominance of Myceneans (ca. 1600–1100 BC) and later transferred to the Classical and Hellenistic times (ca. 500–67 BC) when they were further improved [23].

During the Roman occupation (ca. 67 BC–330 AD), Crete was characterized by technological innovations in infrastructure development including public buildings, fine mosaics, toilets, sewers, drains, and other hydraulic works, in a number of cities (Gortys, Chersonissos, Ierapytna, Aptera, Lyttos, and Lebena), which were heavily influenced by Hellenic philosophy. Furthermore, large-scale aqueducts and cisterns were constructed (e.g., Gortys' aqueduct and cisterns in Aptera) [24].

From 961 to 1204 AD, technologies for water supply were implemented in big cities of Crete, which was then a part of the Byzantine Empire. Large-scale water projects were also developed and implemented during Venetian governance (ca. 1204–1668 AD) to address water supply and management issues. Indicative are the Morozini's aqueduct, as well as the rainwater harvesting cisterns in Rethymnon and Grambousa [25]. During the Ottoman (ca. 1669–1898 AD) and Egyptian (ca. 1830–1840 AD) eras, and until the beginning of 20th century, all of water infrastructure developed by Romans and Venetians was maintained and kept in operation [26]. In the 20th century, new modern water technologies were introduced and applied, such as the development of deep wells, pumps, pipes, etc.

Collectively, the water management and engineering in the island of Crete passed through a number of different stages but maintained a characteristic continuity over the centuries and even through the Dark Ages. Bridges from the past to the future are still present, albeit oftentimes, they are invisible to those who cross them, including ancient constructions that have been in operation continuously or intermittently to this day. The survival of different types of cisterns developed in Minoan and Mycenaean times, which are still in operation in many anhydrous Aegean islands [27], is another example. These constructions constituted the basis for their later development into the modern technological age [28].

Current-day engineers typically use a design period for hydro-structures operation of about 40 to 50 years, which is related to economic considerations. However, it is notable that several ancient Cretan hydraulic works have been operating efficiently for millennia, in some cases until the contemporary times, e.g., sewerage and drainage systems [29]. A. Mosso (1910), an Italian writer who visited the area in the early 20th century, during a heavy rain, noticed that the drains/sewers functioned perfectly and recorded the incident saying *"I doubt if there is other case of stormwater drainage system that works 4000 years after its construction"* [30]. Additionally, the American H. F. Gray (1940), reported that *"you can enable us to doubt whether the modern sewerage and drainage systems will operate at even a thousand years"* [31].

3. The Physical Setting Demographics, Activities and Land Use

3.1. Location Morphology Population and Economic Activities

The island of Crete is located in southern Europe, bordering the Aegean Sea, Ionian Sea, and the Mediterranean Sea (Figure 1). Crete has an elongated shape, of about 260 km from west to east and at its widest 60 km and a coastline of 1046 km long. To the south, Crete is bordered by the Libyan Sea, to the west the Myrtoon Sea, to the east the Karpathion Sea and to the north the Sea of Crete. The geographical position of Crete between three continents has influenced its historical course throughout antiquity and modern times [32].

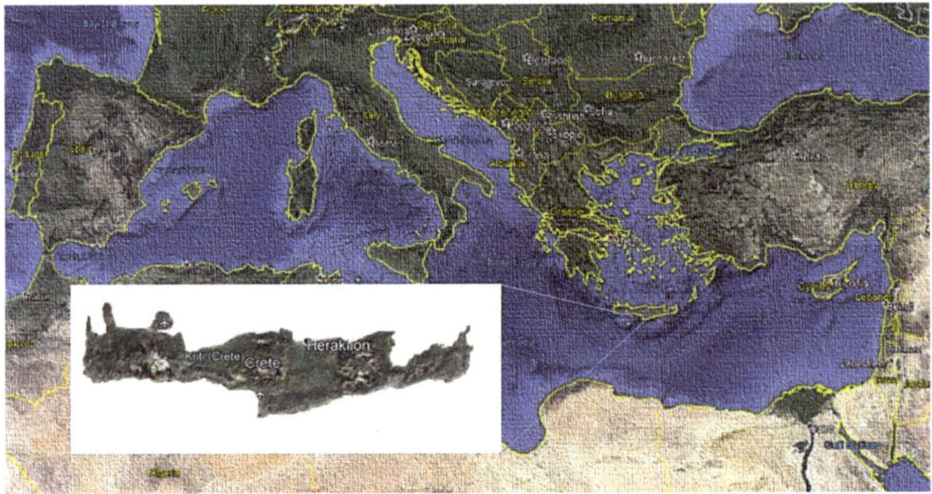

Figure 1. The location of Crete (Google earth, 2019).

The total area of Crete is 8336 km^2, accounting for the 6.36% of the total area of Greece, and is divided into four prefectures, from west to east: Chania (2342 km^2), Rethymnon (1487 km^2), Iraklion (2626 km^2), and Lasithi (1810 km^2). The main mountains running from west to east include the White Mountains (2453 m), the Idis (Pseiloritis) (2456 m) and Asterousian (1280 m) mountains in the central Crete, and the Dikti mountain (2148 m) in the east. Based on its geomorphology, Crete consists of three basic zones: the zone with an altitude of >400 m (mountainous zone), the intermediate zone (200–400 m), and the low zone (<200 m). The mountainous and the intermediate zones occupy approximately 77.3% of Crete's area and extend from the western to the eastern part of the island with some interruptions by valleys and gorges. The island of Crete is considered an independent river basin district (RBD) [14].

The permanent population of Crete is 623,065 following an almost linear increase over the last century until it leveled off in 2011 [33]. It represents 5.76% of the population of Greece [34,35]. The population and has since followed a decreasing trend (Figure 2). Based on the total area and the number of inhabitants, the population density in Crete is estimated to 81.93 inh./km^2, which is close to the country average (82.90 inh./km^2). This number, however, increases significantly in the summer period due to large number of visitors.

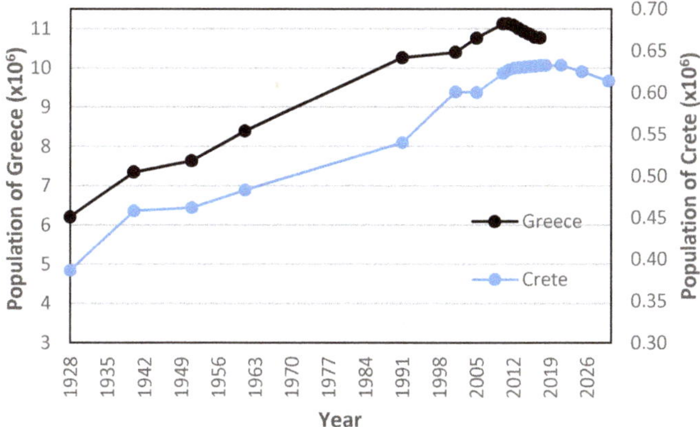

Figure 2. Evolution of the populations of Greece and Crete [33].

In economic terms, Crete produces 4.9% (13,000 M€) of total GNP of the Greece. The critical pillar of the Cretan economy is tourism, based on the accommodation of numerous tourists attracted by Cretan culture, landscapes, and climate as well as by the several archaeological sites throughout the island. The overnight stays per year in Crete are estimated to be about 16,449,065 [36]. There are also 55 Natura 2000 regions in Crete. The agricultural sector is another critical pillar of the Cretan economy based mainly on the production of wine, olive oil, and vegetables. Dairy products also have an important contribution to the Cretan economy.

3.2. Geology and Hydrogeology

Geology exerts considerable influence on surface and subsurface water flow in Crete [37]. The island's geology includes four pre-Neogene major nappes and one autochthonous isotopic zone. The nappes were transported from the north along eastern–western (E-W) trending thrusts and were placed between Late Eocene and Early Miocene times. Shortly after nappe emplacement during the Middle/late Miocene a northern–southern (N-S) directed extensional regime was established in the region, due to the relative plate behavior and the resultant geodynamic condition at the European margin, where the first E-W trending basins were formed [38,39], followed by two main faults generation: (a) in the late Messinian, an arc- parallel extension formed the N-S trending and smaller basins, and (b) in middle Pliocene, a fault development resumed in two normal directions NW-SE and NE-SW [40]. This period was associated with the deposition within the graben of Miocene to Quaternary sediments, which consist mainly of red beds, sandstones, marls, limestones, and evaporites, with overlying depositions of red lacustrine conglomerates and recent alluvium sediments [41].

The hydro-geologic structure of Crete is a result of the nappe emplacement and post emplacement tectonic and depositional history [38]. Shallow aquifers are hosted by the Neogene–Quaternary sediment filled grabens whereas deeper karstic aquifers flow through the carbonates of the pre Neogene nappes [39]. Underground and spring water move through the fault systems, which impede or facilitate the flow depending on structural positions. Major faults with general N-S orientation enable preferential water flow, while faults W-E oriented may act as impermeable boundaries [37]. Approximately 80% of the ground water resources of the island are associated with deep karstic aquifers whereas the remaining is attributed to the Neogene-Quaternary aquifers [39]. In the area of Crete, there are 47 gauged springs divided into freshwater, brackish, and undersea springs [42,43]. Most of springs belong to the karst hydrogeological system (Lefka Ori, Idi, Dikti, and Sitia) discharging around 500 hm^3/year. Brackish springs located at coastal areas have the greater flows, such as Almyros in Iraklion that discharges around 250 hm^3/year while spring in Agios Nikolaos discharge 82 hm^3/year. There are also submarine discharges spread in coastal area of Crete [43]. The geological structure and geomorphology of Crete create many small hydrologic basins (Figure 3). The greater hydrologic basins are these of Anapodiaris (600 km^2) and Geropotamos (525 km^2) found in Messara-Valey (Figure 3).

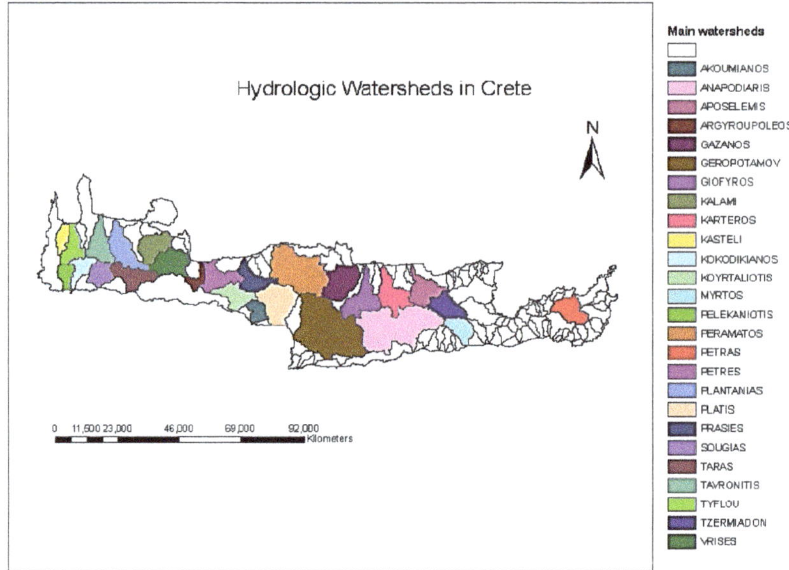

Figure 3. The most important hydrologic basins of Crete (modified from [43]).

3.3. Climate

History. In contrast to earlier ancient civilizations (Egypt, Mesopotamia, Indus) that flourished in water-abundant environments (large river valleys), ancient Greeks and especially Cretans preferred to establish their settlements in dry, water-scarce sites. Thus, all major ancient Cretan cities, during the several phases of the Cretan civilizations, were established in areas that had low water availability. Another characteristic is that, despite the existence of some small-scale rivers and lakes in Crete, no major city has been built close to these resources. It can be argued that climate and health could have been the main criteria, as dry climates are considered healthier and protect the population from water-borne diseases. It is worth noting that the progress in Cretan civilizations over the centuries has been connected closely to hygienic living standards and a comfortable lifestyle. To achieve these, both technological infrastructure and management solutions were developed.

Previous studies on climate variations in the Mediterranean region during the Holocene period have documented the occurrence of distinct climatic periods during the past 5000 years (e.g., cold period, ca. 4500–3000; cold and humid period, ca. 3000–2200; and a warm period, ca. 2200–1400 BC) [44]. Despite the varying climatic conditions during the past 5000 years, it seems that the overall abundance of water resources was never the case for important Cretan cites such as Knossos, Zakros, Phaistos, Kissamos, and Gortys. To address the (occasional) water scarcity in specific areas, ancient Cretans developed innovative technological means to capture, store, and convey water even from long distances [26]. Thus, the main technical and hydraulic innovations were associated with the management of water resources, followed by sewerage and drainage systems, including urinals and toilets, bathrooms with tubs, laundry slabs and basins, as well as by effluent disposal sites. Such operations have been practiced in varying forms since ca. 3000 BC [44].

Tsonis et al. (2010) indicate that wetter conditions during the middle Holocene were followed by drier conditions and that around 1450 BC a long stretch of drier conditions commenced ending around 1200 BC. The authors also presented a synthesis of historical, climatic, and geologic evidence to support the hypothesis that climate, instigated by an intense El Nino activity, contributed to the demise and eventual disappearance of the Minoan civilization. Thereafter, during the Iron period (ca. 1300–600 BC), another cold and humid period prevailed. During classical and Hellenistic times (ca. 600–67 BC), the climate was rather warm and dry. During the Roman period (ca. 67 BC–330 AD), a short colder and more humid period prevailed. Finally, a warm and dry climate prevailed during the Arab period with a peak of high temperatures and drought ca. 800–1000 AD [21]. Taken together, the climatic and hydrologic conditions in Crete have been characterized by high spatial and temporal variability (Figure 4) throughout the long history of the island [32].

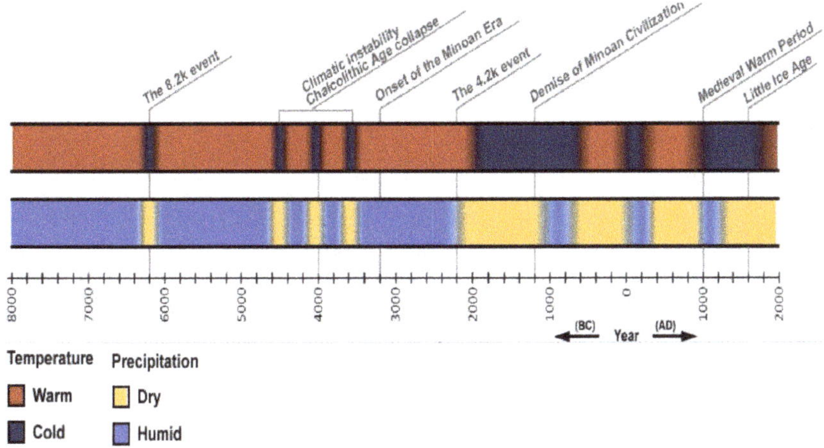

Figure 4. Climate reconstruction of Crete for the last 10,000 years based on proxy and historical data (with permission of [32]).

Today, Crete falls within two major climatic zones, the Mediterranean and the North African zone. As a result, the climate is primarily temperate with relatively high atmospheric humidity and quite mild summer and winters. The average temperature in the summer varies between 15 and 30 °C, with the maximum values up to or exceeding 40 °C. More sunny days and higher temperatures prevail across the south coast, including the Messara valley and Asterousia mountains, driven mainly by the prevailing North African climatic zone [45].

Precipitation. The precipitation in Crete is characterized by spatial and temporal variation increasing towards the western and north parts of the island [11]. The mean annual precipitation in the eastern part has been estimated at 675 mm/year, while in western Crete at 1179 mm/year, with an overall average of 967 mm/year (Table 1). The spatial distribution of mean annual precipitation [39] corelates well with altitude, due to orographic effects [46]. The calculated mean lapse rate is 61 mm/100 m in altitude (range: 25 to 100 mm). Indicative of the precipitation variation along Crete, is the 440 mm/year at the valley of Ierapetra (SE Crete) and the 2000 mm/year at the Askifou upland (NW Crete).

Hydrologic years in Crete are clearly distinguished in wet and dry seasons, with the wet season lasting from October to March and the dry season lasting from April to September. In the prefecture of Iraklion, on average, the 87% of precipitation occurs during the wet period. The average number of rainy days ranges from 15 in December and January to almost null in July and August. Extreme daily precipitation events of 107 mm in Iraklion and 250 mm in Chania (two events on February 2019) have been recorded that exert a strong effect on maximum annual stream flows [49]. The seasonal stability of precipitation pattern in Iraklion, which exhibits a relatively stable wet/dry pattern for more than a century (1909 to 2018), is demonstrated in Figure 5b. Over the past 30 years, the intensity and frequency of daily precipitation maxima have declined slightly according to the time series (Figure 5c), yet precipitation extremes could potentially intensify in the future (see Section 4.1). In Chania, average precipitation is higher than that of Iraklion, but it follows a similar trend over the last century (Figure 5a).

Table 1. Average annual estimations of the hydrologic cycle components in Crete on normal, wet, and dry years. Source: [14].

Hydrologic Conditions	Unit	Precipitation	Actual ET (57.50%)	Run-off (15.00%)	Infiltration (27.50%)
Normal year	mm	967			
	hm^3	7727.47	4443.30	1159.12	2125.05
Wet year	mm	1244			
	hm^3	10,369.98	5962.74	1555.50	2851.74
Dry year	mm	610			
	hm^3	5084.96	2923.85	762.75	1398.36
Year 2017–2018	mm	480			
	hm^3	4001.28	2300.74	600.19	1100.35

Data of 40 years from 90 Meteorological stations.

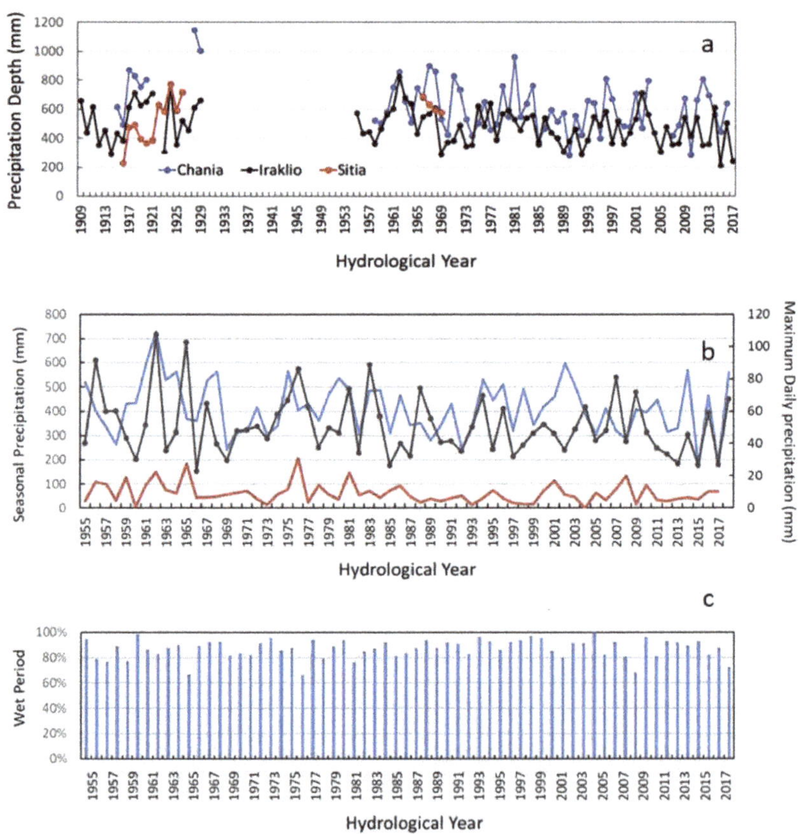

Figure 5. Precipitation in Chania, Iraklion, and Sitia from 1909 to 2017 (**a**). Precipitation in the prefecture of Iraklion, Crete, during the wet (blue line) and dry seasons (red line) (**b**). Maximum daily precipitation is given in black line (left y-axis) (**b**). Blue bars (**c**) above illustrate the wet season precipitation as a percentage of annual rainfall height (data available by the National Oceanic and Atmospheric Administration. Sources: [32,47,48].

Air temperature. The mean annual temperature ranges from 17 to 20 °C. Air temperature increases from West (16.96 °C Alikianos station) to east (18.33 °C Siteia station) and decreases from south (19.55 °C Ierapetra station) to north (18.55 °C Siteia station).

Humidity. The driest months of the year are June and July (mean relative humidity 48.9% in Souda and 59.9% in Iraklion). The most humid month is December (72% in Souda and Siteia and 67% in Iraklion).

Potential evapotranspiration. Potential evapotranspiration (ETo), estimated with Penman–Monteith method, varies from 1240 to 1570 mm/year. The intra-annual monthly ET ranges from about 25 mm (winter) to 225 mm (summer). The annual actual ET accounts from 75% to 85% of the annual precipitation in low elevation areas (< 300 m absl) and 50% to 70% in high elevation areas.

Hydrological water balance. The estimated hydrological water balance of Crete for three hydrological conditions, namely a normal year with a return period equal to or exceeding 50%, a wet year with a return period equal to or exceeding 10% and a dry year with a return period equal to or exceeding 90% is shown in Table 1. Estimations were derived from the analysis of a 40-year time series and on surface models of major catchments and ground water models of aquifers [14].

3.4. Land Use

Land use and habitat characterization has been studied extensively [50] as a tool of guidance for regional planning and policy. An overview of categorized land uses, including cropland, drylands, forests, mountains, and urban areas is given in Figure 6. According to the latest water management plan of Crete [15], forests/semi-natural areas and agricultural land account for 55.56% and 42.38% of the total area, respectively. The distribution of the different land uses across the three drainage basins of Crete (north (EL1339), south (EL1340), and east (EL1341) parts) are summarized in Table 2.

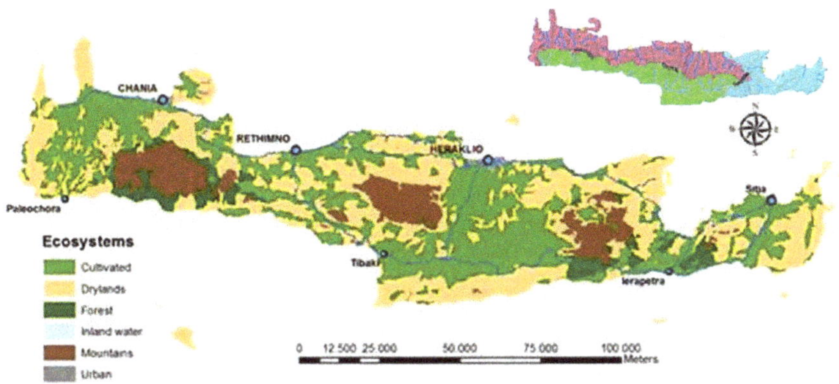

Figure 6. Ecosystems in Crete (modified from [51]). Small figure illustrates the three drainage basins of Crete (north (EL1339) (purple), south (EL1340) (green), and east (EL1341) (blue) (modified from [15]).

Table 2. Drainage basin coverage of Crete by different land uses. Source: [15]).

Drainage Basin	Agriculture (%)	Forest and Semi-Natural Areas (%)	Artificial Areas (%)	Water Lands (%)
EL1339	45.65	51.06	3.22	0.07
EL1340	42.54	56.90	0.50	0.05
EL1341	35.68	62,51	1.76	0.05
Average	42.38	55.59	1.98	0.06

4. Water Management

Sustainable water management is a challenge for Cretan regional authorities to meet water requirements driven by different economic activities (agricultural, domestic, livestock, and industrial use) and to preserve the water resources of the island. Sustainability is not a straightforward task considering the complexity of the existing legislative framework involving several stakeholders from multidisciplinary sectors. Given the instability in the current economic environment and the challenges arising from and the need for adaption to climate variability, sustainability becomes even more challenging. In this context, available administrative structure and legislative framework, spatial and temporal variations water resources availability, and challenges in water management are presented and interpreted to identify critical topics, options, and alternatives.

4.1. Administrative Structure and Principal Legislation

In Crete, water resources management is covered by the 3199/2003 law and the 5/2007 presidential decree, established to achieve synchronization with 200/60/EC WFD of the EU [52]. That Directive, among others, settled for each of the 14 river basin districts (RBD) of Greece (Crete represents the 13th RBD) water management plans, as way to address critical issues and challenges in current and future water management planning. For Crete, the first water management plan was released in 2015 [14]; the first revision of the plan was presented in 2017 [15], covering the period of 2016 to 2021. In the latter, the administrative structure and the responsible authorities in water resources management in Crete are identified and described, including the National Water Committee, National Water Council, and General Secretariat for Water and Environment, at national level, and the Regional Water Council of the Decentralized Administration of Crete, Electoral Region of Crete, and Municipalities [15]. The Decentralized Administration is responsible for the development of national strategic planning (water protection measures), while the Electoral Region and Municipalities are responsible for its implementation. Furthermore, Electoral Region and Municipalities are responsible for the monitoring and control of water resources (level and quality of the ground and surface water) as well as for execution of projects related to water resources exploitation [15].

Besides the WFD, EU via the new EU CAP 2014–20 and other supporting EU directions and measures, such as circular economy concept [53] and measures to mitigate climate change [54], provides a comprehensive institutional umbrella on issues related to water management across the member states to tackle water scarcity in sensitive areas, mitigate climate change impacts, ensure necessary adaptations, and protect water resources. Reuse of treated effluent is among the actions that have been considered and promoted by EU being fully compatible with other (e.g., circular economy and climatic change) policies as aforementioned. To support its policies, the EU provides a variety of financial tools to strengthen knowledge in critical water management issues and to enhance networking and dissemination of knowledge across stakeholders.

4.2. Water Availability and Climate Variability Impacts

The average yearly precipitation on Crete (969 mm) corresponds to approximately 6109 hm^3 [55] (Table 3). However, less than 36% of the precipitation is stored in the soil or percolates to deeper horizons. By contrast, ET and runoff to the sea account for 73% and 19% of the precipitation, respectively. As a result, the theoretical total water reserves are estimated to be 3284.17 hm^3/year (Table 3), accounting for 54% of precipitation, without considering the potential contribution of non-conventional water recourses.

Table 3. General hydrological data (annual average values of a normal year) for the river basin districts (RBDs) of the island of Crete.

Parameter	Unit	RBD of Crete
Area	km^2	8315.00
Precipitation	mm	969.00
Volume of precipitation	hm^3	6109.00
Evapotranspiration	hm^3	4443.30
Percolation	hm^3	2172.31
Surface runoff	hm^3	1159.12
Theoretical water potential	hm^3	3284.17

Although precipitation theoretically satisfies water requirements (consumption accounts for the 7% of the total precipitation) [12,56], water imbalances have been experienced across the island. These imbalances have been driven by temporal and spatial variations in precipitation, terrain characteristics, vegetation distribution, urban water needs, distribution of water infrastructure (e.g., artificial lagoons and dams), local economy potential and seasonal water demands, and water transportation constraints [12].

Due to its geographical location, the island of Crete is subjected to great vulnerability to climatic conditions [5,56,57]. This vulnerability will become more challenging in the upcoming years due to climate variability [6,57,58], impacting further water availability [56,59] and likely crop productivity [60,61]. Among climatic extremes, in Crete, intense precipitation events, increased frequency of flooding, longer and more intense droughts have been projected [56,57,62]. However, the validity of climate change projections is still under investigation due to assumed initial conditions and assumptions and methodological constraints related to the downscaling of the climatic projections [63,64].

4.3. Water Uses and Critical Topics

The major water uses in Crete are irrigation and domestic use, with relatively small volumes of water used for livestock, landscape irrigation, and industrial applications (Table 4). Agriculture relies on the groundwater (about 93%), whereas domestic and livestock use are equally dependent on surface and subsurface water (Table 4).

Table 4. Withdrawals from surface and underground waters and overall water use in Crete. Source: [15].

Source	Water Uses in 2016 (hm^3/Year)				
	Domestic	Agriculture	Livestock	Industry	Total
Surface water	39.40 (30.87%)	34.60 (7.23%)	2.10 (50.48%)	0.27 (36.00%)	76.37
Sub-surface water	88.21 (69.10%)	443.81 (92.77%)	2.08 (50.00%)	0.48 (64.00%)	534.58
Total	127.65	478.39	4.16	0.75	610.94
Consumption index (%)	20.89	78.30	0.68	0.12	NA

Domestic applications. Domestic use has been estimated to 127.65 hm^3/year (Table 4) driven by permanent population and tourism. Water consumption follows a seasonal pattern with the highest water demands in the summer period due to tourism and the lowest ones in winter. Currently, non-revenue water (NRW) may be the major problem in managing potable water in Greece and especially in Crete. The actual amount of NRW is one of the highest among the EU countries; in some cases, it exceeds 60% of the potable water due to losses in the networks, illegal connections, and non-payments from no-payers/consumers. Mitigation of NRW must be considered in collaboration of water companies with the private sector, under transparent, clean, and controlled processes.

Agricultural Use. In Crete, 87,040 recorded land holdings exist, consisting of mixed (agricultural and livestock), agricultural, and livestock farms, which occupy an area of 364,095 ha (Table 5). Cultivated land occupies 280,075 ha, of which 151,550 ha (about 54%) is irrigated (Table 6). Tree crops (mostly olive trees) dominate agricultural land covering 203,946 ha, of which 119,216 ha (about 58%)

are irrigated. Vineyards occupy 18,962 ha, of which 72% (13,590 ha) are irrigated. Overall, based on records from the Hellenic Statistical Agency [65], the irrigated area has been increasing since 1995 (Figure 7). It has been estimated that water from public irrigation networks is utilized for about 30,300 ha; the rest are irrigated with private water systems. It is worth noting that most of the water used for irrigation is derived from groundwater (Table 4). The irrigation efficiency has been estimated at about 80% [66], due to the wide use of drip or micro-sprinkler irrigation systems.

Table 5. Holdings and areas in mixed, agricultural, and livestock areas of Crete. Source: [65].

	Total		Mixed		Agricultural		Livestock	
Holdings/Areas	**Hold-ings**	**Area (ha)**	**Hold-ings**	**Area (ha)**	**Hold-ings**	**Area (ha)**	**Hold-ings**	**Area (ha)**
Region of Crete	87,040	364,096	14,590	225,867	71,619	125,197	831	13,032
Iraklion	41,162	144,098	4787	70,068	36,184	70,348	191	3682
Lasithi	12,981	41,063	1168	19,903	11,729	19,507	84	1653
Rethimnon	13,024	93,703	4517	76,238	8131	14,133	376	3332
Chania	19,873	85,232	4118	59,658	15,575	21,209	180	4366

Table 6. Total and irrigated agricultural areas of Crete. Source: [65].

	Total (Incl. Fallow Land)	**Crops on Arable Land**	**Garden Area**	**Vineyards**	**Tree Crops**	**Fallow Land**
			Total area (ha)			
Region of Crete	280,075	20,774	6965	18,962	203,946	29,427
Iraklion	129,046	10,173	2814	15,012	90,588	10,459
Lasithi	52,517	3008	1953	1459	36,043	10,054
Rethimnon	41,956	4377	609	934	29,348	6688
Chania	56,556	3217	1590	1558	47,966	2226
			Irrigated area (ha)			
Region of Crete	151,550	11,550	7195	13,590	119,216	
Iraklion	87,643	6821	3090	12,519	65,212	
Lasithi	24,731	1851	1915	567	20,398	
Rethimnon	8761	374	594	85	7708	
Chania	30,416	2504	1596	419	25,898	

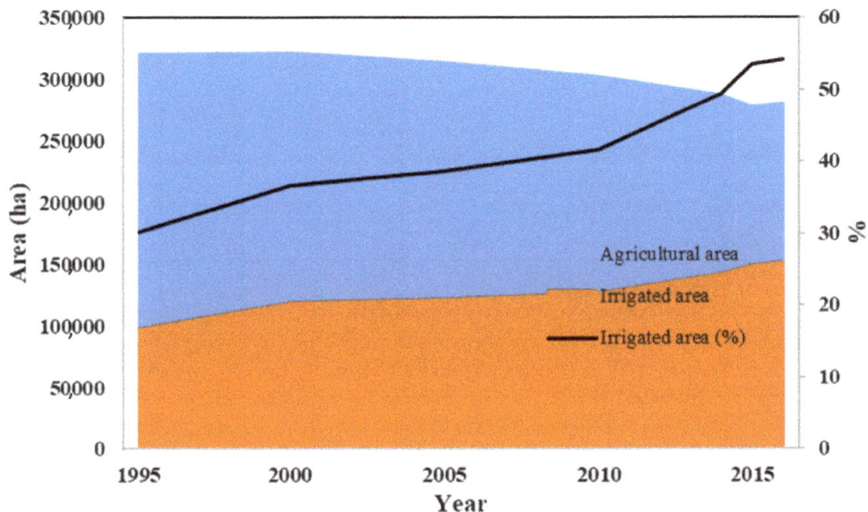

Figure 7. Change of the irrigated area from 1995 until the present (modified from [65]).

Agriculture is by far the greater consumer of water on the island (Table 4) exerting significant pressure on water resources, especially on subterranean water [67–69]. Over-pumping of groundwater to support agricultural production has led to significant depletion of the aquifers, a problem that is being exacerbated over time, mainly in areas of the southern (e.g., valley of Messara) and eastern part of the island [15,69–71]. These areas are characterized by low precipitation and high temperatures and are highly vulnerable to climate variability. These areas are also subject to other serious risks, such as sea intrusion, soil degradation, and desertification [72–74]. To address the problem of groundwater overexploitation, specific strategies and measures should be applied enhancing primarily water availability in the area by using other water sources as well as measures to improve water use efficiency or availability (transportation, dams) [75,76]. For instance, adoption of appropriate field practices, such as deficit irrigation, new irrigation technologies, use of tolerant to water deficits crops, and/or more water efficient crops, will result is significant water savings eliminating concurrently the adverse effects on crop performance [77,78]. Furthermore, measures that improve monitoring and control of water from the source to farms, along with provision of disincentives (high pricing) for those with high water consumption, will contribute to the efficient use of water. According to recent records, water price in Crete varies greatly across or even within the catchment area, depending on responsible agency, between 0.05–0.65 €/m^3 [13,79,80]. Finally, measures that promote knowledge transfer and capacity building across stakeholders may enhance farmers' ability to apply sustainable practices and increase public awareness. Both pricing and non-pricing measures have been supported by the legislative basis of the EU for water management.

Water demand for stock raising and industrial applications. The water demand for stock raising is relatively low but not negligible. The bulk volume of water is used for raising free range stock (sheep and goats) and is estimated to be 4.16 hm^3/year (Table 4) [15]. Industrial activities on Crete are limited. The main water consuming industries are wineries and olive oil mills, which are scattered throughout the island. The annual quantities of water used by the industrial sector is estimated to be 0.75 hm^3/year (Table 4).

Total water demand. The total annual water needs for Crete are estimated to be about 610.94 hm^3/year (Table 4) [15], which correspond to approximately 7.91% of the annual precipitation and to 18.60% of the annual theoretical water potential. Based on these data, it can be inferred that Crete is characterized by high availability of water reserves that may be utilized within an optimized water management plan.

4.4. Conventional Water Resources

4.4.1. Surface

Surface aquatic bodies on the island of Crete include 128 rivers, 1 lake, and 29 transitional and coastal systems distributed throughout Crete. Aquatic bodies also include artificial lagoons and reservoirs (dams). Most of these dams have been built since 1990 onwards and are located all over the area of Crete with a total volume of 280 hm^3 (Table 7). Large-scale dams (e.g., the Valsamiotis dam in Chania) are designed to meet existing demands, while others are designed to meet existing and/or future water demands, such as dams of Roumatianos and Derianos in Chania, dam of Plakiotissa in Iraklion, and dam of Amari in Rethymnon [56]. Moreover, the potential of some of the dams, mainly those in the eastern part of the island (e.g., Aposelemis dam) to meet the design expectations, is challenged by the prevalence of consecutive dry years or by shifts in precipitation patterns arising from climate variability [56,81]. The surface aquatic systems of Crete are assessed in the latest river basin water management plan [15] that focuses mainly on the anthropogenic activity induced impacts and on the compatibility with EU 2000/60 Directive criteria. Most of the surface water bodies are classified at the category "is not at risk" (40%) or "probably not at risk" (56%) [15].

Table 7. Major dams in Crete. Source: [14,46,82,83].

Name	Location	Period of Contraction	Type	Total Volume (hm^3)	Usable Volume (hm^3)	Comments
			Regular dams			
Potamon	Amari, Rethymnon	1995–200 2003–2008	Earth dam	22.50	17.50	Water supply and irrigation
Aposelemis	Avdou, Iraklion	2006–2012	Earth dam	25.27	24.36	Irrigation
Valsamiotis	Vatolakos, Chania	2005–2014	RCC [a] (FSHD)	6.00	5.90	Irrigation
Faneromeni	Western Messara	2005	Earth dam	19.67		Irrigation
Mpramianon	Ierapetra	1986	Earth dam	16.00		Irrigation
Ini	Iraklion	2002	Earth dam	1.75		Irrigation
Damanion	Iraklion	2003	Earth dam	1.50		Irrigation
Amourgeles	Iraklion	2004		1.56		Irrigation
Plakiotisas	Iraklion		Earth dam	18.60		Irrigation/ Under construction
Chalavrianou	Iraklion	2018	Earth dam	1.20		Irrigation and Water supply
Partiron	Iraklion	2000	Earth dam	1.50		Irrigation
Armanogion	Iraklion	2004	Earth dam	1.50		Irrigation
Agias	Chania	1929	Earth dam	0.13		Energy, Water supply and Irrigation
Gerakari	Rethymnon		Earth dam	1.75	1.45	Irrigation
			Under planning (Major)			
Plati Potamou	Agia Galini, Rethymnon			51.00		Irrigation
System of 3-dams	Chania			45.00		Irrigation
Dematiou	Iraklio			30.00		Irrigation
Agiou Ioanni	Lasiyhi			18.50		Irrigation
Lithinon	Lasithi			9.00		Irrigation
			Small dams (limnodecsamenes)			
Vizariou	Rethimnon	1994		0.66		Irrigation
Agiou Georgiou	Lasithi plateau	2008–2012		2.15		Irrigation
Chavga	Lasithi plateau	1995		0.86		Irrigation
Karavado	Iraklion	1996		0.11		Irrigation
Agion Theodoron	Chania	1998		0.65		Irrigation
Anogion	Rethymnon	2001		0.75		Irrigation and Water supply
Gergeri	Iraklion	2001		0.26		Irrigation
Chrisoskalitisa	Chania	2005		0.56		Irrigation
Arkadiou	Rethymnon	2006		0.60		Irrigation
Thrapsanou	Iraklion	2006		0.21		Irrigation
Kountouras	Chania	2008		0.65		Irrigation
Skinia	Iraklion	1997		0.38		Irrigation
Elous	Chania			0.35		Irrigation/ Under construction
Omalou	Chania			0.50		Irrigation/ Under construction
Zou	Lasithi			0.30		Irrigation/ Under construction

[a] Roller-Compacted Concrete.

Thermal springs. Out of about 750 recorded thermal springs in Greece, almost 100 are found in Crete [84]. Thermal springs were known in Crete since Classical times when most of them were associated with *Asclepieia* (ancient hospital, e.g., *Asclepieia* in Levina and in Lissos in Crete [85]). There are indications that the water spring in the *Asclepieion* of Levina was saline and was thought to have healing properties. The spring water of Levina, at a temperature of 22 °C, continued to be used for healing purposes during the Historical and Byzantine times [86]. Moreover, analyzed water samples taken from the spring in ancient Lissos in Crete, during the hot season, were found to be in the hypothermic range from 20.3 to 20.7 °C, among the hot and cold seasons, respectively [86]. The spring water, based on chemical analyses, was dominated by a calcium-magnesium-oxycarbonate (Ca-Mg-HCO_3) mineral complex in both periods of time.

4.4.2. Groundwater

The annual underground water supply in Crete is estimated to be 2172.31 × 10^6 m^3/year (Table 8) of which a significant portion is brackish. It is estimated that the total underwater discharges including brackish water amounts to 800–1000 × 10^6 m^3/year. Water potential of the major hydrogeological units of Crete are presented in Table 8 [14,15].

Table 8. Water potential of major hydrogeological units of Crete (table data are based on estimates of over 91 individual hydrogeological units throughout Crete). Source: [14,15].

Hydrogeological Formations	Area (km^2)	Average Annual Precipitation (mm)	Volume of Precipitation (hm^3/Year)	Average Percolation (%)	Volume of Percolated Water (hm^3/Year)
Karstic	3333.07	1300	3549	42.55	1510.24
Neogenic	2950.92	693	1799	27.00	485.66
Others	2031.81	780	761	23.18	176.41
Total/Average	8315.80	969	6109	35.56	2,172.31

The intensive exploitation of groundwater, particularly by agricultural activity over the last 50 years, has led to a continual decline in groundwater level, while several coastal aquifers suffer from seawater intrusion [87]. In addition, the water quality of some aquifers has been degraded due to pollution from agricultural, industrial, and touristic activities. In a recent investigation of 91 aquifers, it was found that nine systems (eight in Iraklion and one in Ierapetra-Sitia) had significant or moderate degradation due to elevated salinity and high NO_3^- and SO_4^{2-} concentrations [15]. Moreover, by applying a new groundwater footprint methodology, one aquifer system in Chania, five in Iraklion, and eleven in Sitia also exhibited elevated rates of deterioration [88]. In Crete, 94 municipal WWTPs, 1549 hotels, and 721 industrial units operate, of which 492 are olive mills and 35 dairies. There are also 25 wineries and 23 cement industries [15]. These units constitute potential sources of pollution through the diffusion of the specific organic and inorganic pollutants to ground and surface water. In total, the surface loadings derived from the point, non-point, and other anthropogenic pressures have been estimated to be 34,291.58 ton/year for BOD, 22,010.91 ton/year for N, and 5347.57 ton/year for P [15].

4.5. Potential for the Use of Non-Conventional Water Sources

Non-conventional sources of water must comply with the principles for sustainable development and should be considered in an integrated water resources management plan [89]. These sources mainly comprise recycled, brackish, and desalinated water. Stormwater and wastewater management are also considered.

4.5.1. Stormwater and Wastewater Management

Stormwater management. Cretan coastal system is characterized by a strong seasonal variability, being typically oligotrophic during the dry summer periods with nutrients delivered primarily during the wet winter periods via pulses associated with high rainfall events [90]. These events are the most important drivers of coastal primary production leading to deterioration of water quality especially near densely populated areas. In Crete, stormwater is dispersed to natural recipients, e.g., land, rivers, and transitional and coastal waters [91]. As all major cities of the island are located in coastal areas, storm-water dispersal directly to the sea is the common practice. Bathing water quality parameters are monitored regularly in accordance with the provisions of Directive 2006/7/EC [92]. Furthermore, the ecological and chemical status of coastal waters in Crete is assessed seasonally in according to the Water Framework Directive 2000/60/EC [93]. In the past years, based on the results from both monitoring programs, the water quality and status of the coastal waters around Crete was very good [94]. Additionally, the contribution of the wastewater treatment to the bathing water quality, especially in coastal urban areas, is significant.

Current status of wastewater. Greece and, of course, Crete have to comply with the EU Urban Wastewater Treatment Directive 271/91/EC [95]. Today, the status of wastewater management in Crete has been improved significantly. The total length of wastewater collection system is estimated to be about 3000 km serving more than 90% of the total population. In contrast to the common practice of Cretan ancient periods, where the wastewater and stormwater networks were combined, separated systems have been dominant throughout the island since the middle of the last century.

The current status of wastewater treatment in Crete, based on the population served, is presented in Table 9. Today, there are about 100 operational WWTPs, most of which serve human settlements of less than 2000 inhabitants. Most of these WWTPs are in the eastern part of Crete. As noted in Table 9, most of the remaining WWTPs to be implemented in the future are for small settlements with a capacity of less than 2000 population equivalents (pe) [91]. It has been estimated that more than 80% of the island's population will be served after the completion of all plants with a capacity of more than 2000 pe.

Regarding the technology applied, a number of different WWT technologies have been adopted for use in Crete. Among the WWTPs serving more than 2000 inhabitants, 95% are conventional activated sludge and/or extended aeration systems. For populations of less than 2000 inhabitants, the predominant technologies are gravel and sand filters, textile filters, and constructed wetlands. Furthermore, it should be noted that 5%–10% of the population resides in villages of less than 500 ep for which on-site sanitation technologies are used [91].

4.5.2. Water Reuse

Because of the large number of operating WWTPs in Crete, the potential for water reuse is high (Table 9) [15]. However, at present, the major proportion of the treated effluent is discharged to the sea, instead of being reused for crop irrigation [96], to replenish aquifers [97] and/or to hinder seawater intrusion [67,98]. Furthermore, recycled water can be an extra source of nutrients (mainly nitrogen and phosphate) for the existing cultivations reducing the consumption of conventional fertilizers and, hereby, the overall production cost [17,18]. It is estimated that the use of recycled water could reduce the use of commercial nitrogen fertilizers by 5–7 kg N/year.ha. Of the 99 operating WWTPs (Table 9) serving about 80% of the island's population, only 10.06% of the treated effluent (Table 9) is currently used for crop irrigation.

Several factors are responsible for the limited use of recycled water in Crete. Among them, barriers arising from the strict regulations set by the national and EU legislation and the low social acceptance are considered the most important [99,100]. A typical example of strict regulations is the required monitoring of heavy metals and metalloids. Based on the capacity of the WWTP, monitoring frequency varies from 2 (<10,000 pe) to 12 (>200,000 pe) times per year. Another example is the need to monitor 40 organic compounds at least twice per year in WWTPs serving more than 100,000 ep [99,101]. Recently, the EU proposed guidelines for effluent reuse [102], defining minimum requirements for crop irrigation and aquifer recharge along with monitoring needs. The proposed criteria define water quality classes depending on the intended use of crops and irrigation method based on BOD/COD thresholds and *E. coli* population (10 up to 10,000 cfu/100 mL). Furthermore, it has been proposed to consider limits for specific parameters, including heavy metals, pesticides, disinfection by-products, pharmaceuticals, and other substances of emerging concern, and anti-microbial resistance [102]. Finally, it is required for operators to establish a risk management plan to ensure addressing of the potential additional dangers [103].

Reuse has low public acceptability in Crete mainly due to pricing and environmental/public health issues [100]. Raising the awareness of the stakeholders and farmers is probably among critical options for authorities [100]. However, given the presence of harmful substances in the effluents, further steps should be taken to revise the water quality criteria list. Moreover, it is important to develop and introduce a new pricing policy for non-conventional waters, as well as new certification processes along the product chain and a uniform labeling policy to ensure that products in the market are safe.

Certification is in agreement with current framework covering the production delivery and marketing of the conventional products and the new EU policy.

Table 9. Current status of wastewater treatment in Crete (adapted from region of Crete) [104].

Population Served	WWTPs (no)	Capacity (hm^3/Year)	Reused (hm^3/Year)	Reuse Opportunities	Comments
<2000	67	3.90	0.75	Agricultural irrigation.	Numerous additional small projects (more than 650[a]) serving less than 2000 persons are in various stages of planning and development. When completed, these treatment plants will serve 15%-20% of the total population of Crete.
2000–5000	15	4.65	0.90	Agricultural irrigation and landscape irrigation.	Two more plants are under implementation and three are under construction.
5000–15,000	10	8.90	2.25	Agricultural irrigation, landscape irrigation, and groundwater recharge.	One more plant remains under implementations and another one is under construction. When those treatment plants (including the above) are completed, the total population served will rise above 80%.
15,000–100,000	5	12.00	0.55	Agricultural irrigation, landscape irrigation, groundwater recharge, and indirect and direct potable reuse.	
100,000–150,000	1	10.20		Agricultural irrigation and landscape irrigation.	
>150,000	1	14.50	1.00	Agricultural irrigation, landscape irrigation, groundwater recharge, and indirect and direct potable reuse.	
Total	99	54.15[b]	5.45[c]		

[a] WWTPs under implementation are not included. [b] The potential for agricultural use is about 10.1% of the total water used for agricultural irrigation [15]. [c] About 1.10% of the water is now used for agricultural purposes.

Based on the above discussion, it can be argued that Crete has the potential to increase significantly the use of recycled water in the coming years, by providing further services and benefits to the economy, natural resources, and the environment. Achieving the above goal, however, presupposes the resolution of key issues, such as those that meet the quality standards of outflows (revised criteria), pricing policy of their use, as well as certification issues for the products produced. It should be emphasized that all actions as a whole should be compatible with current national and EU policies and requirements for sustainable resource management, sustainable agriculture, the development of the circular economy, and adaptation to climate change. Overall, the expansion of water reuse in Crete is expected to have a number of positive impacts for the residents, the local economy, and the environment that can be summarized as follows:

- Improve water availability and strengthen the adaptation potential to climate variability;
- Support crop production and reduce production cost by water and nutrients supply;
- Enhance the transition toward circular economy agricultural practices;
- Reduce pollution risks for water resources;
- Protect groundwater, the major current source of irrigation water, from overexploitation and degradation.

4.5.3. Brackish Waters

Brackish water is a significant source of water in Crete that could be exploited potentially for variety of different applications. Only the well-known sources of brackish springs, developed in karsic formation, known as Almyroi (e.g., Almyros of Iraklion, Almyros Agios Nikolaos, Almyros Georgioupolis, and Malavra spring) around Crete exceed the 1000 hm^3/year [105]. For instance, the Almyros in the Iraklion city releases on an average basis 250 hm^3/year (ranging from 5 to 7 m^3/s in dry and wet period, respectively), which exceeds 50% of total annual water needs of Crete. To date, these sources of water remain unexploited. The use and exploitation of water resources in karstic formation through the centuries, especially in the Mediterranean area, has been studied widely in the literature [106]. The knowledge of the historical techniques of karst water exploitation is significant for better management and planning of water resources under scarcity in the island. A brief synthesis of the numerous investigations carried out in the Almyros spring of Iraklion, obtained with a view to determining practical methodologies for capturing fresh water, follows.

(a) The major hydrological characteristic of the Almyros spring is that the water becomes brackish under low flow rates (less than 12 m^3/s), i.e., about nine to ten months per year. The total dissolved solids (TDS) decrease from a maximum concentration of 5 g/L in October to less than 0.30 g/L during the following maximum discharge flow rate. Water from Almyros spring could serve Iraklion city with potable water for 35–45 day/year when the concentration of TDS remains below 0.30 g/L. The possibility of constructing a reservoir outside of Almyros basin (e.g., Taveronas basin) for storing the produced fresh water, when the TDS concentration is less than 0.30 g/L, and using it for water supply of Iraklion municipality has been evaluated [107]. An exploitable volume of 31.10 hm^3/year of fresh water, to be stored during the 30 day/year (conservative forecast) when the maximum discharge flow rate is greater than 12 m^3/s).

(b) Several investigations have been carried out since 1964 to understand the spring function mechanism and reduce the seawater intrusion in the Almyros karstic system [108,109]. A small dam was constructed in mid 1970s (Figure 9) based on a previous study [110]. The objective of the dam to raise the water up to 10 m above the sea level to increase the hydraulic pressure in the karstic system and to reduce the seawater intrusion. The experiment was carried out at the end 1977, just for a few days, due to a sudden flood event. Based on the data obtained, no improvement was found in the spring water quality. However, in another experiment, carried out from 12 February to 15 September 1987, significant TDS concentration reductions were assessed [106]. In an earlier study, the operation of brackish karst springs was simulated with the MODKARST model. The simulations revealed that sea water intrusion depends on the difference between the freshwater and seawater density. Moreover, with regard to chloride concentration during the depletion period, the difference is due to the lower pressure in the freshwater channel compared to the channel carrying the seawater [111]. Another study reported that the sea intrusion could be prevented by raising the spring water outlet, through the construction of a new dam of an estimated elevation of 26 m above the sea level [112].

(c) Recently, it was reported that increasing the height of the dam up to 25 m would minimize sea water intrusion [113]. It should be noted that none of the coastal brackish springs in Crete are 25 m above sea level [114]. Ntaskas (2018) correlates the flow rate of the Almyros spring with rainfall in the Idi (Psiloritis) mountain and the TDS concentration. Specifically, at a flow rate of less than 5 m^3/s, the TDS concentration is greater than 5 g/L and at a flow rate above 12 m^3/s, it is less than 0.4 g/L. Moreover, the hydraulic pressure in the karstic system under the high flow rates should increase significantly. Furthermore, Ntaskas (2018) found that a dam at an elevation of 25 m, estimated to cost 4 million €, would support a small hydroelectric power plant of 2.4 MW with an annual energy output of 11 million kWh.

(d) Finally, in another earlier study, the construction of an underground infiltration gallery inland, upstream of the salinization zone, to optimize the exploitation the karstic spring's aquifer (Figure 8) was recommended [115]. Such a project is believed to be a long-term and definitive solution to the water supply problem of the Iraklion municipality and the adjacent villages.

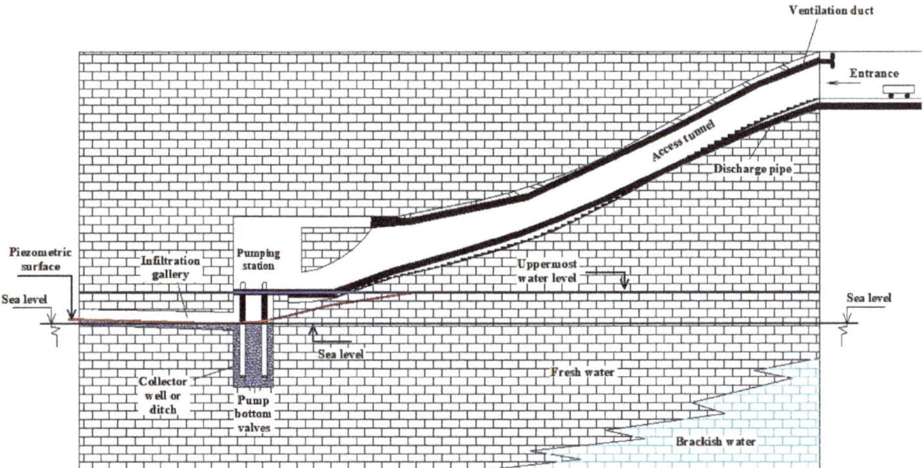

Figure 8. Cross section of proposed underground infiltration gallery (adapted by [87] from [115]).

Figure 9. Almyros spring and the dam with a height of 10 m.

4.5.4. Desalination

Desalination of seawater for the production of potable water has become a popular option, especially in water limited regions [116]. In the past few years, the cost of desalinated water has decreased considerably, while further decreases are expected [117,118] due to advances in membrane technology and improvements in energy conversion coefficient of desalination processes [119].

A new desalination plant has been operated by the Municipal Enterprise for Water Supply and Sewerage of Malevizi, to the west of the Iraklion city, since 2014. Its capacity is 2500 m^3/day and includes ultrafiltration and reverse osmosis. The total operational cost for delivered water is 0.24 €/m^3. The plant uses brackish water from Almyros spring with a TDS concentration of less than 10 g/L. An upgrade of its capacity to 6000 m^3/day was completed recently. Another plant with a capacity of 3500 m^3/day is currently under implementation by the Municipal Enterprise for Water Supply and Sewerage of Iraklion. The plant will use groundwater with a concentration of TDS less than 3 g/L composed mostly of calcium sulfate. Desalination plants should be considered seriously as an option in the regional development plans to avoid unruly development of coastal areas [120].

5. Reorganization of Water Management at Local Level and Water Safety Plans

5.1. Municipal Water Supply and Sewerage Enterprises (DEYA)

The Municipalities (Figure 10) and Municipal Water Supply and Sewerage Enterprises (DEYA) of Crete are responsible for water management at local level. Today, 82.50% of the permanent population is served by the existing DEYA, and the remaining 17.50% is served by the technical services of 12 Municipalities (Table 10). Following the practice of other EU states, either DEYA should be established in the latter, and/or the existing neighbor DEYA should be reformed as Inter-Municipal Water Supply and Sewerage Enterprises (DDEYA) for serving the remaining 17.50% of Crete's permanent population. Such a proposed scheme is shown in Table 10.

Figure 10. Municipalities in Crete.

Table 10. The existing 24 Municipalities, the 12 Municipal Water Supply and Sewerage Enterprises (DEYA), the proposed 9 DEYA (including the existing ones), and the 7 Inter-Municipal Water Supply and Sewerage Enterprises (DDEYA).

Municipalities	Existing	Proposed
Chania	DEYA Chania	DEYA Chania
Platanias	DEYA North Axis	DEYA North Axis
Kantanos-Selino	DEYA Kantanos-Selino	DDEYA Kissamos-Kantanos-Selino
Kissamos		
Apokoronas		DDEYA Apokoronas-Sfakia
Sfakia		
Gavdos		DEYA Gavdos
Rethymno	DEYA Rethymnon	DEYA Rethymnon
Mylopotamos	DEYA Mylopotamos	DDEYA Mylopotamos-Anogeia
Anogeia		
Amari		DDEYA Amari-AgiosVasilios
AgiosVasilios		
Iraklion	DEYA Iraklion	DEYA Irakliou
Malevizi	DEYA Malevizi	DEYA Malevizi
Hersonisos	DEYA Hersonisos	DEYA Hersonisos
Phaistos	DEYA Phaistos	DDEYA Phaistos-Gortys
Gortys		
Arhanes-Asterousia		
Minoa	DEYA Minoa	DDEYA Minoa- Arhanes-Asterousia-Viannos
Viannos		
Agios Nikolaos	DEYA Agios Nikolaos	DDEYA Agios Nikolaos-Plateau Lasithi
Plateau Lasithi		
Ierapetra		DEYA Ierapetras
Sitia	DEYA Sitias	DEYA Sitias

5.2. Water Safety Plans

The SARS-CoV-2 virus causing COVID-19 has not yet been detected in drinking water resources; its transmission by water has not been confirmed, and there is no accepted evidence on the survival times in either drinking water or wastewater and under what conditions. However, if another hydrophilic microorganism is present, it potentially could transmit it. Thus, water agencies should prepare for the worst, "even if it is unnecessary." As the COVID-19 is an invisible enemy, every effective means should be used to deal with it such as:

The virus is not a living organism; it consists of a protein and RNA molecule covered by a protective layer of lipid (fat). The survival time depends on external factors, such as temperature, humidity, exposure to sunlight, and type of hosting material. The virus is controlled by soapy water, which should be available to all citizens as the best protective medicine against COVID-19.

a. Drinking Water Services should prepare Water Safety Plans in each DEYA to ensure the quality of drinking water in accordance with the guidelines of Directive 98/83/EU [121] and those of the World Health Organization (WHO) [122] for the quality of drinking water. These plans are considered to be the most effective means for continuously ensuring the safety and acceptance of drinking water supply. They require a risk assessment, which includes all the necessary steps in water supply from the catchment to the consumer, followed by the implementation and monitoring of risk management control measures, with emphasis on high-risk hazards.
b. Moreover, plans must be developed for water management, mainly for operation of water supply infrastructures under emergency conditions. Such an Emergency Plan has been recently prepared by the Development Organization of Crete (OAK. SA), which can be considered as pioneering, if not at European, certainly at a national level [123]. Well done to those who designed and implemented it, even if it may be unnecessary; κάλλιον το προλαμβάνειν ή το θεραπεύειν, i.e., "it is better to prevent than to cure," Hippocrates (460–370 BC).

6. Water and Energy

Generally, energy and water are inextricably linked. Energy production and generation require water and water use, e.g., pumping, treatment, and distribution require energy. In Crete, awareness of the close relationship between water and energy is raising. The number of new approaches being proposed that will lead to energy security and water sustainability is significant. Discussions and preliminary studies are on the way for further improvement and exploitation of existing water bodies, mainly dams for touristic and energy purposes, as it happens abroad, e.g., in the famous Hoover Dam in California; now, this is the focus of a distinctly 21st-century challenge: turning the dam into a vast reservoir of excess electricity fed by the solar farms and wind turbines that represent the power sources of the future [124].

In Crete, such an energy exploitation of the Potamon dam Pumped Hydro-storage (PHS) system is under implementation. The project consists of a PHS (Figure 11) with a guaranteed power of 50 MW. The renewable energy sources consist of two wind farms with a total installed capacity of 89 MW in the municipality of Sitia in the eastern part of Crete and a water turbine production unit consisting of three reversible fixed speed units with a production power of 50 MW and pumping power of 108 MW [125,126]. The lower reservoir is the reservoir of the existing dam in the municipality of Amari, while the upper reservoir with a capacity of 1.15 million m^3 is located in the area named Gargani in the municipality of Rethymnon, with a height difference of 450 m and a distance of 2.5 km [125]. The PHS will produce 227 GWh hydro-energy annually delivered to the isolated electrical grid of Crete. It will contribute greatly to the further integration of renewable energy resource in Crete and to the stabilization of the electrical network, as the electricity generation is guaranteed. Reductions in pollution from existing conventional Public Power Corporation power plants in Crete are significant [126,127].

Given the current experience and knowledge gained by Potamon dam PHS, there is the option to expand it to other existing water infrastructures, e.g., Aposelemis or other dams, water network, etc. Aposelemis dam has been already the subject of a feasibility study regarding the implementation of a hydroelectric project, indicating, however, the need to consider and address critical socio-economic and environmental issues [128]. Beside dams, specified points of the main irrigation network of Crete, which satisfy the technical requirements (e.g., height difference and water speed and pressure), can be exploited for energy production by using the pumping technique commonly used in small hydroelectric systems. [125]. Recently, Katsaprakakis et al. (2019) reported that Crete has both the renewable energy sources (RES) potential (both wind and solar radiation) and the appropriate land morphology for PHS installation required for the development of the fundamental electricity production and storage plants for high RES annually, for turning itself into an energy independent island [129].

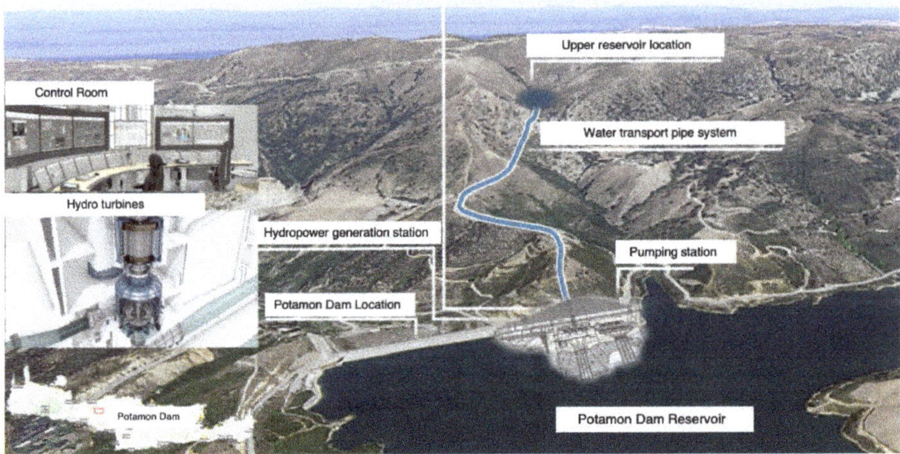

Figure 11. Schematic representation of Pumped Hydro-storage (PHS) in Potamon Dam (modified from [130].

7. Water Sustainability Issues and Opportunities

In Crete, the sustainable use of water resources seems to have its roots in the ancient civilizations and Minoans, as evidenced by their advanced technological water and wastewater achievements [21,131]. Today, increased water demands to meet the needs of agriculture and tourism and the pressure of climate variability, underline the need to re-consider existing water management plans and practices by adopting an integrated water management plan (IWMP) for the Island of Crete and governance tools that consider all the available sources of water. The need for the sustainable use of water resources is still a critical issue for Cretan, Greek, and EU governances aiming at a water efficient society for the benefit of water resources and environment, safe water supply, and sustainable (circular) economy [15,94,132]. Topics that should be considered in the development of an IWMP are as follows.

7.1. Uneven Availability of Water Resources

Despite the overall water sufficiency of Crete, there are regions (Messara valley, Sitia area) that experience severe water shortages, especially during the summer, due to increased residential and agricultural needs [13,133]. To address these water imbalances, short-term and long-term strategies and appropriate measures are necessary to close the gap between water availability and consumption enabling long-term protection and sustainable use of water resources. For example, to increase water availability, use of other water resources (e.g., distant or neighboring lagoons or dams, recycled water, rain harvest water, etc.) should be under consideration ensuring, however, that it satisfies specific social-economic, environmental, and climate prerequisites and requirements [132]. The former means that, beside environmental and climate, issues and measures depicting the cost-benefit value and public acceptance and compliance should also be considered and be taken into account [134]. An example of exploitation of distant or neighboring lagoons or dams, despite the planning and management problems, is the current use of water of Faneromeni and Aposelemis dams located in areas characterized by intensified agricultural production. To protect groundwater aquifers from overexploitation and deterioration, aquifer protection measures should be applied (e.g., reduced abstraction, replenishment with treated effluent or rainwater to increase their capacity or protect them from seawater intrusion). Moreover, it is necessary to establish limitations and prohibitions, particularly in vulnerable areas, relying on an intensified monitoring plan against depletion and chemical pollution (e.g., nitrates, organic compounds, etc.) [15,132,135].

7.2. *Efficient Water Use in Agriculture*

Reduction of water consumption in agriculture seems to be crucial to reduce or even reverse negative imbalance events and eliminate the pressure on groundwater resources in Crete (Table 4). So far, water pricing (e.g., tariffs) and/or non-pricing (e.g., field practices, awareness campaigns, education, etc.) measures are proposed to achieve minimum possible use of water in parallel with the minimum possible losses to the environment (illustrated by increased water use efficiency). To enhance the efficiency of water use in the agricultural sector, a number of actions should be taken including: the modernization of irrigation systems, the adoption of modern methodologies of ET estimation, adoption of smart agriculture and/or deficit irrigation concepts, testing of crop adaptations to meet water-efficiency and climate variability goals, improved soil management (e.g., reduced tillage, residue recycling), and develop practices increasing the small farms resilience under water scarcity.

7.3. *Low Agricultural Water Use Efficiencies*

In addition to the above, water volumetric pricing policies should be used by local agencies to improve water use efficiency in the fields, by controlling the use of water, crop adaptation, and field practices. However, volumetric pricing requires strict control, metering, and a pricing policy to provide incentives or disincentives (based on pricing levels) for those with low or high-water consumption, respectively. Current records indicate that volumetric pricing policy (0.05–0.65 €/m^3) is quite common in water management in Crete. However, there are variations across the island depending on the managing agency policy, geographical region, origin, and the quality and quantity of water [13]. Moreover, there has been significant mismatch between the water use history and the corresponding applied pricing policy attributed mostly to inadequate or imposed estimates by managing agencies. Finally, the proposed measures and practices may require regulatory adjustments, development and adoption of new technological innovations (e.g., new technology for accurate irrigation, effective monitoring of plants water status and soil moisture, overall water management by managing agencies and authorities, etc.), and cooperation schemes as well as support by diverse finance tools. EU (and national agricultural policy) either via Common Agricultural Policy (CAP 2014-020) or other relevant policies (e.g., circular economy concept) provides the legislative framework and the wide spectrum of financial tools aiming at directed knowledge and innovation via collaborations among universities' research institutions, private sector, and (groups of) farmers in agri-food sector.

7.4. *Non-Revenue Water Losses*

Non-revenue water supply is another important issue that should be considered and properly faced by responsible agencies and authorities describing mainly the losses via supply and distribution networks, the illegal connections, and the unbilled authorized use. It is worth noting that non-revenue water in Crete may exceed 60% of the availability potable water supply, highlighting it among the critical issues of water management for the decision makers and authorities.

7.5. Limited Water Reuse

Water reuse in Crete can increase water availability in many problematic areas of the island and contribute to a far more efficient use of water, mitigating in parallel the negative impacts from over-exploitation of subterranean water. Water reuse can also be seen as an additional way to cope with climate variability, as well as to comply with the circular economy concept [136], which are rapidly expanded across EU economies. The geographical distribution of the WWTPs across the island is appropriate for water reuse. Despite the current legislative framework, more work needs to be done with respect to food safety, public acceptability and marketability. Indeed, current EU and national policy supports this direction in many different ways either by providing the legislative framework [101,102] or the necessary economic tools; however, there are still certain issues that need to be addressed, such as legislative requirements (e.g., quality criteria) governing wastewater treatment and reuse, inadequate legislation covering the product value chains, lack of incentives (low pricing levels) for the users, consideration of the potential capital and operating costs of switching to reclaimed wastewater, uncertainties in the quality of the effluents and products; all are considered critical drivers of safe and economically efficient reuse of reclaimed wastewater. All these factors are drivers of public acceptance.

Additional issues can be the available and continuity of the wastewater supply, quality of services by involved agencies, reliability of supply, and adoption of the proper pricing systems; these should be described by agreements between the supplier of reclaimed wastewater and the customers. To date, most of the above issues that interact inevitably with each other have not been outlined or identified fully. It appears that what is needed is targeted research (including environmental and social-economic assessments) and effective corporations and synergies among stakeholders, supported by national or EU funding tools. An example of a critical issue could be the regulatory gap on specific emerging pollutants/contaminants, such as disinfection byproducts, pharmaceuticals, or anti-microbial resistance that are commonly found in municipal wastewater effluents [137,138]. These substances could be moved through the trophic chain and threaten terrestrial and aquatic biodiversity and humans' health [136,139]. Altogether, reuse in Crete clearly lags behind other regions or countries due to weak regulatory framework and governance of water and sanitation sectors, lack of incentives (low pricing policy), and low public acceptance. By solving the regulatory issues and providing incentives (mostly focused on low wastewater effluent prices) public acceptance and private sector would probably move towards water reuse, favoring in parallel the application of decentralized/small onsite treatment systems. These systems could be a bosting leverage for water reuse in the island, allowing the exploitation of wastewater on the site of their production [140].

7.6. Limited Use of Alternative Water Sources

Water storage, desalination, water and rainwater harvesting could be seen as alternative complementary water sources in Crete, particularly when other, more cost-effective resources are insufficient. European experience [135] indicates that these alternatives can have a role in water management and may help in local water balances; however, detailed planning strategies and measures are required, supported by a regulatory framework, investments, and relevant research. In Crete, there is lack of experience on rainwater harvesting, an issue that remains unexplored; however, several proposals like greenhouse rainwater collection are considered [134]. Moreover, the use of Almyros River water particularly in winter months, when the salt concentration is low, should be considered and supported. Previous and recent investigation have highlighted options and solutions for decision makers, such as the construction of dam, to increase the water availability and reduce its degradation by seawater intrusion [113,114]. Moreover, another option may include desalination of salty waters which is now more feasible due to new technological advances [119], probably, small-scale desalination plans for low population areas and higher-scale ones in northern urban areas. Beside the above, for non-conventional water resources, it is also speculated that potable water reuse might be a critical element in the development of sustainable strategies for water supply in the future, particularly in areas with high density population (i.e., those existing in northeastern part of the island). Such an option may seem more viable over the withdrawal of transportation and final discharge of the water of the inland areas.

7.7. Water and Energy Production Nexus

Finally, efforts should be done towards the connection of water with energy production initially by exploitation of existing infrastructure (i.e., dams and water networks) for energy purposes. Of particular importance is to document and highlight the potential benefits and implications for energy reserve, economy and environment over the current situation that relies on operation of Public Power Corporation power plants. The implementation of the Potamon dam PHS, described in detail above, could be a good example and opportunity for the robust assessment of the benefits and drawbacks from such an alternative and may help to expand and enhance water-energy connection along the island, e.g., exploitation of Aposelemis or others dams.

7.8. Local Water Management

Local water management is undoubtedly the concern for all decision makers to ensure a fair settlement of the complex water issues issues. However, it is important that any solutions be a part of a broader regional and integrated strategic plan that serves primary goals and objectives according the needs of the island. The current water management plans for the RBD of Crete [14,15] could be seen as such, however, additional strategies, systematic recording policy of water resources, estimates on water demands and research findings by regional and local authorities, academic institutions, and private enterprises as well as conclusions by public consultation processes should further be incorporated in the scope of an integrated plan construction. It is remarkable that even though there is solid body of literature and knowledge on water issues of Crete, some of which is presented within this paper, it is yet to be used appropriately to identify and address important current or future problems, in the context of improved and more sophisticated water management. A schematic representation of water uses, proposed measures and key actors and managers of water management in Crete is illustrated in Figure 12.

7.9. Knowledge Gaps

Records in knowledge gaps, particularly those addressing current sustainability issues, are also valuable and can provide the basis for future initiatives and directions. From that point of view and for the purposes of this study what has been highlighted is the need to substantially enrich our knowledge on the links between current water resources status (quality and availability) with the economic, cultural and environmental, and human health performances of the areas, in the context of increasing the public awareness and develop more "clean" and water efficient societies. In this line it is also important, to provide links between water status and the terrestrial and aquatic life and biodiversity, including the protected species and areas, due to their multifaceted role in ecosystems and human well-being [141]. The critical topics and impacts by the current water management along with proposed measures and funding sources, discussed previously are summarized in Appendix A.

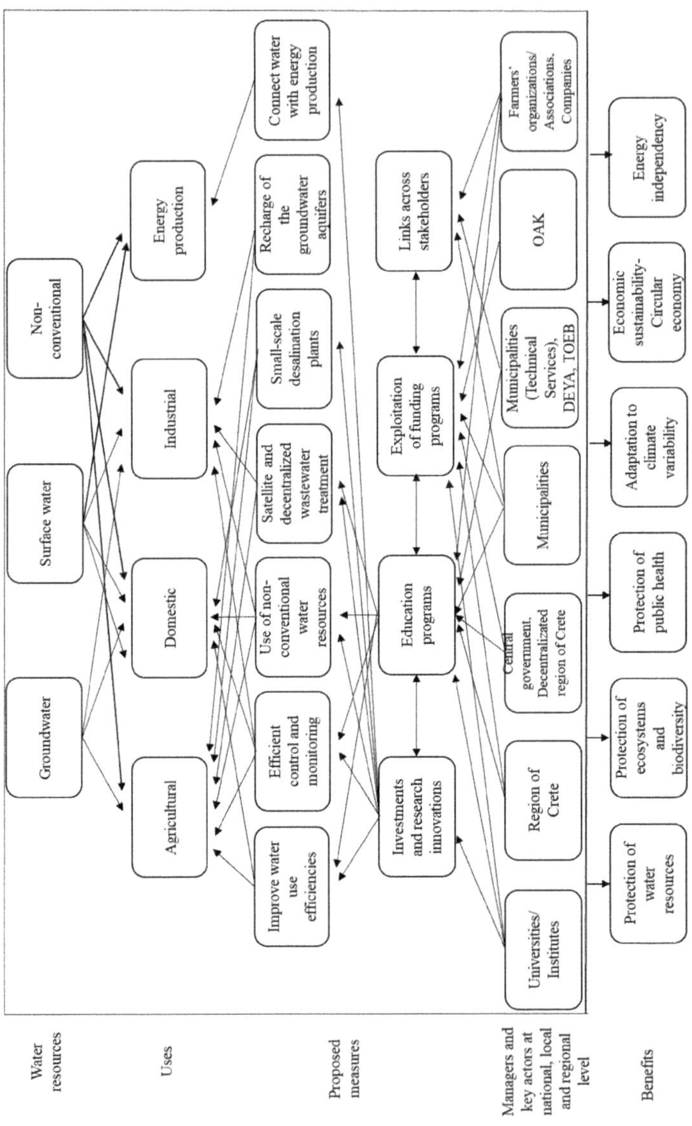

Figure 12. Schematic representation of water uses, proposed measures, and key actors and managers of water management in Crete.

8. Epilogue

Through the long history of the island of Crete, the climatic and hydrologic conditions have been characterized by high variability both spatially and temporally. This variability has had a clear impact on the water availability as well as on the human responses to the observed variability. As a result, a number of piecemeal solutions have evolved over time to deal with local water issues. As the impacts of population growth, tourism, agricultural production, other water uses, and climate variability are now becoming understood more fully, it is clear that the implementation of an IWMP for the Island of Crete is crucial if long-tern water sustainability is to be achieved. Important areas of water management that must be addressed in any IWMP, as discussed in this paper, include:

1. Uneven availability of water resources;
2. Over-exploitation of groundwater;
3. Low agricultural water use efficiencies;
4. High non-revenue water losses;
5. Limited water reuse;
6. Limited use of alternative water sources;
7. Water energy production nexus;
8. Local management concerns.

To address the above issues, the development of an IWMP must be based on the availability of reliable data (systematic recording of water resources and water uses), must take into account contemporary and emerging trends, be in harmony with the guiding principles of sustainable development, must consider the implications of the circular economy, and must include a response plan for the uneven impacts of climate variability. It is hoped that the discussion of these issues, along with the discussion of the islands' water resources presented in this paper will be of value to those charged with the responsibility of developing an IWMP for the island of Crete.

Author Contributions: V.A.T. coordinated the project development, prepared the manuscript and contributed to the most of the sections, A.N.A. had the original idea and mainly contributed to Prolegomena, History, Physical Settings, Water management, Non-conventional water sources, Reorganization of local management, and Water and energy Sections, N.V.P. reviewing and editing the manuscript, Y.G.D. contributed to the Physical settings and G.T. reviewing and editing the manuscript. All authors have read and agreed to the published version of the manuscript.

Funding: This research received no external funding.

Acknowledgments: Information provided by Evangelos Mamagkakis, Organization for the Development of Crete, s.a.; Demetris G. Cristakis, Hellenic Mediterranean Univ., Iraklion, Greece; Emmanouil Steiakakis, Technical Univ. of Crete, Chania, Greece; Demetris Papamastorakis, Iraklion, Greece; and George Chatzakis, Iraklion, Greece, all of whom are gratefully acknowledged.

Conflicts of Interest: The authors declare no conflict of interest.

Appendix A

Table A1. Critical topics, impacts-risks, and proposed measures and funding sources for sustainable water management in Crete.

Critical Topics	Impacts-Risks	Major Goals	Proposed Measures
• Droughts and water scarcity events [13,56,57,62,142] • Mismatches between water demand and water availability (negative water balances) [13,15,39,46,62] • Over-consumption of water in the field [13,143] • Over exploitation of groundwater [14,15,88] • Low performance of water reuse (low use or reclaimed and saline water and rain-harvesting [12,13,143] • Inadequate control and monitoring [11,15,41]	• Reduction of available water resources [87] • Depletion of underground water [14,15,88] • Increased competition for water • High water pricing • Reduction in agricultural and livestock productions • climate variability impacts-low adaptation potential to [56,59,144–146] • Increased water losses	• Sustainable water management-Protection of water (and soil) resources [143] • Environmental sustainability- Protection of biodiversity, ecosystems and public health [9,147–150] • Adaptation to climate variability [143,144,151,152] • Sustainable development and economy (strengthen circular economy practices) [53,136,153]	• Integrated water management plan [132,152] • Detailed records of water resources and accurate estimations (short and long-term) for water demand (including the minimum water quality) for each sector included in a unified platform (including mapping of climate data, water resources and problematic areas) [11,73,154] • Efficient control and monitoring framework by farmers, managing agencies and public authorities • Adoption of practices that increase water use efficiency in fields (efficient irrigation methods such as accurate or deficit irrigation, crop adaptations, reduced soil tillage, preservation and enhancement of soil organic matter etc.) [77,78,143,155,156] • Improved water pricing policy (use of disincentives) [13,79,143] • Improved re-distribution of water across different sectors and users. • Increase in use of alternative water resources-Adoption of specified measures (low pricing policy, development of criteria and expansion of legislation to ensure the quality and marketability of the products) [13,99,157–160] • Establishment of satellite and decentralized wastewater treatment management [140,161]
• Lack of systematic records of water resources and estimates in water use efficiency and water demands • Low linkage among stake holders (e.g., farmers, enterprises, corporations, institutes, universities)-targeted scientific research • Adaptation to climate change [56,59,77,143]	• Pollution of groundwater [162] • Increased risk for sea intrusion risk [72] • Increased risk for the terrestrial and aquatic biodiversity [146,163] • Increased risk for human health [139,164]		• Expansion of small-scale desalination plants (large-scale plants should be implemented particularly in coastal urban areas in the eastern Crete for production potable water, e. g. Iraklion and Agios Nikolaos cities) [117,118,120] • Recharge of the groundwater aquifers targeting either the enrichment of the groundwater supplies, or their protection/restoration from seawater intrusion should be seriously considered particularly at the coastal areas of the island [72,87,97,98] • Managed aquifer recharge in association to reservoir management • Connect water with energy production [125–128] • Investments and research innovations in [99]: i. water management and monitoring, ii. development of efficient water use practices in the field iii. study the links between water management practices and impacts on environmental, economic and cultural performances and impacts on biodiversity and ecosystem services), iv. Development of criteria for non-conventional water resources • Strengthen the links across stakeholders (e.g., farmers, enterprises, corporations, institutes, universities) to address critical issues in water management and increase the use of non-conventional water resources [134] • Education programs for the stakeholders [134] • Exploitation of EU, national and regional funding tools

References

1. Chen, B.; Han, M.Y.; Peng, K.; Zhou, S.L.; Shao, L.; Wu, X.F.; Wei, W.D.; Liu, S.Y.; Li, Z.; Li, J.S.; et al. Global land-water nexus: Agricultural land and freshwater use embodied in worldwide supply chains. *Sci. Total Environ.* **2018**, *613–614*, 931–943. [CrossRef] [PubMed]
2. Godfray, H.C.J.; Beddington, J.R.; Crute, I.R.; Haddad, L.; Lawrence, D.; Muir, J.F.; Pretty, J.; Robinson, S.; Thomas, S.M.; Toulmin, C. Food Security: The Challenge of Feeding 9 Billion People. *Science* **2010**, *327*, 812–818. [CrossRef] [PubMed]
3. Schneider, U.A.; Havlík, P.; Schmid, E.; Valin, H.; Mosnier, A.; Obersteiner, M.; Böttcher, H.; Skalský, R.; Balkovič, J.; Sauer, T.; et al. Impacts of population growth, economic development, and technical change on global food production and consumption. *Agric. Syst.* **2011**, *104*, 204–215. [CrossRef]
4. Gosling, S.N.; Arnell, N.W. A global assessment of the impact of climate change on water scarcity. *Clim. Chang.* **2016**, *134*, 371–385. [CrossRef]
5. Diffenbaugh, N.S.; Giorgi, F. Climate change hotspots in the CMIP5 global climate model ensemble. *Clim. Chang.* **2012**, *114*, 813–822. [CrossRef]
6. Jacob, D.; Kotova, L.; Teichmann, C.; Sobolowski, S.P.; Vautard, R.; Donnelly, C.; Koutroulis, A.G.; Grillakis, M.G.; Tsanis, I.K.; Damm, A.; et al. Climate Impacts in Europe Under +1.5 °C Global Warming. *Earth's Future* **2018**, *6*, 264–285. [CrossRef]
7. Guiot, J.; Cramer, W. Climate change: The 2015 Paris Agreement thresholds and Mediterranean basin ecosystems. *Science* **2016**, *354*, 465–468. [CrossRef]
8. Sivapalan, M.; Konar, M.; Srinivasan, V.; Chhatre, A.; Wutich, A.; Scott, C.A.; Wescoat, J.L.; Rodríguez-Iturbe, I. Socio-hydrology: Use-inspired water sustainability science for the Anthropocene. *Earth's Future* **2014**, *2*, 225–230. [CrossRef]
9. Orlove, B.; Caton, S.C. Water Sustainability: Anthropological Approaches and Prospects. *Annu. Rev. Anthropol.* **2010**, *39*, 401–415. [CrossRef]
10. Cosgrove, W.J.; Loucks, D.P. Water management: Current and future challenges and research directions. *Water Resour. Res.* **2015**, *51*, 4823–4839. [CrossRef]
11. Varouchakis, E.A.; Corzo, G.A.; Karatzas, G.P.; Kotsopoulou, A. Spatio-temporal analysis of annual rainfall in Crete, Greece. *Acta Geophys.* **2018**, *66*, 319–328. [CrossRef]
12. Tsagarakis, K.P.; Dialynas, G.E.; Angelakis, A.N. Water resources management in Crete (Greece) including water recycling and reuse and proposed quality criteria. *Agric. Water Manag.* **2004**, *66*, 35–47. [CrossRef]
13. Chartzoulakis, K.S.; Paranychianakis, N.V.; Angelakis, A.N. Water resources management in the Island of Crete, Greece, with emphasis on the agricultural use. *Water Policy* **2001**, *3*, 193–205. [CrossRef]
14. CMD. *Common Ministerial Decision. Management Plan as a River Basin of the Water Region of Crete*; No 163. ΦΕΚ 570/B'/2015; Ministry of Environment, Energy and Climate Change: Athens, Greece, 2015. (In Greek)
15. CMD. *Common Ministerial Decision. Management Plan as a River Basin of the Water Region of Crete*; No 896, ΦΕΚ B/4666/2017; Ministry of Environment, Energy and Climate Change: Athens, Greece, 2017. (In Greek)
16. Tsagarakis, K.P.; Paranychianakis, N.V.; Angelakis, A.N. *Water Supply and Wastewater Services in Greece. European Water Management between Regulation and Competition, Aqualibrium Project, EU Directorate General for Research, Global Change and Ecosystems*; Mohajeri, S., Ed.; B-1049; European Union: Brussels, Belgium, 2003; pp. 151–170.
17. Gori, R.; Lubello, C.; Ferrini, F.; Nicese, F. Reclaimed municipal wastewater as source of water and nutrients for plant nurseries. *Water Sci. Technol.* **2004**, *50*, 69–75. [CrossRef]
18. Papadopoulos, I.; Savvides, S. Optimisation of the use of nitrogen in the treated wastewater reused for irrigation. *Water Sci. Technol. Water Supply* **2003**, *3*, 217–221. [CrossRef]
19. Lal, K.; Minhas, P.S.; Yadav, R.K. Long-term impact of wastewater irrigation and nutrient rates II. Nutrient balance, nitrate leaching and soil properties under peri-urban cropping systems. *Agric. Water Manag.* **2015**, *156*, 110–117. [CrossRef]
20. Lu, S.; Zhang, X.; Liang, P. Influence of drip irrigation by reclaimed water on the dynamic change of the nitrogen element in soil and tomato yield and quality. *J. Clean. Prod.* **2016**, *139*, 561–566. [CrossRef]
21. Angelakis, A.N.; Koutsoyiannis, D.; Tchobanoglous, G. Urban wastewater and stormwater technologies in ancient Greece. *Water Res.* **2005**, *39*, 210–220. [CrossRef]

22. Angelakis, A.; Dialynas, M.; Despotakis, V. *Evolution of Water Supply Technologies in Crete, Greece through the Centuries*; IWA Publishing: London, UK, 2012; pp. 227–258.
23. Koutsoyiannis, D.; Zarkadoulas, N.; Angelakis, A.N.; Tchobanoglous, G. Urban water management in Ancient Greece: Legacies and lessons. *J. Water Resour. Plan. Manag.* **2008**, *134*, 45–54. [CrossRef]
24. Davaras, K. *Guide to Cretan Antiquities*; Noyes Press: Park Ridge, NJ, USA, 1976.
25. Strataridakis, A.; Chalkiadakis, E.; Gigourtakis, N. *Water Supply of Iraklion City (Greece) through the ages. Evolution of Water Supply Technologies through the Mellenia*; IWA Publications: London, UK, 2012; pp. 467–496.
26. Angelakis, A.N.; Mays, L.W.; Koutsoyiannis, D.; Mamassis, N. *Evolution of Water Supply through the Millennia*; IWA Publishing: London, UK, 2012.
27. Enriquez, J.; Tipping, D.; Lee, J.-J.; Vijay, A.; Kenny, L.; Chen, S.; Mainas, N.; Holst-Warhaft, G.; Steenhuis, T. Sustainable Water Management in the Tourism Economy: Linking the Mediterranean's Traditional Rainwater Cisterns to Modern Needs. *Water* **2017**, *9*, 868. [CrossRef]
28. Antoniou, G.; Xarchakou, R.; Angelakis, A. Water cistern systems in Greece from Minoan to Hellenistic period. In Proceedings of the 1st IWA International Symposium on Water and Wastewater Technologies in Ancient Civilizations, National Fountation for Agricultural Research, Institute of Iraklion, Iraklion, Greece, 28–30 October 2006; pp. 28–30.
29. Angelakis, A. Hydro-technologies in the Minoan Era. *Water Sci. Technol. Water Supply* **2017**, *17*, 1106–1120. [CrossRef]
30. Mosso, A. *Escursioni nel Mediterraneo e gli scavi di Creta*; Treves: Milano, Italy, 1910; Volume 1.
31. Gray, H.F. Sewerage in ancient and mediaeval times. *Sew. Work. J.* **1940**, *12*, 939–946.
32. Markonis, Y.; Angelakis, A.; Christy, J.; Koutsoyiannis, D. Climatic variability and the evolution of water technologies in Crete, Hellas. *Water Hist.* **2016**, *8*, 137–157. [CrossRef]
33. HSA. Hellenic Statistical Authority. 2019. Available online: http://www.statistics.gr/el/statistics/-/publication/SAM03/ (accessed on 15 May 2020).
34. HSA. Hellenic Statistical Authority. Real Population of Greece. Athens, Greece. 2002. Available online: http://www.statistics.gr (accessed on 15 May 2020).
35. HSA. *Hellenic Statistical Authority. Statistic of Tourism for the Years 1994–1996*; HAS: Athens, Greece, 1999.
36. HSA. Hellenic Statistical Authority. 2011. Available online: https://www.statistics.gr/en/provision-of-statistical-data (accessed on 15 May 2020).
37. MEE. Ministry of Environment and Energy. 2015. Available online: http://www.ypeka.gr/Default.aspx?tabid=245&language=en-US (accessed on 10 May 2020).
38. Gaki-Papanastassiou, K.; Karymbalis, E.; Papanastassiou, D.; Maroukian, H. Quaternary marine terraces as indicators of neotectonic activity of the Ierapetra normal fault SE Crete (Greece). *Geomorphology* **2009**, *104*, 38–46. [CrossRef]
39. Tzoraki, O.; Kritsotakis, M.; Baltas, E. Spatial Water Use Efficiency Index towards resource sustainability: Application in the island of Crete, Greece. *Int. J. Water Resour. Dev.* **2015**, *31*, 669–681. [CrossRef]
40. Fassoulas, C. The tectonic development of a Neogene basin at the leading edge of the active European margin: The Heraklion basin, Crete, Greece. *J. Geodyn.* **2001**, *31*, 49–70. [CrossRef]
41. Papanikolaou, D. *The Geology of Greece*; Publication of the University of Athens: Athens, Greece, 1986; 240p.
42. Malagò, A.; Efstathiou, D.; Bouraoui, F.; Nikolaidis, N.P.; Franchini, M.; Bidoglio, G.; Kritsotakis, M. Regional scale hydrologic modeling of a karst-dominant geomorphology: The case study of the Island of Crete. *J. Hydrol.* **2016**, *540*, 64–81. [CrossRef]
43. MEDIWAT. State of the Art of Water Resources in Mediterranean Islands. 2013. Available online: http://www.mediwat.eu/sites/default/files/D.3.1.1.pdf (accessed on 15 May 2019).
44. Angelakis, A.N.; Spyridakis, S.V. The status of water resources in Minoan times: A preliminary study. In *Diachronic Climatic Impacts on Water Resources*; Springer: Berlin/Heidelberg, Germany, 1996; pp. 161–191.
45. Angelakis, A. Evolution of rainwater harvesting and use in Crete, Hellas, through the millennia. *Water Sci. Technol. Water Supply* **2016**, *16*, 1624–1638. [CrossRef]
46. Voudouris, K.; Mavrommatis, T.; Daskalaki, P.; Soulios, G. *Rainfall Variations in Crete Island (Greece) and Their Impacts on Water Resources*; Serie: Hidrogeologia y aguas subterráneas; Publicaciones del Instituto Geologico y Minero de Espana: Madrid, Spain, 2006; pp. 453–463.
47. NOAA. National Oceanic and Atmospheric Administration—National Centers for Environmental Information. 2019. Available online: http://www.ncdc.noaa.gov/cdo-web/datasets (accessed on 15 May 2020).

48. Platakis, E. *The Climate of Crete*; Cretan Estia: Iraklion, Greece, 1964. (In Greek)
49. Koutroulis, A.G.; Tsanis, I.K.; Daliakopoulos, I.N. Seasonality of floods and their hydrometeorologic characteristics in the island of Crete. *J. Hydrol.* **2010**, *394*, 90–100. [CrossRef]
50. Sarris, A.; Maniadakis, M.; Lazaridou, O.; Kalogrias, V.; Bariotakis, M.; Pirintsos, S.A. Studying land use patterns in Crete Island, Greece, through a time sequence of landsat images and mapping vegetation patterns. *Wseas Trans. Environ. Dev.* **2005**, *1*, 272–279.
51. Leddra. Crete Socio-Ecological System (SES). 2010. Available online: http://leddris.aegean.gr/index.html (accessed on 15 May 2020).
52. WFD. *European Water Framework Directive. Directive 2000/60/EC of the European Parliament and of the Council Establishing a Framework for the Community Action in the Field of Water Policy*; WFD: Brussels, Belgium, 2000.
53. EC-COM. *European Commission. From the Commission to the European Parliament, the Council, the European Economic and Social Committee and the Committee of the Regions on the Implementation of the Circular Economy Action Plan (190 Final)*; European Commission: Brussels, Belgium, 2019.
54. EEA. *Report No 6. Tracking Progress towards Europe's Climate and Energy Targets until 2020*; Trends and projections in Europe; EEA: Copenhagen, Denmark, 2014.
55. Koutsoyiannis, D.; Andreadakis, A.; Mavrodimou, R.; Christofides, A.; Mamassis, N.; Efstratiadis, A.; Koukouvinos, A.; Karavokiros, G.; Kozanis, S.; Mamais, D.; et al. *National Programme for the Management and Protection of Water Resources, Support on the Compilation of the National Programme for Water Resources Management and Preservation*; Department of Water Resources and Environmental Engineering—National Technical University of Athens: Athens, Greece, 2008; p. 748. [CrossRef]
56. Koutroulis, A.G.; Tsanis, I.K.; Daliakopoulos, I.N.; Jacob, D. Impact of climate change on water resources status: A case study for Crete Island, Greece. *J. Hydrol.* **2013**, *479*, 146–158. [CrossRef]
57. Tsanis, I.K.; Koutroulis, A.G.; Daliakopoulos, I.N.; Jacob, D. Severe climate-induced water shortage and extremes in Crete. *Clim. Chang.* **2011**, *106*, 667–677. [CrossRef]
58. Vautard, R.; Gobiet, A.; Sobolowski, S.; Kjellström, E.; Stegehuis, A.; Watkiss, P.; Mendlik, T.; Landgren, O.; Nikulin, G.; Teichmann, C. The European climate under a 2 C global warming. *Environ. Res. Lett.* **2014**, *9*, 034006. [CrossRef]
59. Koutroulis, A.G.; Grillakis, M.G.; Daliakopoulos, I.N.; Tsanis, I.K.; Jacob, D. Cross sectoral impacts on water availability at +2 °C and +3 °C for east Mediterranean island states: The case of Crete. *J. Hydrol.* **2016**, *532*, 16–28. [CrossRef]
60. Brilli, L.; Lugato, E.; Moriondo, M.; Gioli, B.; Toscano, P.; Zaldei, A.; Leolini, L.; Cantini, C.; Caruso, G.; Gucci, R.; et al. Carbon sequestration capacity and productivity responses of Mediterranean olive groves under future climates and management options. *Mitig. Adapt. Strat. Glob. Chang.* **2019**, *24*, 467–491. [CrossRef]
61. Chartzoulakis, K.; Psarras, G. Global change effects on crop photosynthesis and production in Mediterranean: The case of Crete, Greece. *Agric. Ecosyst. Environ.* **2005**, *106*, 147–157. [CrossRef]
62. Koutroulis, A.G.; Vrohidou, A.-E.K.; Tsanis, I.K. Spatiotemporal Characteristics of Meteorological Drought for the Island of Crete. *J. Hydrometeorol.* **2011**, *12*, 206–226. [CrossRef]
63. Koutsoyiannis, D. Hydrology and change. *Hydrol. Sci. J.* **2013**, *58*, 1177–1197. [CrossRef]
64. Koutsoyiannis, D.; Makropoulos, C.; Langousis, A.; Baki, S.; Efstratiadis, A.; Christofides, A.; Karavokiros, G.; Mamassis, N. HESS Opinions: "Climate, hydrology, energy, water: Recognizing uncertainty and seeking sustainability". *Hydrol. Earth Syst. Sci.* **2009**, *13*, 247–257. [CrossRef]
65. *Change of the Irrigated Area from 1995 until the Present*; Hellenic Statistical Authority: Pireas, Greece, 2016; Available online: https://www.statistics.gr/en/provision-of-statistical-data (accessed on 15 May 2020).
66. EUROSTAT. European Statistics Database [WWWDocument]. Eurostat URL. Available online: http://ec.europa.eu/eurostat/statistics-explained/index.php/Agri-environmental_indicator_-_irrigation (accessed on 11 January 2017).
67. Karatzas, G.P.; Dokou, Z. Optimal management of saltwater intrusion in the coastal aquifer of Malia, Crete (Greece), using particle swarm optimization. *Hydrogeol. J.* **2015**, *23*, 1181–1194. [CrossRef]
68. Papadopoulou, M.; Karatzas, G.; Koukadaki, M.; Trichakis, Y. Modeling the saltwater intrusion phenomenon in coastal aquifers–A case study in the industrial zone of Herakleio in Crete. *Glob. Nest J.* **2005**, *7*, 197–203.
69. Tsanis, I.K.; Coulibaly, P.; Daliakopoulos, I.N. Improving groundwater level forecasting with a feedforward neural network and linearly regressed projected precipitation. *J. Hydroinform.* **2008**, *10*, 317–330. [CrossRef]

70. Varouchakis, E.A.; Theodoridou, P.G.; Karatzas, G.P. Spatiotemporal geostatistical modeling of groundwater levels under a Bayesian framework using means of physical background. *J. Hydrol.* **2019**, *575*, 487–498. [CrossRef]
71. Varouchakis, E.A. Modeling of temporal groundwater level variations based on a Kalman filter adaptation algorithm with exogenous inputs. *J. Hydroinform.* **2017**, *19*, 191–206. [CrossRef]
72. Pappa, A.; Dokou, Z.; Karatzas, G.P. Saltwater intrusion management using the swi2 model: Application in the coastal aquifer of hersonissos, crete, greece. *Desalin. Water Treat.* **2017**, *99*, 49–58. [CrossRef]
73. Morianou, G.G.; Kourgialas, N.N.; Psarras, G.; Koubouris, G.C. Mapping sensitivity to desertification in Crete (Greece), the risk for agricultural areas. *J. Water Clim. Chang.* **2018**, *9*, 691–702. [CrossRef]
74. Croke, B.; Cleridou, N.; Kolovos, A.; Vardavas, I.; Papamastorakis, J. Water resources in the desertification-threatened Messara Valley of Crete: Estimation of the annual water budget using a rainfall-runoff model. *Environ. Model. Softw.* **2000**, *15*, 387–402. [CrossRef]
75. Varouchakis, E.; Palogos, I.; Karatzas, G. Application of Bayesian and cost benefit risk analysis in water resources management. *J. Hydrol.* **2016**, *534*, 390–396. [CrossRef]
76. Varouchakis, E.A.; Yetilmezsoy, K.; Karatzas, G.P. A decision-making framework for sustainable management of groundwater resources under uncertainty: Combination of Bayesian risk approach and statistical tools. *Water Policy* **2019**, *21*, 602–622. [CrossRef]
77. Georgopoulou, E.; Mirasgedis, S.; Sarafidis, Y.; Vitaliotou, M.; Lalas, D.P.; Theloudis, I.; Giannoulaki, K.D.; Dimopoulos, D.; Zavras, V. Climate change impacts and adaptation options for the Greek agriculture in 2021–2050: A monetary assessment. *Clim. Risk Manag.* **2017**, *16*, 164–182. [CrossRef]
78. Udias, A.; Pastori, M.; Malago, A.; Vigiak, O.; Nikolaidis, N.P.; Bouraoui, F. Identifying efficient agricultural irrigation strategies in Crete. *Sci. Total Environ.* **2018**, *633*, 271–284. [CrossRef] [PubMed]
79. Giannakis, E.; Bruggeman, A.; Djuma, H.; Kozyra, J.; Hammer, J. Water pricing and irrigation across Europe: Opportunities and constraints for adopting irrigation scheduling decision support systems. *Water Supply* **2015**, *16*, 245–252. [CrossRef]
80. Varouchakis, E.; Apostolakis, A.; Siaka, M.; Vasilopoulos, K.; Tasiopoulos, A. Alternatives for domestic water tariff policy in the municipality of Chania, Greece, toward water saving using game theory. *Water Policy* **2018**, *20*, 175–188. [CrossRef]
81. Chifos, C.; Doxastakis, Z.; Romanos, M.C. Public discourse and government action in a controversial water management project: The damming of the Aposelemis River in Crete, Greece. *Water Policy* **2019**, *21*, 526–545. [CrossRef]
82. Voudouris, K.S.; Tsatsanifos, C.; Yannopoulos, S.; Marinos, V.; Angelakis, A.N. Evolution of underground aqueducts in the Hellenic world. *Water Supply* **2016**, *16*, 1159–1177. [CrossRef]
83. Nikolaou, T.; Christodoulakos, I.; Piperidis, P.; Angelakis, A. Evolution of Cretan, Greece Aqueducts. In Proceedings of the 4th IWA International Symposium on Water and Wastewater Technologies in Ancient Civilizations, Coimbra, Portugal, 22–24 March 2014; pp. 17–19.
84. Lekkas, N. *The 750 Mineral Springs of Greece*; IGME: Athens, Greece, 1938. (In Greek)
85. Angelakis, A.N.; Antoniou, G.P.; Yapijakis, C.; Tchobanoglous, G. History of Hygiene Focusing on the Crucial Role of Water in the Hellenic Asclepieia (ie, Ancient Hospitals). *Water* **2020**, *12*, 754. [CrossRef]
86. Manutsoglou, E. The Thermal Springs of Asklepieion in Crete, Greece. In Proceedings of the 11th Hydrogeological Conference of Greece, Athens, Greece, 4–6 October 2017; Hellenic Hydrogeology Committee of International Association of Hydrogeologists. IAH: Athens, Greece, 2017.
87. Steiakakis, E.; Vavadakis, D.; Kritsotakis, M.; Voudouris, K.; Anagnostopoulou, C. Drought impacts on the fresh water potential of a karst aquifer in Crete, Greece. *Environ. Earth Sci.* **2016**, *75*, 507. [CrossRef]
88. Kourgialas, N.N.; Karatzas, G.P.; Dokou, Z.; Kokorogiannis, A. Groundwater footprint methodology as policy tool for balancing water needs (agriculture & tourism) in water scarce islands—The case of Crete, Greece. *Sci. Total Environ.* **2018**, *615*, 381–389.
89. Gikas, P.; Tchobanoglous, G. Sustainable use of water in the Aegean Islands. *J. Environ. Manag.* **2009**, *90*, 2601–2611. [CrossRef]
90. Karydis, M.; Lesvos, I. Eutrophication assessment of coastal waters based on indicators: A literature review. In Proceedings of the International Conference on Environmental Science and Technology, Crete, Greece, 3–5 September 2009.

91. Angelakis, A. Urban waste-and stormwater management in Greece: Past, present and future. *Water Sci. Technol. Water Supply* **2017**, *17*, 1386–1399. [CrossRef]
92. EU. *Concerning the Management of Bathing Water Quality and Repealing Directive 76/160/EEC*; Official Journal L. 64/37; EU: Brussels, Belgium, 2006.
93. EU. *Directive 2000/60/EC of the European Parliament and of the Council of 23 October 2000 Establishing a Framework for Community Action in the Field of Water Policy*; Official Journal L 327, 22/12/2000 P. 0001–0073; EU: Brussels, Belgium, 2000.
94. EEA. European Environment Agency. Sustainable Water Management. 2018. Available online: https://ec.europa.eu/info/news/sustainability-at-the-water-source_en (accessed on 15 May 2020).
95. EU. Urban Wastewater Council Directive of 21 May 1991 concerning urban waste water treatment (91/271/EEC). *J. Eur. Commun.* **1991**, *34*, 40.
96. Paranychianakis, N.V.; Angelakis, A.N.; Leverenz, H.; Tchobanoglous, G. Treatment of Wastewater with Slow Rate Systems: A Review of Treatment Processes and Plant Functions. *Crit. Rev. Environ. Sci. Technol.* **2006**, *36*, 187–259. [CrossRef]
97. De Feo, G.; Galasso, M.; Belgiorno, V. Groundwater recharge in an endoreic basin with reclaimed municipal wastewater. *Water Sci. Technol.* **2007**, *55*, 449–457. [CrossRef] [PubMed]
98. Masciopinto, C.; Carrieri, C. Assessment of water quality after 10 years of reclaimed water injection: The Nardò fractured aquifer (Southern Italy). *Groundw. Monit. Remediat.* **2002**, *22*, 88–97. [CrossRef]
99. Paranychianakis, N.V.; Salgot, M.; Snyder, S.A.; Angelakis, A.N. Water Reuse in EU States: Necessity for Uniform Criteria to Mitigate Human and Environmental Risks. *Crit. Rev. Environ. Sci. Technol.* **2015**, *45*, 1409–1468. [CrossRef]
100. Menegaki, A.N.; Hanley, N.; Tsagarakis, K.P. The social acceptability and valuation of recycled water in Crete: A study of consumers' and farmers' attitudes. *Ecol. Econ.* **2007**, *62*, 7–18. [CrossRef]
101. CMD. *Common Ministerial Decision. Measures, Limits and Procedures for Reuse of Treated Wastewater*; No. 145116; Ministry of Environment, Energy and Climate Change: Athens, Greece, 2011. (In Greek)
102. EC-COM. *European Commission. Proposal for a Regulation of the European Parliament and of the Council on Minimum Requirements for Water Reuse (337 Final)*; European Commission: Brussels, Belgium, 2018.
103. Shoushtarian, F.; Negahban-Azar, M. Worldwide Regulations and Guidelines for Agricultural Water Reuse: A Critical Review. *Water* **2020**, *12*, 971. [CrossRef]
104. RC. *National Business Plan: Urban Wastewater Management*; Region of Crete: Iraklion, Greece, 2019. (In Greek)
105. Alexopoulos, A. *Brackish Water Resources in Crete: Quantities and Characteristics. Desalination Technologies for the Production of Desalinated Water*; Karamouzis, D., Alexopoulos, A., Angelakis, A.N., Eds.; Hellenic Union of Municipal Enterprises for Water Supply and Sewerage: Larisa, Greece, 2008; pp. 37–41. (In Greek)
106. Stevanović, Z. *Karst Aquifers-Characterization and Engineering*; Springer: Berlin/Heidelberg, Germany, 2015.
107. Vergis, G. Investigation of the Possibilities Construction a Dam for Storage the Water of Iraklion Almyros Spring. Master's Thesis, Interdepart. of Water Sciences and Technology, National Technical Univ. of Athens, Athens, Greece, 2006; p. 148.
108. Arfib, B.; De Marsily, G.; Ganoulis, J. Locating the zone of saline intrusion in a coastal karst aquifer using springflow data. *Groundwater* **2007**, *45*, 28–35. [CrossRef]
109. Tsakiris, G.; Spiliotis, M.; Paritsis, S.; Alexakis, D. Assessing the water potential of karstic saline springs by applying a fuzzy approach: The case of Almyros (Heraklion, Crete). *Desalination* **2009**, *237*, 54–64. [CrossRef]
110. FAO-UNDP. *Study of Water Resources and Their Exploitation for Irrigation in Eastern Crete: Study of Almyros Springo Iraklion*; Technical Report 3; FAO, UN Development Programme: Iraklion, Greece, 1972.
111. Maramathas, A. Simulation of Brakish-Karstic Spring. Ph.D. Thesis, School of Chemical Engineering, National Technical University of Athens, Athens, Greece, 2002.
112. Maramathas, A.; Maroulis, Z.; Marinos-Kouris, D. Blocking sea intrusion in brackish karstic springs. *Case Almiros Spring Heraklion Crete Greece Eur. Water* **2003**, *1*, 14–20.
113. Ntaskas, A. Πρόταση για την αξιοποίηση της πηγής Αλμυρού (Ύδρευση, Άρδευση, Ενέργεια). Patris Newspaper, 30 April 2018 (In Greek). Publisher A. Mykoniatis A.E. Available online: https://www.patris.gr/2018/04/30/protasi-gia-tin-axiopoiisi-tis-pigis-almyroy-ydreysi-ardeysi-energeia (accessed on 20 May 2020).
114. Papamastorakis, D. (Region of Crete, Direction of Water Resources, Iraklion, Greece). Personal communication, 2020.

115. Monopolis, D.; Klidopoulou, M. *Exploitation of the Brackish-Water Spring of Almyros*; DEYA Iraklion: Iraklion, Greece, 2002.
116. Garcia-Rodriguez, L. Seawater desalination driven by renewable energies: A review. *Desalination* **2002**, *143*, 103–113. [CrossRef]
117. Shannon, M.A.; Bohn, P.W.; Elimelech, M.; Georgiadis, J.G.; Marinas, B.J.; Mayes, A.M. Science and technology for water purification in the coming decades. In *Nanoscience and Technology: A Collection of Reviews from Nature Journals*; World Scientific: Singapore, 2010; pp. 337–346.
118. Khawaji, A.D.; Kutubkhanah, I.K.; Wie, J.-M. Advances in seawater desalination technologies. *Desalination* **2008**, *221*, 47–69. [CrossRef]
119. Fritzmann, C.; Löwenberg, J.; Wintgens, T.; Melin, T. State-of-the-art of reverse osmosis desalination. *Desalination* **2007**, *216*, 1–76. [CrossRef]
120. UNEP/MAP. *Sea water desalination in the Mediteranean: Assessment and guidelines*; MAP Technical Reports Series; UNEP/MAP: Nairobi, Kenya, 2003.
121. EU. *Council Directive 98/83/EC of 3 November 1998 on the Quality of Water Intended for Human Consumption*; Official Journal L 330, 05/12/1998 P. 0032–0054; EU: Brussels, Belgium, 1998.
122. WHO. *Water Safety Plan Manual: How to Develop and Implement a Water Safety Plan*; WHO and IWA: Genova, Switzerland, 2009.
123. Development Organization of Crete (OAK, SA). *The no 3278/16/03/2020 Decision Approving and Implementing of an Emergency Plan for Continuous Operation of Water Supply Infrastructure Managed by OAK. SA during the Risk of Appearance and Dissemination of the Corona SARS-CoV-2*; Development Organization of Crete (OAK, SA): Chania, Greece, 2020. (In Greek)
124. Penn, I. The $3 Billion Plan to Turn Hoover Dam into a Giant Battery. *New York Times*. 24 July 2018. Available online: https://www.nytimes.com/interactive/2018/07/24/business/energy-environment/hoover-dam-renewable-energy.html (accessed on 15 May 2020).
125. Nikolaou, T.G.; Christodoulakos, I.; Piperidis, P.G.; Angelakis, A.N. Evolution of Cretan aqueducts and their potential for hydroelectric exploitation. *Water* **2017**, *9*, 31. [CrossRef]
126. Nikolaou, T.; Tsamoudalis, K.; Stavrakakis, G. Modeling and Optimal Design of a Pump–Storage System for the Exploitation of the Rejected Energy in Non—Interconnected Electrical Power Systems. Case—Study in Potamon Dam, Rethymno of Crete, Greece. In Proceedings of the 11th ICOLD European Club Symposium, Chania, Crete, 2–4 October 2019.
127. Nikolaou, T.; Piperidis, P.; Kopasis, L.; Vardoulaki, E.; Bazdanis, G. Systems and methods of best management and saving energy in hydro projects in Crete, Greece. In Proceedings of the Jointed Conference of HHA, HWA, and HCoMWR, Athens, Greece, 10–12 December 2015; pp. 10–12.
128. Kavvalakis, A. Study of Viability of Energy Efficiency of the Drainage Dam. Bachelor's Thesis, Technical University of Crete, Chania, Greece, 2017.
129. Katsaprakakis, D.; Ioannis, A.; Dakanali, E.; Christakis, D. Turning Crete into an energy independent island. In Proceedings of the 4th International Hybrid Power Systems Workshop, Crete, Greece, 22–23 May 2019.
130. Tsiknakou, Y. *A Hybrid Power Station in Rethymnon and Lasithi of Crete*; Terna Enregy: Athens, Greece, 2014.
131. Tzanakakis, V.; Koo-Oshima, S.; Haddad, M.; Apostolidis, N.; Angelakis, A.; Angelakis, A.; Rose, J. *The History of Land Application and Hydroponic Systems for Wastewater Treatment and Reuse. Evolution of Sanitation and Wastewater Technologies through the Centuries*; IWA Publishing: London, UK, 2014; p. 457.
132. EEA. *European Environment Agency. Water Management in Europe: Price and Nonprice Approaches to Water Conservation*; European Environment Agency: København, Denmark, 2017; Available online: https://www.eea.europa.eu/themes/water/european-waters/water-management/water-management-in-europe (accessed on 15 May 2020).
133. Varouchakis, E.A. Integrated water resources analysis at basin scale: A case study in Greece. *J. Irrig. Drain. Eng.* **2016**, *142*, 05015012. [CrossRef]
134. Demetropoulou, L.; Lilli, M.A.; Petousi, I.; Nikolaou, T.; Fountoulakis, M.; Kritsotakis, M.; Panakoulia, S.; Giannakis, G.V.; Manios, T.; Nikolaidis, N.P. Innovative methodology for the prioritization of the Program of Measures for integrated water resources management of the Region of Crete, Greece. *Sci. Total Environ.* **2019**, *672*, 61–70. [CrossRef]
135. EC-SWD. *153 Final. Commision Staff Working Document*; Agriculture and Sustainable Water Management in the EU; EC-SWD: Filderstadt, Germany, 2017.

136. Voulvoulis, N. Water reuse from a circular economy perspective and potential risks from an unregulated approach. *Curr. Opin. Environ. Sci. Health* **2018**, *2*, 32–45. [CrossRef]
137. Karkman, A.; Do, T.T.; Walsh, F.; Virta, M.P.J. Antibiotic-Resistance Genes in Waste Water. *Trends Microbiol.* **2018**, *26*, 220–228. [CrossRef]
138. Rizzo, L.; Manaia, C.; Merlin, C.; Schwartz, T.; Dagot, C.; Ploy, M.C.; Michael, I.; Fatta-Kassinos, D. Urban wastewater treatment plants as hotspots for antibiotic resistant bacteria and genes spread into the environment: A review. *Sci. Total Environ.* **2013**, *447*, 345–360. [CrossRef]
139. Xie, J.; Jin, L.; He, T.; Chen, B.; Luo, X.; Feng, B.; Huang, W.; Li, J.; Fu, P.; Li, X. Bacteria and Antibiotic Resistance Genes (ARGs) in PM2.5 from China: Implications for Human Exposure. *Environ. Sci. Technol.* **2019**, *53*, 963–972. [CrossRef]
140. Capodaglio, A.G.; Callegari, A.; Cecconet, D.; Molognoni, D. Sustainability of decentralized wastewater treatment technologies. *Water Pract. Technol.* **2017**, *12*, 463–477. [CrossRef]
141. Haines-Young, R.; Potschin, M. The links between biodiversity, ecosystem services and human well-being. *Ecosyst. Ecol. New Synth.* **2010**, *1*, 110–139.
142. Vrochidou, A.-E.; Tsanis, I. Assessing precipitation distribution impacts on droughts on the island of Crete. *Nat. Hazards Earth Syst. Sci.* **2012**, *12*, 1159–1171. [CrossRef]
143. Chartzoulakis, K.; Bertaki, M. Sustainable Water Management in Agriculture under Climate Change. *Agric. Agric. Sci. Procedia* **2015**, *4*, 88–98. [CrossRef]
144. Olmstead, S.M. Climate change adaptation and water resource management: A review of the literature. *Energy Econ.* **2014**, *46*, 500–509. [CrossRef]
145. Papadopoulou, M.P.; Charchousi, D.; Tsoukala, V.K.; Giannakopoulos, C.; Petrakis, M. Water footprint assessment considering climate change effects on future agricultural production in Mediterranean region. *Desalin. Water Treat.* **2016**, *57*, 2232–2242. [CrossRef]
146. Vozinaki, A.E.K.; Tapoglou, E.; Tsanis, I.K. Hydrometeorological impact of climate change in two Mediterranean basins. *Int. J. River Basin Manag.* **2018**, *16*, 245–257. [CrossRef]
147. Becerra-Castro, C.; Lopes, A.R.; Vaz-Moreira, I.; Silva, E.F.; Manaia, C.M.; Nunes, O.C. Wastewater reuse in irrigation: A microbiological perspective on implications in soil fertility and human and environmental health. *Environ. Int.* **2015**, *75*, 117–135. [CrossRef]
148. Gordon, L.J.; Finlayson, C.M.; Falkenmark, M. Managing water in agriculture for food production and other ecosystem services. *Agric. Water Manag.* **2010**, *97*, 512–519. [CrossRef]
149. Richter, B.D.; Mathews, R.; Harrison, D.L.; Wigington, R. Ecologicaly sustainable water management: Managing river flows for ecological entegrity. *Ecol. Appl.* **2003**, *13*, 206–224. [CrossRef]
150. Johnson, N.; Revenga, C.; Echeverria, J. Managing Water for People and Nature. *Science* **2001**, *292*, 1071–1072. [CrossRef] [PubMed]
151. Iglesias, A.; Garrote, L. Adaptation strategies for agricultural water management under climate change in Europe. *Agric. Water Manag.* **2015**, *155*, 113–124. [CrossRef]
152. Gain, A.K.; Rouillard, J.J.; Benson, D. Can integrated water resources management increase adaptive capacity to climate change adaptation? A critical review. *J. Water Resour. Prot.* **2013**, *5*, 11–20. [CrossRef]
153. Sgroi, M.; Vagliasindi, F.G.A.; Roccaro, P. Feasibility, sustainability and circular economy concepts in water reuse. *Curr. Opin. Environ. Sci. Health* **2018**, *2*, 20–25. [CrossRef]
154. Agou, V.D.; Varouchakis, E.A.; Hristopulos, D.T. Geostatistical analysis of precipitation in the island of Crete (Greece) based on a sparse monitoring network. *Environ. Monit. Assess.* **2019**, *191*, 353. [CrossRef]
155. Kourgialas, N.N.; Karatzas, G.P. A modeling approach for agricultural water management in citrus orchards: Cost-effective irrigation scheduling and agrochemical transport simulation. *Environ. Monit. Assess.* **2015**, *187*, 462. [CrossRef]
156. Kourgialas, N.N.; Koubouris, G.C.; Dokou, Z. Optimal irrigation planning for addressing current or future water scarcity in Mediterranean tree crops. *Sci. Total Environ.* **2019**, *654*, 616–632. [CrossRef]
157. Agrafioti, E.; Diamadopoulos, E. A strategic plan for reuse of treated municipal wastewater for crop irrigation on the Island of Crete. *Agric. Water Manag.* **2012**, *105*, 57–64. [CrossRef]
158. Angelakis, A.N.; Asano, T.; Bahri, A.; Jimenez, B.E.; Tchobanoglous, G. Water reuse: From ancient to modern times and the future. *Front. Environ. Sci.* **2018**, *6*, 26. [CrossRef]
159. Gikas, P.; Angelakis, A.N. Water resources management in Crete and in the Aegean Islands, with emphasis on the utilization of non-conventional water sources. *Desalination* **2009**, *248*, 1049–1064. [CrossRef]

160. Monopolis, D.; Lambrakis, N.; Perleros, B. The brackish karstic spring of Almiros of Heraklion. *EUR* **2005**, *21366*, 321–328.
161. Asano, T.; Burton, F.; Leverenz, H.; Tsuchihashi, R.; Tchobanoglous, G. *Water Reuse-Issues, Technologies, and Applications;* McGrow-Hill: Enfield, NH, USA, 2007.
162. Kourgialas, N.N.; Karatzas, G.P. Groundwater contamination risk assessment in Crete, Greece, using numerical tools within a GIS framework. *Hydrol. Sci. J.* **2015**, *60*, 111–132. [CrossRef]
163. Tan, D.T.; Shuai, D. Research highlights: Antibiotic resistance genes: From wastewater into the environment. *Environ. Sci. Water Res. Technol.* **2015**, *1*, 264–267. [CrossRef]
164. Adegoke, A.A.; Amoah, I.D.; Stenström, T.A.; Verbyla, M.E.; Mihelcic, J.R. Epidemiological Evidence and Health Risks Associated with Agricultural Reuse of Partially Treated and Untreated Wastewater: A Review. *Front Public Health* **2018**, *6*, 337. [CrossRef] [PubMed]

Article

Addressing the Scarcity of Traditional Water Sources through Investments in Alternative Water Supplies: Case Study from Florida

Tatiana Borisova [1,*], Matthew Cutillo [2], Kate Beggs [2] and Krystle Hoenstine [2]

[1] Food and Resource Economics Department, Institute of Food and Agricultural Sciences, University of Florida, Gainesville, FL 32611, USA

[2] Economic and Demographic Research, Florida Legislature, Tallahassee, FL 32399, USA; cutillo.matthew@leg.state.fl.us (M.C.); beggs.kate@leg.state.fl.us (K.B.); krystle.hoenstine@gmail.com (K.H.)

* Correspondence: tborisova@ufl.edu

Received: 1 June 2020; Accepted: 17 July 2020; Published: 23 July 2020

Abstract: This paper examines the capital costs for alternative water supply projects in Florida, the third most populous state in the United States. The increasing scarcity of fresh groundwater in Florida has led to investments in alternative water supply sources, including brackish groundwater, surface water capture and storage, reclaimed water, and stormwater. Expenditures to meet the growing water demand for the 20-year planning horizon are estimated using water demand projections and existing supply estimates from Florida's five water management districts. In the regions where demand projections exceed the existing supply, the districts are required to identify project options to meet the growing water demand while protecting the natural systems. This study uses the database of 645 projects implemented in the past or considered for the future. The Ordinary Least Squares regression model shows that project implementation costs depend on project capacity, type, implementation status, and implementation region. Given the most common project types and project sizes, the total investments to meet the state's future water demand could reach almost $2 billion in the next 20 years. The expenditures necessitate more cost-effective options (such as expanding stormwater use and water conservation).

Keywords: long-term; regional water supply planning; alternative water supply; projects; expenditures; investments; conservation

---

1. Introduction

Expenditure forecasting is an integral part of long-term regional water supply development and water resource management [1]. In many regions, a growing population, expanding economy, and propagating groundwater depletion have increased water scarcity and impacted water-dependent ecosystems [2–5]. To address communities' water supply and environmental protection priorities, water suppliers and government agencies seek to diversify water supply sources, invest in demand management, and implement environmental protection strategies [1,6]. Due to the high costs of such initiatives, they require funding from a mix of local, regional, state, and federal sources [7,8]. Government cost-share and private–public partnerships have become common funding strategies for water infrastructure projects, supplementing funding from service fees, bonds, and other funding sources available to water utilities. Infrastructure funding from the general government revenue has been justified as a method to promote economic development, given that infrastructure is a prerequisite of economic growth [7,8]. Despite the seemingly diverse funding opportunities, funding sources are becoming less available. Economic recessions and utility revenue shrinkages due to per capita water use reduction translate into a funding deficit for water supply development and environmental protection. Therefore, forecasting expenditures and funding needs have grown in importance in long-term regional water supply planning on the local, regional, and state levels.

This paper aims to estimate the expenditures required to develop alternative water supplies and meet the water demand increase in Florida, the third most populous state in the United States. Florida's population is expected to grow from almost 22 million in 2020 to over 26 million in 2040 [9]. The increase in population is projected to escalate water demand unless water use efficiency significantly improves [10]. The state's traditional water source—fresh groundwater—is becoming scarce, with withdrawals causing or projected to cause harm to the water resources or ecology in several areas [11]. "Rethinking water supply" [12] (p. 21302) to include the sources that have previously been ignored, along with water conservation, is seen as a solution to water challenges. Assessing the costs and evaluating funding sources for such supply projects become increasingly important in Florida, mirroring the trend observed in other states and countries.

2. Study Area, Materials, and Methods

2.1. Study Area: Florida's Long-Term Water Supply Planning Framework

Florida has been primarily relying on groundwater, which is referred to as "traditional" water supply source in most of the state. One of the leading groundwater users in the United States, Florida reports more than four billion gallons (that is more than 15 million m^3) of groundwater withdrawals daily [13]. In 2015, groundwater accounted for 63.0 percent of total freshwater withdrawals, and it was particularly prominent in the public supply (86.2% of the total freshwater withdrawals in this category) [13]. The groundwater withdrawals exceed the recharge rates in many Florida areas, with reductions in the flows and levels documented for selected aquifers, springs, streams, and lakes. Minimum flows or minimum water levels (MFLs) are defined as the limits at which further water withdrawals would be significantly harmful to the area's water resources or ecology (§ 373.042, Fla. Stat.). As of 1 March 2019, out of 426 adopted MFLs, 105 are in recovery status, implying that their water flows or water levels are below their adopted MFLs [14], and groundwater withdrawals in the relevant basins should be carefully managed or reduced.

To address and prevent conflicts among water users (e.g., see [15,16]), Florida has developed a rigorous framework for long-term water supply planning. The state is divided into five water management districts (WMDs) based on surface water hydrologic boundaries (Figure 1): Northwest Florida (NWFWMD), Suwannee River (SRWMD), St. Johns River (SJRWMD), Southwest Florida (SWFWMD), and South Florida (SFWMD). Each WMD's governing board is required to develop a district water management plan (§ 373.036, Fla. Stat.) for at least a 20-year planning horizon. A principal element of a district's water management plan is a water supply assessment (WSA). The WSA determines whether existing and reasonably anticipated sources of water and conservation efforts are adequate to sustain the water resources and related natural systems into the future. In a region where a WMD has determined that existing water sources are inadequate to meet projected demand, a more in-depth regional water supply plan (RWSP) must be developed for that region. Both districtwide WSAs and RWSPs are required to be updated at least once every five years (§ 373.036, Fla. Stat.).

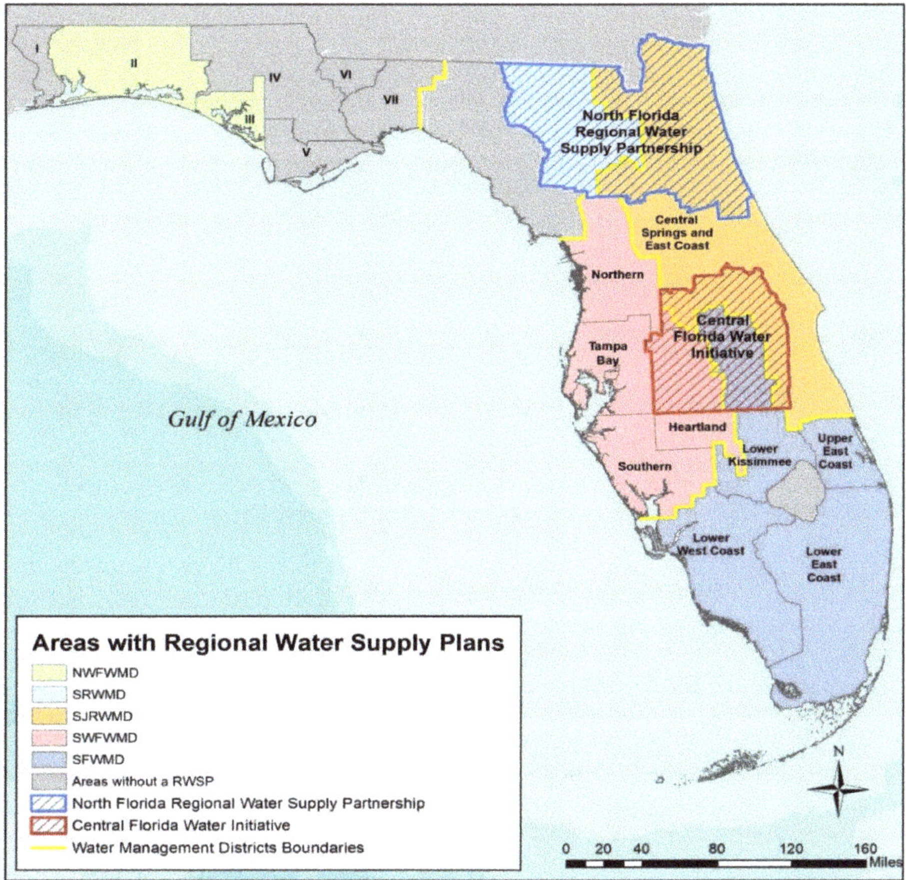

Figure 1. Florida's water management districts and water supply planning regions (Source: Originally developed by FDEP, Office of Water Policy, for [17]).

As Figure 1 indicates, RWSPs have been developed for many areas, implying that the projected demand exceeds the existing supplies in most of the state. In two regions, WMDs have cooperatively developed RWSPs that span across WMD boundaries. These are the North Florida Regional Water Supply Partnership (NFRWSP) and the Central Florida Water Initiative (CFWI) established for the regions surrounding Jacksonville (in northeast Florida) and Orlando (in central Florida). In these two areas, populations are booming, and water withdrawals from the shared aquifer are particularly high.

Each RWSP contains a list of water supply development project options and water resource development programs. According to the Florida Statutes, water resource development project options must include projects that support water supply development for all uses and the natural systems in the region (§ 373.709, Fla. Stat.). Examples of such project options are regional model development and data collection. In turn, the water supply development project options include traditional and alternative water supply projects (§ 373.709, Fla. Stat.). The water that can be made available from the projects in an RWSP must exceed the water supply needs in the water supply planning region for all existing and future reasonable-beneficial uses within the 20-year planning horizon (§ 373.709, Fla. Stat.).

The water supply development project options should be technically and financially feasible. Note that unless otherwise exempt, all water withdrawals in Florida are regulated through a system of consumptive use permits granted for a fixed time (e.g., 20 years) by WMDs. The water supply development projects included in RWSPs are referred to as "options:" local governments, public and private utilities, self-suppliers, and other water users may choose from these project options when applying for the new water use permits or an extension or modification of the existing permits. Alternatively, water users can propose different projects. The WMD shall determine whether such proposed projects meet the goals of the RWSP. If so, the projects shall be included in the list of projects supporting RWSP.

An RWSP should also account for water conservation and other demand management measures, as well as water resources constraints. In addition, RWSPs must include projects identified in an applicable recovery or prevention strategy (RPS) for established minimum flow (for rivers, streams, estuaries, and springs) or minimum water levels (for lakes, wetlands, and aquifers) (§ 373.0421, Fla. Stat.).

2.2. Florida's Long-Term Water Demand and Supply Projections

The most substantial increase in the statewide demand is expected for the public supply category, representing residential, commercial, and industrial customers served by public and private utilities. The demand is also forecasted to grow in the other water use categories that are not supplied by the water utilities and therefore classified as "self-supplied." Based on [10], these categories are referred to as domestic (e.g., wells providing for both indoor and outdoor household water needs), agriculture, landscape/recreational (e.g., golf courses and parks), commercial/industrial/institutional, and power generation categories (see Figure 2). The demand is generally projected from the five-year average per capita water use for counties or utility service areas and relevant population forecast [17]. For agriculture, an econometric water demand forecasting model utilizes past water withdrawal data, land use trends, and projected agricultural commodity prices [18]. These demand projections follow guidelines for the regional water supply planning [19]. They do not account for future improvements in water use efficiency, potential adjustments in urban development, and other changes that can reduce the future per capita use and per acre irrigation rates. Therefore, these demand projections represent conservative, "status quo" water use forecast.

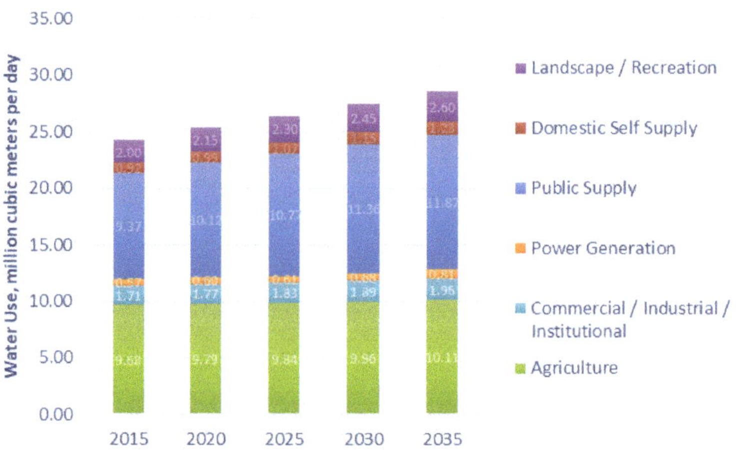

Figure 2. Total Statewide Water Use Projections for 2015–2035 Developed by the water management districts (WMDs) for Planning Purposes, assuming average rainfall (Source: based on data from [10]).

Overall, the statewide water demand increase cannot be met solely with the existing sources. WMDs evaluate existing water supplies by using hydrogeological models, examining water utilities' planned projects, comparing existing and permitted water withdrawals, and other methods [17]. The differences between the projected water demand and existing supply are expected to be met through alternative water supplies and conservation.

"Alternative water supplies" are statutorily defined as salt water, brackish surface and ground water, surface water captured predominately during wet-weather flows, surface and groundwater storage, reclaimed water, stormwater, and any other water supply source that is designated as nontraditional in the applicable regional water supply plan (§ 373.019, Fla. Stat.). Reclaimed water refers to "the water that has received at least secondary treatment and basic disinfection and is reused after flowing out of a domestic wastewater treatment facility" (§ 373.019, Fla. Stat.). Alternative water supply sources will play a key role in meeting the increase in water demands. A significant number of options for water supply development are identified in the RWSPs.

Water conservation will also play an essential role in meeting or offsetting the increase in water demand. For planning purposes, water conservation is defined as "the prevention and reduction of wasteful, or unreasonable uses of water to improve the efficiency of use" [19] (p. 30). Districts' conservation projections are intended to represent "reasonably expected demand reduction at the end of the planning period due to conservation activities" [19] (p. 30). The projections are based on agricultural irrigation efficiency trends, residential end-use surveys, statistical evaluation of actual billing data, information from the Water Conservation Tracking Tool [20], and other available data and methods [17].

WMDs consider both passive and active conservation. In the public supply category, passive conservation occurs when fixtures and appliances reach the end of their service life, and they are replaced by new (usually, more efficient) designs. Passive conservation also occurs when newly built homes are more efficient in their water use than the existing homes [21]. Passive conservation can result in substantial per capita water use reduction while having zero costs to the water providers and government agencies. In turn, active conservation accounts for the additional water use reduction due to conservation programs administered by public suppliers, government agencies, Extension, non-governmental groups, and other entities. These programs can include cost-share funds for installing more efficient indoor fixtures, higher water prices, residential irrigation restrictions, outreach strategies, and certification for low water users. Potential reductions in water use due to water conservation are estimated by the WMDs separately from the water demand projections shown in Figure 1. Given that the conservation projections can depend on funding for conservation initiatives, WMDs continue to focus on the "status quo" water demand forecast shown in Figure 2.

The Florida Department of Environmental Protection (FDEP) is charged with providing an annual status summary of regional water supply planning (§ 373.709, Fla. Stat.). The following definitions are used in the summary. "Net Demand Change" is the change in total water demands projected over the planning horizon [19]. "Net Demand Change of which Additional Projects and/or Conservation Must Surpass" is defined as the total water volume needed to meet future demands. It is calculated as the difference between "Net Demand Change" and "Estimated Existing Sources Available to Meet Future Demands" [19]. In turn, "Conservation Projection to Meet Future Demands" refers to the projected conservation savings that could be achieved during the planning horizon [19]. This conservation projection can exceed the values reported in "Net Demand Change of which Additional Projects and/or Conservation Must Surpass." The 2018 status summary, published in August 2019 [10], is compiled in Table 1.

Table 1. Projections of the water demand change, existing water sources, and additional water needed for Florida regions (based [10]).

Water Management Districts and Planning Regions	Net Demand Change (Million Cubic Meters Per Day)	Net Demand Change of which Additional Alternative Water Supply or Conservation Must Surpass (Million Cubic Meters Per Day)	Conservation Projection to Meet Future Demands (Million Cubic Meters Per Day) [1]
Areas planning for 2015–2035			
NWFWMD	0.15	0.01	0.07
SRWMD & SJRWMD [2]	0.81	0.53	0.42
CFWI	0.88	0.88	0.14
SWFWMD [2]	0.66	0.13	0.37
Total for regions planning to 2035	2.50	1.55	1.01
Areas planning for 2020–2040			
SFWMD [2]	1.76	0.24	0.54
Statewide			
Total Statewide	4.26	1.78	1.55 [1]

[1] This column reports the higher of the values provided in [10]. For the lower estimate, the total statewide conservation potential is 1.45 million m^3 per day over the 20-year planning period. [2] Excludes areas in CFWI.

As discussed above, the statewide water demand increase cannot be met solely with the existing sources. In the areas with a planning horizon to 2035, the net demand is projected to exceed the existing water supplies by 1.55 million m^3 per day (or 408.90 million gallons per day, mgd) by 2035. In the regions planning to 2040, the water demand is expected to surpass the existing water supplies by 0.24 million m^3 per day (or 62.58 mgd) by 2040 (Table 1). At the same time, the total water conservation potential statewide is 1.55 million m^3 per day (or 408.3 mgd) over the 20 years. To assess the offset for the alternative water supply costs that the water conservation can provide, we compared the water conservation potential, demand increase, and existing supplies for each of the water supply planning regions identified in Figure 1. Conservation potential can offset the requirement to develop water supplies for all planning regions except the northern region in SWFWMD, as well as CFWI, and NFRWSP. The remaining additional water supply needs for these three planning regions are 0.99 million m^3 per day (or 260.80 mgd). In other words, water conservation can reduce the necessity for alternative water supplies by almost 45 percent (subject to funding availability for the active water conservation initiatives).

2.3. *Database of Water Supply and Water Resource Development Projects*

This study utilizes a dataset assembled by the FDEP from past project funding information and data from the RWSPs [10]. FDEP annually publishes a project list summarizing the options from current RWSPs along with the projects funded in the past [10]. For implementation costs of completed projects and projects in design or construction, the FDEP dataset includes the "Project Total" column with information about the total project funding by the state (if any), district, or cooperating entity (e.g., county, city, water utility, farm, homeowner association, or golf club). This information is not always reflective of the total implementation cost of the project since it generally does not include information about land purchases or the costs of project components ineligible for cooperative funding from regional or state sources. This information also excludes any funding provided by federal agencies. It is assumed, however, that the funding from the state, district, or cooperating entity accounts for most of the implementation costs. For the projects that are listed as RWSP or RPS options, the "Projected Total Funding (for RWSP/RPS Options Only)" column summarizes information about potential funding requirements (i.e., planning-level cost estimates). This "Projected Total Funding" is an estimate only and is not verified until the project is submitted for cost-share funding to begin the design or implementation. Still, this projected funding represents the best available information regarding the future funding needs and, therefore, we include it in the analysis. Below, the combined "Project Total" and "Projected Total Funding (for RWSP/RPS Options Only)" is referred to as the "project total ($)". These values are indexed to 2019 USD using the consumer price index [22]. It is important to note that these implementation costs primarily include capital investment needs, while not accounting for operation and maintenance costs [17].

The FDEP list also includes information about project capacity (expressed as water or reuse flow, in mgd). Further, the project status is identified as completed, in design, under construction, on hold, canceled, or "RWSP/RPS Options Only." The projects are classified into 15 types, for example, brackish groundwater, agricultural conservation, reclaimed water (for potable offset), and data collection and evaluation. In addition to the project type, a narrative description is available for each project. Finally, for some of the completed projects, the funding shares of the state, regional, and local funding sources are described.

2.4. Regional Project Implementation Scenarios

To estimate statewide expenditures to meet a projected increase in demand, the criteria for identifying the project types and capacities for different planning regions should be defined. Existing studies offer a range of optimization models to assist in the project selection; however, optimization may not serve water planning agencies' needs. Optimization models can consider either a single objective—such as minimizing the total discounted infrastructure investment—or multiple objectives—such as reducing the expenditures while maximizing resilience (e.g., see [6,23–26]). Nevertheless, the number of objectives built into optimization is usually limited, and it may not reflect the full diversity of the stakeholder goals. Water supply planning typically aims at integrating the perspectives of various stakeholder groups, including government agencies, water suppliers, agricultural interests, and environmental nongovernmental organizations, that bring a myriad of considerations to the planning discussions [26,27]. Further, the recommendations based on an optimization model may be perceived by the stakeholders as prescriptive, which may impede reaching an agreement among the diverse interests involved in the planning. For example, in Florida, WMDs rely on water suppliers to identify a range of feasible water supply options. The districts do not have the authority or local knowledge to prescribe specific projects to the suppliers. Therefore, an optimization model may not be suitable for the districts to make water supply recommendations and develop funding estimates on the regional scale. Optimization-based models may also provide inadequate guidance to adaptive planning approaches when the water supply development is expected to evolve as uncertainties are resolved, and additional information becomes available [28].

In this paper, the scenario is based on the description of the potential future water supply sources and the number of projects of each type in the regional water supply plans, as well as informal discussions with the WMDs described in [17]. Specifically, in the FDEP project database [10], the number and the capacity of projects of different types can be calculated, focusing on the projects that have not yet been completed (and therefore, can be used to meet the future water demand). The project number and total capacity vary among the Florida regions. In the expenditure analysis we assume that the share of the different water supply sources should correspond to the share of the projects of different types in the total project capacity for each region. The mean project size is assumed for every project type. The project costs are estimated using a regression model, as discussed below.

2.5. Regression Analysis of Project Implementation Costs

The project cost analysis reported in [29] implies a linear relationship between the natural logarithm of the project implementation costs and the natural logarithm of the project capacity. For this study, we assumed a similar linear relationship between the natural logarithms of the costs and capacities. Because factors other than the project capacity can influence implementation costs, a multivariate regression analysis is used to explore the potential relationship between the project costs and project characteristics:

$$\ln(Cost) = \alpha + \beta \times Type + \gamma \times Region + (\delta \times Status + \zeta \times Type + \eta \times Region) \times \ln(Capacity) + \theta \times \ln(Capacity) + \varepsilon$$

where $\beta, \gamma, \delta, \zeta$ and η are vectors of model coefficients; *Status*, *Type*, *Region*, and *Capacity* are vectors of respective project characteristics (i.e., independent variables); and ε is the error term representing factors other than those captured in the independent variables that influence the dependent variable [28]. In the model, α refers to the intercept, and θ refers to the model coefficient capturing the relation between the natural logarithms of the project size and project costs. Project size's effect can depend on the project type, location, and status, as captured by the coefficients δ, ζ and η. The independent variables *Status*, *Type*, and *Region* are dummy variables that describe the projects' qualitative characteristics; therefore, the intercept α includes information for the benchmark category of *Status*, *Type*, and *Region* used for the comparison with the other categories [30].

The model is first estimated using the ordinary least squares method [30] implemented in the *glmselect* procedure in the SAS 9.4 software (© 2002–2012, SAS Institute, Inc., Cary, NC, USA). Backward selection method with select and stop criteria being the significance level of 0.1 is used to test the selected model specification against the specifications with some of the variables omitted. Furthermore, the model coefficients were compared between the *glmselect* and *robustreg* procedures to account for the potential effect of outliers and high-leverage observations on the model estimation coefficients. Below, the results are reported for the *robustreg* procedure only.

3. Results

3.1. Water Supply Projects

The original FDEP project dataset included 1623 projects or project phases. After linking all the phases for multi-phased projects, 1417 unique projects were identified. We further removed 302 projects that were canceled or lacked information about the implementation costs or project capacity (leaving 1115 projects in the dataset). Next, 426 projects were excluded since these projects did not directly contribute to the development of new water supply. These projects were described as "data collection and evaluation" (1 project), "flood control works" (4 projects), "agricultural conservation" (63 projects), and "public supply and commercial-industrial-institutional conservation" (358 projects). Discussions with WMD staff allowed us to also exclude desalination (5 projects) and reclaimed water for groundwater recharge or natural system restoration (36 projects) from the expected project types for meeting the future water demands. Finally, three projects were excluded because their type information was not clearly identifiable. The remaining database included 645 projects of seven alternative water supply types (see Table 2).

Table 2. Project number, size, and total capacity, by project types.

Project Type	Number of Projects	Project Size (m^3 per day)				Total Water or Reuse Flow for All Projects (m^3 per day)
		Mean	Median	Minimum	Maximum	
Aquifer Storage and Recovery	19	12,583.90	11,356.24	624.59	53,942.12	239,094.18
Brackish Groundwater	119	14,980.88	11,356.24	37.85	113,562.35	1,782,724.54
Groundwater Recharge	20	13,979.53	10,712.72	832.79	37,854.12	279,590.51
Other Non-Traditional Source and Projects [1]	23	11,250.90	2839.06	113.56	57,159.72	258,770.75
Reclaimed Water (for potable offset)	377	8331.70	2536.23	18.93	124,161.51	3,141,049.52
Stormwater	23	16,598.21	9463.53	454.25	46,939.11	381,758.78
Surface Water and Storage [2]	64	35,264.78	18,927.06	7.57	463,334.40	2,256,945.78

[1] Combines "Other Non-Traditional Sources" and "Other Project Type" defined in [10]. [2] Combines "Surface Water" and "Surface Water Storage" defined in [10].

Reclaimed water (for potable offset), brackish groundwater, and surface water and storage were the most numerous project types, accounting for 86.8 percent of the projects in the database (Table 2). The narrative descriptions provided for the projects were summarized for illustrative purposes and displayed as WordClouds (see Figure 3). Reclaimed water (for potable offset) projects included a broad range of projects, primarily construction or expansion of transmission capacity to provide reclaimed water for urban irrigation. In fact, in 2015, out of 2.31 million m^3 per day (or 609.0 mgd) of reclaimed water used to meet water demand in Florida (i.e., excluding reclaimed water for groundwater recharge), 68.9 percent was applied for residential irrigation, golf course irrigation, and other public access areas. Reclaimed water use for these three reuse activities had been growing steadily since 2000, while the reuse for agricultural irrigation had been declining (as summarized in [10] from the annual FDEP reuse reports [31]). Reclaimed water is expected to be a main source of alternative water supply for urban uses statewide.

Brackish groundwater projects were primarily described in the database as brackish groundwater development or well construction (Figure 3). Such projects were especially important for SFWMD, where this category accounts for the largest proportion of the projects, based on both the number and the total project capacity. Brackish groundwater was also important in the CFWI area; however, this source was less available in the other parts of the state.

Surface water and storage was the third most prominent project type in the dataset. Many surface water projects included recharge components (Figure 3), increasing aquifer recharge rates and benefiting both natural systems and groundwater users. Projects of this type were proposed or implemented in all WMDs, and their number and capacity was especially large in SWFWMD, where surface water and storage accounted for 44.7 percent of the capacity of all alternative water supply projects.

The other types included a relatively small number of projects, likely because of regional (as opposed to statewide) significance of the respective sources. In NFRWSP—the planning region shared by SRWMD and SJRWMD—groundwater recharge accounted for 65.48 percent of projects' total capacity. In turn, in NWFWMD, projects listed as "other non-traditional sources and projects" accounts for 47.04 percent of the entire capacity for all the listed projects. Almost all ASR projects were from SFWMD. Finally, stormwater projects were listed by all the WMDs (except NWFWMD); however, their total number and capacity was small.

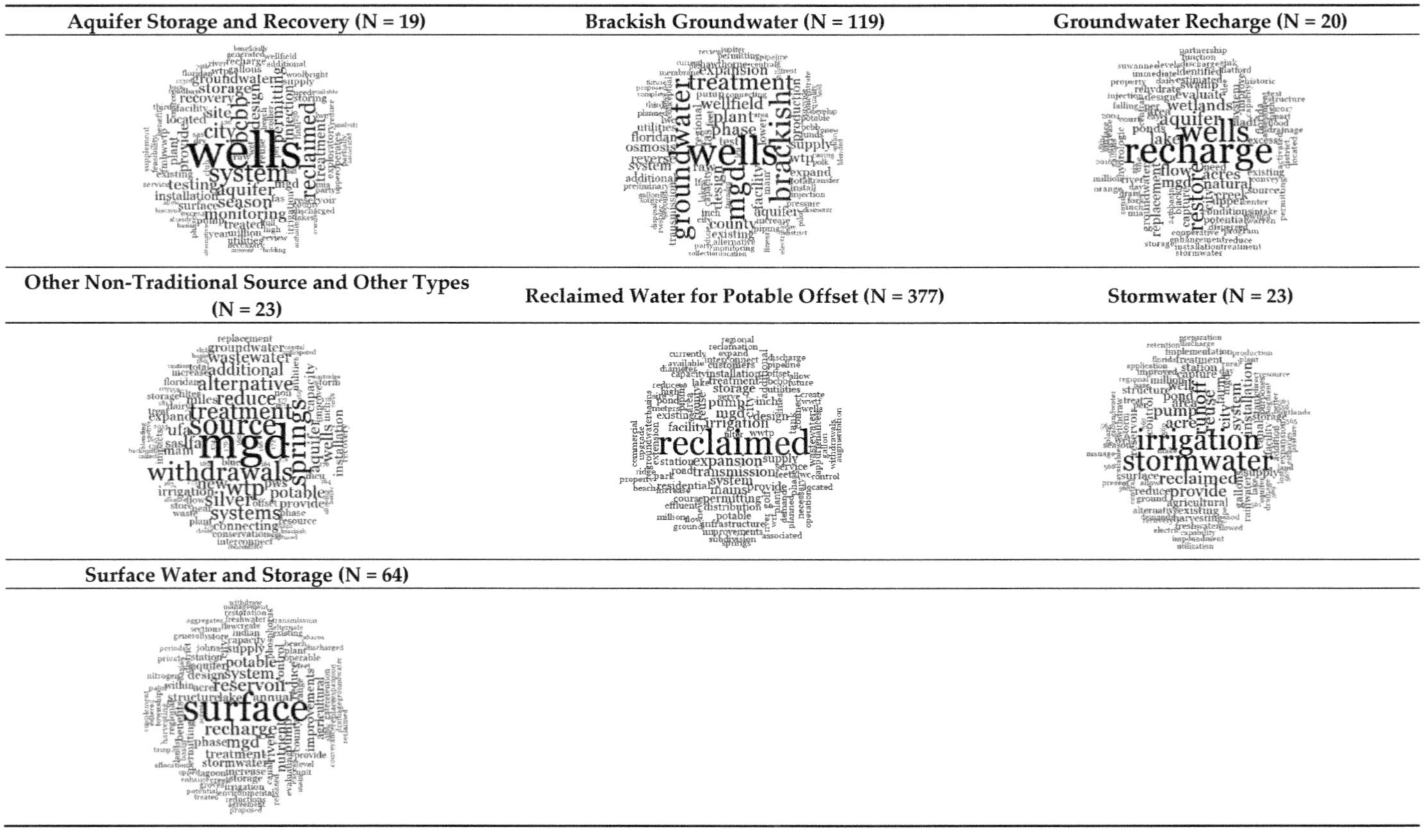

Aquifer Storage and Recovery (N = 19)	Brackish Groundwater (N = 119)	Groundwater Recharge (N = 20)
Other Non-Traditional Source and Other Types (N = 23)	**Reclaimed Water for Potable Offset (N = 377)**	**Stormwater (N = 23)**
Surface Water and Storage (N = 64)		

Figure 3. Project description: WordCloud summary, by project type (100 most common words; created in NVivo Software, version 12, using "with stemmed words" option).

The project number varied significantly by region. Given that existing supplies were projected to meet most of the demand increase in NWFWMD, the database included only 12 projects from the WMD (i.e., less than 2% of the database). We combined NWFWMD, SRWMD, and SJRWMD (outside CFWI) into the same region representing north Florida for the analysis below. In contrast, the fast-growing CFWI area accounted for approximately 20 percent of all projects in the database (Table 3). In terms of the project sizes, surface water and storage projects tended to be exceptionally large (Table 2). The projects of all types tended to be larger in SFWMD. Overall, regional variation in the project sizes and types was expected to impact the expenditures for satisfying the state's growing water demand.

Table 3. Project number, by geographical region.

Geographical Regions	Number of Projects	Percent
NWFWMD, SRWMD, and SJRWMD [1]	161	24.96
CFWI	132	20.47
SWFWMD [1]	182	28.22
SFWMD [1]	170	26.36
Total	645	100.00

[1] Excluding CFWI.

The majority of the projects (52.9%) had not yet been completed (Table 4), and therefore, they can provide a guide of potential future water supply sources in various regions. Based on the total project capacity for such projects, reclaimed water, brackish groundwater, and surface water and storage were expected to be keys to meeting the future water demand, with other project types also being prominent in specific regions (Table 5).

Table 4. Project number, size, and total capacity, by project types.

Project Status	Number of Projects	Percent
Completed	304	47.13
Design, construction/underway, or on hold	98	15.19
RWSP/RPS Options Only	243	37.67
Total	645	100.00

Table 5. Total capacity of the projects that had not yet been completed, by type and geographic region (million cubic meters per day).

Project Type	Geographic Regions				
	NWFWMD, SRWMD, and SJRWMD [1]	SWFWMD [1]	SFWMD [1]	CFWI	Total
Aquifer Storage and Recovery	0.00	0.01	0.06	0.00	0.07
Brackish Groundwater	0.00	0.10	0.42	0.37	0.90
Groundwater Recharge	0.20	0.03	0.00	0.00	0.23
Other Non-Traditional Source and Projects [2]	0.11	0.00	0.04	0.00	0.15
Reclaimed Water (for potable offset)	0.08	0.80	0.39	0.14	1.41
Stormwater	0.04	0.19	0.04	0.00	0.28
Surface Water and Storage [3]	0.26	0.85	0.58	0.14	1.83
Total	0.69	1.98	1.53	0.66	4.86

[1] Excluding CFWI. [2] Combines "Other Non-Traditional Sources" and "Other Project Type" defined in [10]. [3] Combines "Surface Water" and "Surface Water Storage" defined in [10].

3.2. Regression Analysis of the Project Costs

The median project implementation costs were \$3.54 million (with the mean of \$24.01 million). The project cost increased with the project size; the natural logarithm of the implementation costs and the natural logarithm of the project capacity were highly correlated (Figure 4). The regression analysis showed that, as expected, the project implementation costs increased with the project sizes (Table 6). Further, the projects identified as options to meet the future increase in water demand (i.e., RWSP and RPS options) tended to be more expensive than the projects of the other statuses. This result could be due to the planning-level cost estimates provided for the future projects, or because the opportunities to implement the low-cost projects had been exhausted. Among water sources, stormwater projects were significantly less expensive as compared with surface water and storage projects (i.e., the benchmark). Finally, projects from north and central Florida (i.e., NWFWMD, SRWMD, and SJRWMD and CFWI) tended to be less expensive than the same projects implemented in SWFWMD (i.e., the benchmark region). Overall, the model explained approximately 54 percent of the variability of the independent variable.

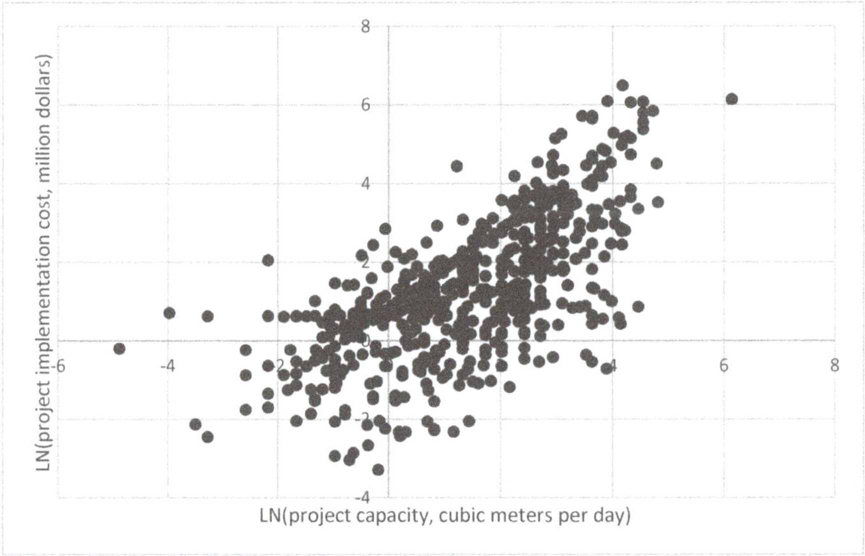

Figure 4. Scatter Plot of the Natural Logarithms of the Project Implementation Costs and Capacities.

Table 6. Estimated coefficients for the variables used in the regression model (with the coefficients that are statistically different from zero highlighted in bold font).

Parameter		Estimate	Standard Error	95% Confidence Limits		Chi-Square	Pr > χ^2
Intercept		**−8.560**	0.797	−10.121	−6.998	115.470	<0.0001
Natural logarithm of project capacity		**1.256**	0.083	1.093	1.419	227.700	<0.0001
Project Type	Aquifer Storage and Recovery	**6.051**	2.036	2.060	10.042	8.830	0.003
	Brackish Groundwater	**3.263**	0.972	1.358	5.168	11.260	0.001
	Groundwater Recharge	**−4.766**	2.044	−8.772	−0.760	5.440	0.020
	Other Nontraditional Sources and Project Types	**3.293**	1.317	0.712	5.875	6.250	0.012
	Reclaimed Water (for potable offset)	**3.053**	0.788	1.508	4.597	15.010	0.000
	Stormwater	**5.242**	1.528	2.246	8.237	11.760	0.001
	Surface Water and Storage (benchmark)	0.00					
Region	NWFWMD, SRWMD, & SJRWMD	**2.387**	0.600	1.212	3.562	15.850	<0.0001
	CFWI	**2.246**	0.605	1.060	3.431	13.780	0.000
	SFWMD	−0.328	0.709	−1.717	1.061	0.210	0.644
	SWFWMD (benchmark)	0.00					
Interaction between project status and natural logarithm of project capacity	Completed	**−0.070**	0.012	−0.093	−0.047	35.790	<0.0001
	Design, construction/underway, or on hold	**−0.074**	0.016	−0.106	−0.043	20.960	<0.0001
	RWSP/RPS Options (benchmark)	0.00					
Interaction between project region and natural logarithm of project capacity	NWFWMD, SRWMD, & SJRWMD	**−0.368**	0.074	−0.513	−0.222	24.420	<0.0001
	CFWI	**−0.334**	0.073	−0.477	−0.190	20.770	<0.0001
	SFWMD	−0.057	0.079	−0.213	0.099	0.510	0.473
	SWFWMD (benchmark)	0.00					
Interaction between project type and natural logarithm of project capacity	Aquifer Storage and Recovery	**−0.674**	0.226	−1.116	−0.231	8.910	0.003
	Brackish Groundwater	**−0.313**	0.104	−0.516	−0.109	9.080	0.003
	Groundwater Recharge	**0.446**	0.224	0.007	0.884	3.960	0.047
	Other Nontraditional Sources and Project Types	**−0.341**	0.152	−0.639	−0.043	5.020	0.025

Table 6. *Cont.*

	Parameter	Estimate	Standard Error	95% Confidence Limits		Chi-Square	Pr > χ^2
	Reclaimed Water (for potable offset)	**−0.329**	0.084	−0.492	−0.165	15.500	<0.0001
	Stormwater	**−0.748**	0.166	−1.073	−0.424	20.410	<0.0001
	Surface Water and Storage (benchmark)	0.00					
R-Square	0.539						
AICR	830.192						
BICR	930.593						
Deviance	640.206						

3.3. Regional Project Implementation Scenarios and Statewide Expenditure Projections

The relation between the project implementation costs and project capacity, type, and location was used to estimate the costs to implement the projects in various regions. In this analysis, we used the average capacity for each project type, with the assumption that the projects design or implementation had not yet started (i.e., project status is "RWSP/RPS Options"). The estimated project costs were then divided by the mean project capacity to assess the implementation cost per unit of the project capacity (Table 7). As discussed above, stormwater projects tended to be the least expensive option, while brackish groundwater was relatively more expensive. Also, projects implemented in SWFWMD tended to be more expensive, especially when compared with north Florida (i.e., NWFWMD, SRWMD, and SJRWMD).

Table 7. Estimated implementation costs, assuming mean projects capacity. [1]

Project Types	Mean Project Capacity (Cubic Meters Per Day)	Implementation Costs (dollars Per Cubic Meter Per Day of Capacity)			
		NWFWMD, SRWMD, and SJRWMD [2]	SWFWMD [2]	SFWMD [2]	CFWI
Aquifer Storage and Recovery	12,583.90	534.71	1578.60	664.77	640.02
Brackish Groundwater	14,980.88	919.88	2895.41	1207.27	1107.58
Groundwater Recharge	13,979.53	429.96	1319.38	552.30	516.48
Other NTS&PT	11,250.90	826.83	2342.59	992.81	985.90
Reclaimed Water (for potable offset)	8331.70	829.20	2103.77	906.96	978.68
Stormwater	16,598.21	92.66	302.85	125.54	111.96
Surface Water and Storage	35,264.78	648.19	2794.61	1109.84	803.51

[1] Note that these costs include capital investment only, and do not account for operation and maintenance costs. While all these project types are expected to provide water for growing water demand, the water quality provided by the projects can differ; and therefore, the project capacity reported in this table should not be the only metrics to consider. [2] Excluding CFWI.

The statewide expenditures to meet the water demand increase in the next 20 years is presented in Table 8. The estimate utilizes the implementation costs reported in Table 7; we also assume that the mix of water supply sources used to meet the projected water demand should mimic the mix of various sources reported in Table 5. In other words, the share of various project types observed in the database (excluding the completed projects) is assumed to correspond to the share of the alternative sources in the future water supply mix in each region. The total statewide expenditures are projected to approach $2 billion over the next 20-year planning horizon.

Table 8. Estimated implementation costs to meet the increase in water demand in the next 20 years, million dollars [1].

	NWFWMD, SRWMD, and SJRWMD [2]	SWFWMD [2]	SFWMD [2]	CFWI	Total Statewide
Total project implementation costs	322.50	287.10	246.11	897.14	1752.85

[1] Note that these costs include capital investment only, and does not account for operation and maintenance costs.
[2] Excluding CFWI.

4. Discussion

The analysis shows that over the next 20 years, Florida must invest almost $2 billion into the development of alternative water supplies to meet the growing water demand. On average, approximately 65 percent of expenditures for such water supply projects are funded by local partners. In comparison, regional and state funding shares account for about 35 and 5 percent of the total, respectively [17]. Therefore, the local entities must plan for these expenditures accordingly. These investments can also be particularly high for the urbanizing regions with significant population growth, requiring such communities to explore appropriate funding mechanisms.

The actual investments to meet the water demand increase can be even higher than reported in Table 8. The average beneficial offset provided by the reclaimed water projects is determined at 0.55 [17,32]. Such a beneficial offset ratio implies that approximately two cubic meters per day of water should be created by reclaimed water projects to meet the rise in the future water demand of one cubic meter per day. The offset values vary among the types of reuse activities. For example, the offset provided by reclaimed water for fire protection is estimated at one [32]. For various urban irrigation activities, the offset ratio is between 0.25 and 0.75 [32]. This low ratio reflects the poor irrigation efficiency pertinent specifically to reclaimed water use. This low ratio also captures an expansion of irrigated areas due to reclaimed water availability (e.g., irrigation of road medians if reclaimed water is available, but no irrigation otherwise) [32]. If the offset ratio is significantly lower than 1, the costs of reclaimed water projects needed to meet the future demand will dramatically increase.

Further, the expenditures reported in Table 8 include capital investments only and do not account for the operation and maintenance (O&M) costs. O&M expenses will further increase the total funding required to satisfy the water demand increase. This study also does not consider the expenditures for data collection and evaluation (such as water quality monitoring, or hydrogeologic modeling), which can also be expensive, and which are critical for selecting water resource use and management strategies [33].

This study relies on the demand projections that do not explicitly account for the potential demand increase during droughts. For the last year of their planning horizons, Florida's WMDs forecast the water demand for both normal and drought rainfalls. The drought event is defined as "a year in which below normal rainfall occurs with a 10 percent probability of occurring in any given year" [19] (p. 4). Drought demand coefficients, historical water use analysis, and crop irrigation requirement models are used to develop the drought demand forecast [34]. The compilation of WMDs' most recent drought demand projections is not available; however, a 2018 compilation shows that the statewide drought demand can exceed the demand given normal rainfall by approximately 20 percent [34]. Such a significant increase in water demand during droughts and potential impacts of the drought on water supply can require additional investments in the alternative water supplies.

At the same time, considerable opportunities exist in the state to reduce funding needs for alternative water supplies. First, implementation costs differ significantly among the water supply sources, and more substantial utilization of inexpensive sources, such as stormwater, can reduce the regional and statewide cost. For example, stormwater is expected to provide approximately 3 percent of the water to meet future demand (based on the project database utilized in this study). If, in every region, stormwater projects are used to satisfy 10 percent of the future water demand (with a corresponding reduction of surface water and storage), the total statewide implementation costs become $1662.11 million over 20-year period, or 5.18 percent lower than reported in Table 8.

Second, the statewide expenditure can be reduced by varying the sizes of the projects implemented to meet the future water demand, if feasible. The regression model shows a statistically significant effect of the project size on the project implementation costs, even though a massive increase in the project sizes is needed to noticeably alter the statewide expenditure estimate. For example, for a 10 percent increase in the project size assumed for each project type, the estimated total statewide cost reduces by only 1.73 percent.

Third and most importantly, water conservation can provide a cost-effective alternative to water supply development. This study utilizes the water demand projections that do not account for continuous water use efficiency improvement. Meanwhile, in the public supply sector, the statewide average per capita use declined sharply in 2000–2010 and then remained unchanged in 2010–2015 [13]. While no statewide data are available for the 2015–2020 period, potential reductions in the per capita use are discussed in all RWSPs. To assess the offset for the alternative water supply costs that the water conservation can provide, we compared the water conservation potential, demand increase, and existing supplies for each of the water supply planning regions identified in Figure 1. Conservation potential can offset the requirement to develop water supplies for all planning regions except the Northern region in SWFWMD, as well as CFWI and NFRWSP. The remaining additional water supply needs for these three planning regions is 987,235.39 cubic meters per day (or 260.80 mgd). In other words, water conservation can reduce the necessity of developing alternative water supplies by 44.68 percent. The statewide water supply development investments for this scenario become $931.53 million over 20 years, showing more than 45 percent reduction in the water supply development costs as compared with Table 8.

Water conservation can be a more cost-efficient strategy for addressing the projected gap between water demand and existing supplies than developing alternative water supplies. For example, in central Florida, 11 active water conservation programs for the public supply category are considered, including such strategies as high-efficiency toilet installations, residential low-flow showerheads, irrigation and landscape evaluations, residential irrigation controllers, and irrigation enforcements. These 11 programs are estimated to result in 17,185.77 cubic meters per day (or 4.54 mgd) of water use reduction, costing $8.1 million over the 20-year planning horizon [21], or approximately $475 per m^3 per day (or $1.8 million per mgd). These average costs are significantly lower than that for widely used brackish groundwater or reclaimed water alternative water supply sources (see Table 8). Note that as mentioned above, the project database used in this study initially included 358 projects classified as "public supply and commercial-industrial-institutional conservation." For these projects, the median costs were $1207.27 per cubic meter per day of water conserved (or $4.57 million per mgd), which is relatively high and comparable to the costs of alternative water supply projects. These high costs are likely because the low-cost conservation initiatives, such as conservation water pricing or outreach strategies, are not eligible for WMDs' cost-share funding and are not included in the project database. Ultimately, both "rethinking water supply" and "rethinking water demand" [12] (p. 21302) should be considered when addressing water scarcity challenges.

5. Conclusions

In 2010, Dr. Peter H. Gleick, co-founder and director of the nonprofit research think-tank Pacific Institute, identified rethinking water supply and demand, reforming water management institutions, and analyzing climate change impacts as the solutions to global water crisis [12]. These suggestions reaffirm and extend the proposals expressed in other manuscripts (e.g., [35–37]), and they remain relevant in 2020. Moreover, the recommendations become applicable to even the historically water-rich regions like the eastern United States.

In this paper, we estimate the capital investments for developing new water supply sources alternative to the traditionally used groundwater in one of the largest and fastest-growing states in the United States – Florida. In line with Gleick's recommendations, Florida has been rethinking water supply and developing sources that have been ignored in the past, such as reclaimed water, brackish groundwater, surface water storage, and stormwater. Based on a database of projects assembled by FDEP, this study shows that over the next 20 years, Florida must invest almost 2 billion dollars into the development of alternative water supplies to meet the growing water demand. The expenditures projections depend on the mix of projects considered, with scenarios relying on such commonly used water sources as reclaimed water and brackish groundwater being relatively expensive. Less expensive project options include groundwater recharge and stormwater projects.

The scarcity of traditional water supply and the high costs of alternative water supply projects call for combining "rethinking water supply" with "rethinking water demand" [12] (p. 21302). In Florida, water conservation is defined as reduction in wasteful and inefficient water uses [19]. Water conservation can reduce the need for alternative water supplies by more than 40 percent, based on WMDs' estimates. This drop can translate into a reduction of more than 45 percent in the capital investments for alternative water supplies. While water conservation programs considered by the WMDs also require investments, many of them are less expensive (per m^3 or mgd of water) than the alternative water supply projects.

The solutions of "rethinking water supply" and "rethinking water demand" should be integrated and coordinated (e.g., [37]). Gleick suggests that "a key to improving efficiency is understanding where, when, and why we use water" [12] (p. 21302). A better understanding of our water use may allow tailoring the water supply treatment levels to the water use requirements. This adaptation of supply sources has been partially accomplished by separating the water applied for irrigation and providing reclaimed water specifically for this use. More specific water quality requirements for different use types may be developed in the future, with various supply mixes appropriate for each application type. In this case, the statewide total water demand projections may become irrelevant, as the demand, supply, and expenditure projections will need to be developed for each type of use. Reduction in the expenditures could be possible if treatment needs vary significantly among the use types, with many uses requiring reduced treatment levels.

Gleick [12] also emphasizes climate change implications for water resource management decisions. Potential impacts of climate change on both water demand and supply sources are discussed in Florida's RWSPs; however, no quantitative assessment is available. Similarly, this study includes no analysis of establishing better institutions to address potential groundwater quality or the implications of storage, recovery, and recharge activities [38].

Overall, climate change impacts, natural systems restoration needs, and maintenance and replacement requirements for the aging infrastructure will make the total expenditures far exceed the estimates reported in this study [17]. For example, the United States Environmental Protection Agency's (US EPA's) Drinking Water Infrastructure Needs Survey provides the most comprehensive infrastructure cost estimate in the United States [39]. The US EPA survey focuses on existing infrastructure maintenance, and it largely excludes new infrastructure necessary to meet the increasing demand due to future growth. The study estimates that $21.89 billion is needed for Florida's existing infrastructure in the next 20 years [39]. The expenditures to continue providing water to communities and to support future growth are real needs for many U.S. states and water providers [40]. Overall, expenditure analysis should continue to be an essential part of the long-term water supply planning and water resource management.

Author Contributions: Conceptualization, T.B., M.C., K.B., and K.H.; methodology, T.B.; formal analysis, T.B.; Writing—Original draft preparation, T.B.; writing—review and editing—original draft, M.C., K.B., and K.H.; Writing—Review and Editing—Revised versions of the manuscript, M.C., and K.B.; supervision, T.B. and M.C.; project administration, M.C. All authors have read and agreed to the published version of the manuscript.

Funding: This study is partially funded through the cooperative agreement between the Office of Economic and Demographic Research, Florida Legislature, and the University of Florida. Any opinions, findings, conclusions, or recommendations expressed in this publication are those of the author(s). They do not necessarily reflect the view of Office of Economic and Demographic Research, Florida Legislature. This paper is also partially supported by the U.S. Department of Agriculture Project FLA-FRE−005565 (PI: Tatiana Borisova).

Acknowledgments: We appreciate insightful feedback provided by two anonymous reviewers that allowed us to improve the manuscript significantly. We also valuable acknowledge assistance from the professional editor, Ms. Carol Fountain (Food and Resource Economics Department, University of Florida). Finally, we wanted to thank Dr. James Colee, Consultant, Statistical Consulting Unit, Institute of Food and Agricultural Sciences, University of Florida, for constructive discussions about the methods used in this study.

Conflicts of Interest: The authors declare no conflict of interest.

References

1. Zeff, H.B.; Kasprzyk, J.R.; Herman, J.D.; Reed, P.M.; Characklis, G.W. Navigating financial and supply reliability tradeoffs in regional drought management portfolios. *Water Resour. Res.* **2014**, *50*, 4906–4923. [CrossRef]
2. Gleick, P.H.; Cohen, M.; Cooley, H.; Donnelly, K.; Fulton, J.; Ha, M.-L.; Morrison, J.; Phurisamban, R.; Rippman, H.; Woodward, S. *The World's Water. The Report on Freshwater Resources*; Pacific Institute for Studies in Development, Environment, and Security: Oakland, CA, USA, 2018; Volume 9.
3. WWAP (United Nations World Water Assessment Programme). *The United Nations World Water Development Report 2015: Water for a Sustainable World*; UNESCO: Paris, France, 2015; Available online: https://unesdoc.unesco.org/ark:/48223/pf0000231823 (accessed on 20 July 2020).
4. Vorosmarty, C.J. Global water resources: Vulnerability from climate change and population growth. *Science* **2000**, *289*, 284–288. [CrossRef]
5. Vörösmarty, C.J.; McIntyre, P.B.; Gessner, M.O.; Dudgeon, D.; Prusevich, A.; Green, P.; Glidden, S.; Bunn, S.E.; Sullivan, C.A.; Liermann, C.R.; et al. Global threats to human water security and river biodiversity. *Nature* **2010**, *467*, 555–561. [CrossRef]
6. Padula, S.; Harou, J.J.; Papageorgiou, L.G.; Ji, Y.; Ahmad, M.; Hepworth, N. Least Economic Cost Regional Water Supply Planning–Optimising Infrastructure Investments and Demand Management for South East England's 17.6 Million People. *Water Resour. Manag.* **2013**, *27*, 5017–5044. [CrossRef]
7. Houlihan, B. Politics, finance, and the environment. *Util. Policy* **1995**, *4*, 243–252. [CrossRef]
8. Emek, U. Turkish experience with public private partnerships in infrastructure: Opportunities and challenges. *Util. Policy* **2015**, *27*, 120–129. [CrossRef]
9. Economic and Demographic Research (EDR). *Population and Demographic Data-Florida Products*; Economic and Demographic Research (EDR): Tallahassee, FL, USA, 2019; Available online: http://edr.state.fl.us/Content/population-demographics/data/index-floridaproducts.cfm (accessed on 20 July 2020).
10. Florida Department of Environmental Protection (FDEP). *Regional Water Supply Planning 2018 Annual Report*; Florida Department of Environmental Protection (FDEP): Tallahassee, FL, USA, 2019. Available online: https://floridadep.gov/water-policy/water-policy/documents/2018-annual-regional-water-supply-planning-report-ada-compliant (accessed on 20 July 2020).
11. Florida Department of Environmental Protection (FDEP). *Minimum Flows and Minimum Water Levels and Reservations*; Florida Department of Environmental Protection (FDEP): Tallahassee, FL, USA, 2020. Available online: https://floridadep.gov/water-policy/water-policy/content/minimum-flows-and-minimum-water-levels-and-reservations (accessed on 20 July 2020).
12. Gleick, P. Roadmap for Sustainable Water Resources in Southwestern North America. *Proc. Natl. Acad. Sci. USA* **2010**, *107*, 21300–21305. [CrossRef]
13. Marella, R. *Water Withdrawals, Uses, and Trends in Florida, 2015*; Scientific Investigations Report 2019–5147; U.S. Geological Survey: Reston, VA, USA, 2019. Available online: https://pubs.usgs.gov/sir/2019/5147/sir20195147.pdf (accessed on 20 July 2020).
14. Florida Department of Environmental Protection (FDEP). *2019 Statewide Annual Report on Total Maximum Daily Loads, Basin Management Action Plans, Minimum Flows or Minimum Water Levels, and Recovery or Prevention Strategies*; Florida Department of Environmental Protection (FDEP): Tallahassee, FL, USA, 2019. Available online: https://floridadep.gov/dear/water-quality-restoration/content/statewide-annual-report (accessed on 20 July 2020).
15. Regan, K. Adverse Environmental Impacts Under Florida Water Law: From Water Wars towards Adaptive Management. *J. Land Use Environ. Law.* **2003**, *19*, 123–184.
16. Hagen, G.; Brown, M. An Experimental Internet Delphi: Florida Water Supply Alternatives. *Fla. Sci.* **1996**, *59*, 163–168.
17. Economic and Demographic Research (EDR). *Annual Assessment of Florida's Water Resources and Conservation Lands-2020 Edition*; Economic and Demographic Research (EDR): Tallahassee, FL, USA, 2020; Available online: http://edr.state.fl.us/Content/natural-resources/index.cfm (accessed on 20 July 2020).

18. Florida Department of Agriculture and Consumer Services (FDACS). *Agricultural Water Supply Planning. Undated*; Florida Department of Agriculture and Consumer Services (FDACS): Florida State Capitol, FL, USA, 2020. Available online: https://www.fdacs.gov/Agriculture-Industry/Water/Agricultural-Water-Supply-Planning (accessed on 20 July 2020).
19. Florida Department of Environmental Protection (FDEP); Northwest Florida Water Management District (NWFWMD); South Florida Water Management District (SFWMD); Southwest Florida Water Management District (SWFWMD), St. Johns River Water Management District (SJRWMD); Suwannee River Water Management District (SRWMD). *Format and Guidelines For Regional Water Supply Planning*; Florida Department of Environmental Protection: Tallahassee, FL, USA, 2019; 42p.
20. Alliance for Water Efficiency (AWE). *Water Conservation Tracking Tool*; Alliance for Water Efficiency (AWE): Chicago, IL, USA, 2020; Available online: https://www.allianceforwaterefficiency.org/resources/topic/water-conservation-tracking-tool (accessed on 20 July 2020).
21. Central Florida Water Initiative (CFWI). *Regional Water Supply Plan 2020, Appendices*; Public Review Draft; Central Florida Water Initiative: Orlando, FL, USA, 2020; 323p, Available online: https://cfwiwater.com/planning.html (accessed on 20 July 2020).
22. U.S. Bureau of Labor Statistics (USBLS). *CPI for All Urban Consumers (CPI-U) 1967=100 (Unadjusted)-CUUR0000AA0*; U.S. Bureau of Labor Statistics (USBLS): Washington, DC, USA, 2019. Available online: https://data.bls.gov/cgi-bin/surveymost?bls (accessed on 20 July 2020).
23. Rosenberg, D.E.; Howitt, R.E.; Lund, J.R. Water management with water conservation, infrastructure expansions, and source variability in Jordan. *Water Resour. Res.* **2008**, 44. [CrossRef]
24. Matrosov, E.S.; Huskova, I.; Kasprzyk, J.R.; Harou, J.J.; Lambert, C.; Reed, P.M. Many-objective optimization and visual analytics reveal key trade-offs for London's water supply. *J. Hydrol.* **2015**, *531*, 1040–1053. [CrossRef]
25. Zeff, H.B.; Herman, J.D.; Reed, P.M.; Characklis, G.W. Cooperative drought adaptation: Integrating infrastructure development, conservation, and water transfers into adaptive policy pathways. *Water Resour. Res.* **2016**, *52*, 7327–7346. [CrossRef]
26. Huskova, I.; Matrosov, E.S.; Harou, J.J.; Kasprzyk, J.R.; Lambert, C. Screening robust water infrastructure investments and their trade-offs under global change: A London example. *Glob. Environ. Chang.* **2016**, *41*, 216–227. [CrossRef]
27. Furlong, C.; De Silva, S.; Guthrie, L.; Considine, R. Developing a water infrastructure planning framework for the complex modern planning environment. *Util. Policy* **2016**, *38*, 1–10. [CrossRef]
28. Giordano, T. Adaptive planning for climate resilient long-lived infrastructures. *Util. Policy* **2012**, *23*, 80–89. [CrossRef]
29. Hertzler, P.C.; Davies, C. The cost of infrastructure needs. *J. Am. Water Work. Assoc.* **1997**, *89*, 55–61. [CrossRef]
30. Wooldridge, J.M. *Introductory Econometrics: A Modern Approach*, 6th ed.; Cengage Learning: Boston, MA, USA, 2016.
31. Florida Department of Environmental Protection (FDEP). *Reuse Inventory Database and Annual Report*; Florida Department of Environmental Protection (FDEP): Tallahassee, FL, USA, 2019. Available online: https://floridadep.gov/water/domestic-wastewater/content/reuse-inventory-database-and-annual-report (accessed on 20 July 2020).
32. Florida Department of Environmental Protection (FDEP). *Report on Expansion of Beneficial Use of Reclaimed Water, Stormwater and Excess Surface Water (Senate Bill 536)*; Florida Department of Environmental Protection: Tallahassee, FL, USA, 2015; 230p. Available online: https://floridadep.gov/water/domestic-wastewater/documents/senate-bill-536-final-report-expansion-beneficial-use-reclaimed (accessed on 20 July 2020).
33. Varady, R.G.; Zuniga-Teran, A.A.; Gerlak, A.K.; Megdal, S.B. Modes and Approaches of Groundwater Governance: A Survey of Lessons Learned from Selected Cases across the Globe. *Water* **2016**, *8*, 417. [CrossRef]
34. Economic and Demographic Research (EDR). *Annual Assessment of Florida's Water Resources and Conservation Lands-2018 Edition*; Economic and Demographic Research (EDR): Tallahassee, FL, USA, 2018; Available online: http://edr.state.fl.us/Content/natural-resources/LandandWaterAnnualAssessment_2018Edition.pdf (accessed on 20 July 2020).
35. Dziegielewski, B. Strategies for Managing Water Demand. *Univ. Counc. Water Resour.* **2003**, *126*, 29–39.

36. Gleick, P.H. The changing water paradigm: A look at twenty-first century water resources development. *Water Int.* **2000**, *25*, 127–138. [CrossRef]
37. Thomas, J.S.; Durham, B. Integrated Water Resource Management: Looking at the Whole Picture. *Desalination* **2003**, *156*, 21–28. [CrossRef]
38. Van der Gun, J.; Aureli, A.; Merla, A. Enhancing Groundwater Governance by Making the Linkage with Multiple Uses of the Subsurface Space and Other Subsurface Resources. *Water* **2016**, *8*, 222. [CrossRef]
39. U.S. Environmental Protection Agency (US EPA). *Drinking Water Infrastructure Needs Survey and Assessment: Sixth Report to Congress*; Office of Water, EPA 816-K-17-002: Washington, DC, USA, 2018; 76p. Available online: https://www.epa.gov/sites/production/files/2018-10/documents/corrected_sixth_drinking_water_infrastructure_needs_survey_and_assessment.pdf (accessed on 20 July 2020).
40. Job, C.; Barles, R. USEPA's 2015 Drinking Water Infrastructure Needs Survey. *J. Am. Water Works Assoc.* **2015**, *107*, 84–89. [CrossRef]

Case Report

Characterizing Supply Variability and Operational Challenges in an Intermittent Water Distribution Network

John J. Erickson [1,2], Yamileth C. Quintero [3] and Kara L. Nelson [1,*]

1 Department of Civil and Environmental Engineering, University of California, Berkeley, CA 94720, USA; jerickson@hazenandsawyer.com
2 Hazen and Sawyer, Dallas, TX 75206, USA
3 Instituto de Acueductos y Alcantarillados Nacionales (IDAAN), Panama City, Panama; yquintero@idaan.gob.pa
* Correspondence: karanelson@berkeley.edu; Tel.: +1-510-643-5023

Received: 20 May 2020; Accepted: 23 July 2020; Published: 29 July 2020

Abstract: Intermittent piped water supply is common in low- and middle-income countries and is inconvenient for users, particularly when supply schedules are unreliable. In this study, supply schedules and operational challenges were characterized in intermittent areas of the Arraiján, Panama distribution network based on one year of pressure and flow monitoring in four study zones, analysis of three years of pipe break data, and observations of system operation. Service quality was found to vary among users and supply schedules were often irregular and unpredictable. Direct causes of unanticipated supply outages included pump failures, chronic pipe breaks in specific parts of the system, transmission main breaks, irregular valve operations, and treatment plant outages. The extent and duration of these outages were often increased by high rates of water loss, insufficient storage capacity, and difficulty detecting and resolving infrastructure failures. Factors associated with intermittent supply, such as intermittent pumping, appeared to be associated with a higher frequency of pipe breaks. However, the analysis did not indicate a strong general correlation between intermittent supply and pipe breaks. Pressure and flow monitoring in intermittent supply areas, similar to that undertaken in this study, could be a valuable tool to improve regular operations as well as longer-term planning and prioritization of system improvements. Water loss reduction and adequate distribution storage capacity could also mitigate the effects of operational failures. Investments in monitoring and data analysis have the potential to improve the reliability of intermittent supply in cases where continuous supply is not immediately feasible.

Keywords: intermittent water supply; pressure monitoring; unreliable water supply; pipe breaks; water distribution system; water system operation

---

1. Introduction

A water utility operations manager arrives at the regional office in the morning to find a group of local residents waiting to voice their frustration with the lack of water service in their part of Arraiján, Panama. They live near the end of the pipe that supplies their area and normally only receive water intermittently. However, it has now been several days since piped water last arrived. In the outskirts of Arraiján, a rapidly growing peri-urban area outside of Panama City, the area where these residents live was initially served by a rural groundwater system before it was connected to the larger Arraiján water network, operated by Panama's Institute of National Aqueducts and Sewers (IDAAN for initials in Spanish). Since then, the utility has struggled to provide reliable service. Several months before this morning, residents had closed a lane of Panama's largest highway to protest poor service quality.

Today, the operations manager tells the residents he does not know what might be causing problems in their area this time. He pleads with them to be patient while he looks into the matter. This is just one of many fires he needs to put out. He and the regional office are overwhelmed by a backlog of customer complaints and broken pipes in the system that serves a quarter of a million people. All of these problems must be addressed with two active repair crews and one working backhoe.

Further investigation reveals that one of the two pumps serving the area in question is out of service. The one working pump is not enough to push water to the far reaches of the leaky pipe network. Operators had not noticed the problem on their daily drive-by inspections when they listen to check if the pumps are running. The faulty pump is quickly repaired and all is well for a few days until an electrical failure shuts down the entire pump station the next Saturday. The weekend operator does not notice the failure, and the residents again close a lane of the highway before the utility is even aware of the problem.

Utilities in low- and middle-income countries around the world often face problems similar to the ones described above. Intermittent drinking water supply (IWS), defined as supply that is available to consumers less than 24 h per day [1], is a common deficiency in piped water systems [2]. Intermittent supply can be caused by a combination of institutional and technical factors, including the unplanned expansion of the distribution network, insufficient system data to inform an optimal operation of the distribution network, excessive water losses, insufficient water resources, and inadequate infrastructure [3–6]. A recent study [7] extrapolated data from the World Bank Water and Sanitation Program's International Benchmarking Network (IBNET) to estimate that approximately 1 billion people living in low- and middle-income countries worldwide are likely exposed to IWS. IWS is an inconvenience for users [8–10], can make it difficult for a utility to provide equitable supply to all customers in the distribution network [3,11,12], is hypothesized to lead to pipe damage [13–15], and is a risk to water quality [4,16,17]. The nature and severity of IWS varies considerably throughout the world, between water systems, and often within water systems. A recent review revealed the complex and diverse factors that cause IWS and identified that there currently is a knowledge gap regarding characteristics of the different types of IWS and their implications for those involved [15].

While predictable supply is important for customer satisfaction in intermittent systems, many factors make it difficult for water utilities to operate intermittent distribution networks predictably and equitably. High peak demands in intermittent networks often result in excessive pressure losses, as large flows are forced through small-diameter pipes, which lead to supply inequities between users at upstream and downstream ends of a pipe [12]. The operation of intermittent networks is often hindered by incomplete knowledge of the distribution system [3], the inapplicability of hydraulic modeling methods developed for continuous systems [18,19], inadequate monitoring of dynamic hydraulic conditions, frequent pipe breaks [13], and high rates of water loss [6].

Real-time monitoring of intermittent distribution networks is rare and often inadequate when it does exist. Hydraulic conditions in intermittent networks are much more variable than in continuous networks, since pipes may be full or empty during intermittent supply; but agencies managing intermittent networks often do not have SCADA (Supervisory Control And Data Acquisition) or other similar sensor systems to monitor their networks [20]. Even if SCADA is available, monitoring equipment is usually not installed at enough locations to provide a complete picture of complex intermittent networks. In a network studied in Hubli-Dharwad, India, >800 valves were operated during a complete supply cycle, and the state of the system was communicated between employees via phone calls and field visits. Some operating data were recorded in written logbooks, but with varying levels of accuracy [20].

In a previous article, we examined the water quality impacts of intermittent supply in four study zones (one with continuous supply and three with intermittent supply) in Arraiján, Panama [17]. Despite sustained low and negative pressures and water quality sometimes being degraded during the first-flush period (when supply first returned after an outage), random grab samples consistently had good quality. These results contrasted with results reported from a previous study in India, where water quality in intermittent zones was highly degraded [21], indicating that water quality may vary greatly among intermittent systems depending on the context.

This paper seeks to build on the previous publication to characterize in the same Arraiján, Panama distribution system: (i) the detailed supply schedule experienced by users in the four different supply zones over a 1-year period; (ii) the infrastructural and operational challenges that contributed to irregular and unreliable supply; and (iii) the occurrence of pipe breaks throughout the network over a 3-year period and their relationship to IWS. While previous studies of IWS networks have described supply schedules and inequities based on general observations or surveys of users [21–23], this study provides a much more detailed characterization based on a full year of continuous flow and pressure monitoring in four different study zones. This study also provides novel insights by linking the measured supply patterns to specific operational events and challenges, observed through extensive informal interactions with and interviews of system operators.

For improved administration and operation of complex sectors of distribution networks like the Arraiján study zones, it is important to understand the reality of their current operation. To this end, we characterized supply patterns in the study zones based on their continuity (i.e., the portion of time that water was supplied) and their regularity (i.e., the extent to which the supply schedule followed a consistent pattern). Based on insights from this detailed year-long picture of supply in the four study zones, we identify opportunities to improve service quality in networks facing similar challenges. It should be noted that at the time of publication, many of the supply deficiencies identified herein have not been resolved, despite recent investments to increase conveyance, pumping and storage capacity in Arraiján to address intermittent supply. We hope that the results of this study will help utility managers, policy-makers, infrastructure finance institutions, and researchers to better understand the technical nature, impacts, and causes of IWS with the aim of developing solutions to make supply more continuous, regular and reliable.

2. Methods

Data collection for this study was conducted in Arraiján, Panamá from October 2013 to August 2015. Arraiján's drinking water network offered a variety of supply situations to study. Most customers received nearly continuous supply, but others were faced with a range of intermittent supply situations, varying in severity and in how they were controlled. Four study areas were selected in the Arraiján distribution network, each with a different supply situation (one nearly continuous and three intermittent). Supply was monitored in these study zones by a variety of methods, including pressure and flow sensors, field visits, and informal interviews with operators. Pressure and flow sensors provided a more detailed, accurate, and objective characterization of supply than would be obtained by interviewing operators or users. Continuous monitoring over 1 year allowed supply variability and anomalies that may not occur frequently to be captured. Informal interviews with operators allowed us to connect the supply observed to the reality of operating a complex intermittent system. In addition to an in-depth study of the four study zones, pipe break records were reviewed for the entire Arraiján network.

2.1. *Arraiján's Drinking Water System*

Arraiján grew rapidly in recent decades, from 60,000 inhabitants in 1990 to an estimated 263,000 in 2014 [24,25]. This growth varied in terms of formality and the degree of planning involved (Figure 1). The rapid pace of residential expansion made it difficult for IDAAN to expand the distribution system accordingly. Frequent changes in the Arraiján system's management structure (between being managed by IDAAN's central office and a regional office) and difficulties with the implementation of infrastructure improvements also affected IDAAN's ability to respond to the challenges of rapid growth.

Summary statistics on Arraiján's complex water distribution network are provided in Table 1. Despite the ample quantity of water entering the system from the treatment plants, rates of water loss were high and many utility customers in Arraiján received deficient service. The entire system was vulnerable to temporary outages due to pipe breaks and treatment plant stoppages. Some areas had chronic supply deficiencies caused by: (1) insufficient local distribution capacity (pipe diameter, storage capacity or pump capacity) to supply the water demand in the area; or (2) drawing supply from parts of the network that frequently lost pressure when the capacity of the entire network was surpassed because of high user demand (for example, a Sunday when many users were at home) or operational failures such as pipe breaks. In addition to those with deficient service, some users did not receive piped water at all and were supplied by tanker trucks contracted by the utility.

Figure 1. Examples of recent residential growth in Arraiján.

Table 1. Arraiján Distribution System Vital Statistics (from [26] and interviews with IDAAN personnel).

Supply:	• 3 surface water treatment plants supplied by water from the Panama Canal watershed
Pipe:	• 431 km PVC, 10-inch or smaller diameter • 73 km ductile iron, 12-inch or larger diameter • Small quantities of cast iron and asbestos-cement • Over half of the network <25 years old. Some portions >35 years old
Pump Stations:	• 27, with approximately 3 to 300 horsepower capacity
Storage Tanks:	• 39 distribution storage tanks, 38,000–5.7 million liter (ML) capacity • 5.7 ML capacity at water treatment plants • 33 ML total capacity; but 12.3 ML out of service • 20.7 ML available storage = 13.4% of 154 ML daily production *
Water balance:	• 154 ML daily production = 585 L/capita (2014) • 310 L/capita not billed to customers = 53% non-revenue water
Service quality:	• 6420 connections (13%) received a monthly discount in 2014 due to deficient service • Many more clients suffered occasional interruptions due to pipe breaks or treatment plant stoppages • According to a 2010 survey: 443 households served by tanker trucks because they do not receive piped supply (the number was likely much higher at time of this research, given that 10 trucks were distributing water fulltime)
Pipe breaks:	• 604 breaks in pipes ≥ 2-inch diameter in 2014; equivalent to 1.46 breaks/km-year • Similar break rate to the average for 13 Latin American utilities in a regional benchmarking report [27] • Much higher break rate than 0.13–17 breaks/km-year reported in studies of U.S. and Canadian utilities [28]

* Available storage did not meet IDAAN's standard that storage capacity should equal one-third of daily demand.

2.2. Study Zones and Monitoring Locations

The four study zones were selected to be as large as possible, while still maintaining a supply regime with similar characteristics within each zone. The locations of the four study zones in Arraiján are shown in the Supporting Information (Figure S1). The four study zones were located in Burunga, Loma Cova, and Arraiján Cabecera, contiguous sectors within the Arraiján network. These zones were supplied by two of the three treatment plants serving Arraiján, referred to here as WTP A and WTP B. Although the neighborhoods and housing developments in these three sectors varied in terms of urban development and water supply, many shared common characteristics that influenced water supply:

- Complex topography created a need for pump stations.
- The ubiquity of informal housing settlements and unplanned and older (more than 30 years old) developments contributed to the complexity of the water network and often to a lack of data about its configuration.

The four zones are detailed in Table 2. A schematic of Zone 2 is shown in Figure 2 as an example, and schematics of the other study zones are included in the Supporting Information (Figures S2, S3 and S4).

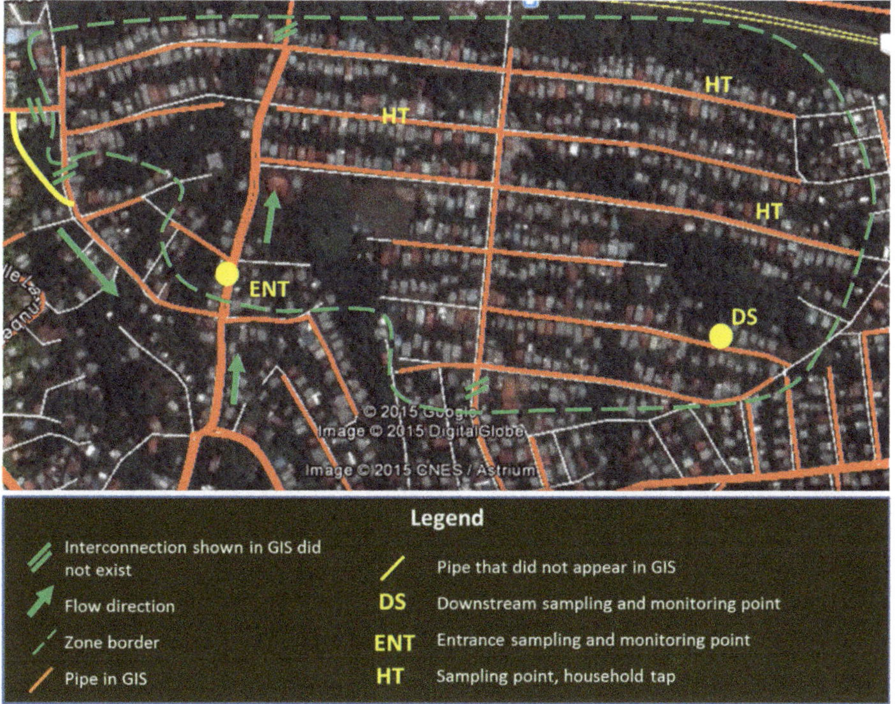

Figure 2. Schematic of Study Zone 2. The entrance (ENT) and downstream (DS) locations at which continuous monitoring stations were installed are indicated, as well as the household taps at which water quality grab samples were collected. (Source of satellite images and study zone schematic: Google Earth and IDAAN's GIS database).

Table 2. Summary of study zones.

Zone (Supply Type)	Approx. No. of Customer Connections	Water Source	Supply
1 (continuous)	348	WTP A	Supplied by the main transmission pipe from WTP A through two entrance locations (ENT 1 and ENT 2). Continuous supply except for eleven outages during the year of monitoring and several houses at high elevation.
2 (tank-fed)	650	WTP A & WTP B	Received most supply by gravity from two 3.8-million-liter storage tanks and some supply from the main transmission pipe. High elevations lost supply when the storage tanks drained, which was most common during afternoon hours and on weekends.
3 (valve-controlled)	232	WTP B	Supplied by a local pump station that supplied water to Zone 3 and two other nearby sectors. Operation schedule called for supplying Zone 3 for three days and then closing a control valve for three days to stop supply to Zone 3 and fill a tank supplying an adjacent area. However, supply often deviated from the schedule due to irregular valve operation, pipe breaks, and pump station failures.
4 (pumped)	368	WTP A & WTP B	Most of the supply was from a local pump station pumping directly to the zone's local network. The pump station stopped frequently due to insufficient supply or power failures, causing most of the zone to lose supply. A small amount of supply also entered the zone through two other small diameter pipes, enabling some parts of Zone 4 to have supply even when the main pump station was off.

2.3. Continuous Pressure and Flow Monitoring

Continuous pressure and flow monitoring at the entrance(s) and a downstream location in each study zone allowed detection of when the supply was on and off. Pressure was monitored with ECO-3 RTUs (remote telemetry units, AQUAS Inc., Taipei, Taiwan). The pressure monitors normally recorded a measurement every 30 seconds (s). They also were programmed to record measurements more frequently when a pressure transient was detected. In addition to measuring pressure, the RTUs received signals from other sensors and sent the data periodically to an internet server. The RTUs also had the capacity to send text messages when pressure or other parameters went out of a programmed range. At the Zone 3 entrance monitoring point, pressure was monitored by an LPR-31i pressure monitor (Telog Instruments Inc., Victor, NY, USA). Data were downloaded from that sensor each week to a laptop computer.

IP80 Paddle-wheel insertion flow meters (Seametrics Inc., Kent, WA, USA) were installed at the entrance(s) to each zone. These sent an electrical pulse signal to the RTUs. Some stations were also equipped with turbidity and free chlorine sensors for a related water quality study (results reported in [17]).

Monitoring equipment was installed in above-ground metal boxes (Figure 3). Each set of equipment was powered by a 12-volt battery charged with a solar panel installed on the top of the box. Each monitoring station was connected to the distribution pipe via a saddle installed on the pipe, a $\frac{1}{2}''$ PVC pipe, and a 3/8″ PVC hose.

Pressure data were smoothed (running average of five nearest data points) before analysis. Zones 1–3 were considered to be without supply when pressure (at ground level) was <2 psi at the downstream monitoring station. Zone 4 (pump-controlled) was considered to be without supply when the Zone 4 pump station was stopped (with both pumps off) because the Zone 4 downstream monitoring station received supply from interconnections with adjacent areas of the network and often had supply even though the pump station was off and much of Zone 4 was without supply. Outages with less than 10 minutes (min) of supply between them were grouped together and considered single outages for analysis purposes, but reported durations only include the time when water was actually off. Outage groups with total duration < 10 min are not reported.

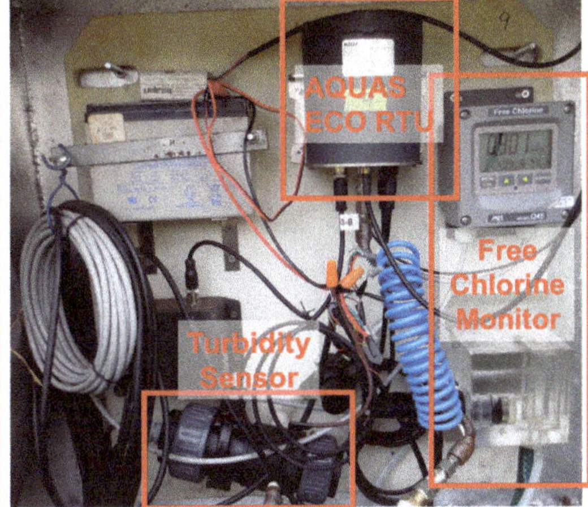

Figure 3. Continuous monitoring station (**top**) and sensors (**bottom**).

2.4. *Qualitative Observation of Water Supply Schedule*

To complement the data from continuous monitoring stations, observations were made at three household taps in each zone whenever grab samples for water quality analysis were collected. This sampling provided the opportunity to observe supply conditions in parts of the study zones that did not have continuous monitoring stations. Each time a household tap sample was collected, the researcher asked the user whether they had experienced any supply interruptions in the last week and, if so, when the last interruption had ended. Sample households were distributed geographically throughout each zone, so qualitative observation of supply in these locations during the approximately 500 household sampling visits enabled a more complete picture of supply in the zones.

This portion of the research was carried out under Protocol 2012-04-4278, approved by the UC Berkeley Committee for Protection of Human Subjects on 17 June 2013.

2.5. *Qualitative Observations of Network Operation*

Hundreds of hours were spent informally observing and interacting with Arraiján system operators, which offered an up-close view of the operation of the network. Those informal observations, when coupled with continuous monitoring data, provided insight into the challenges of operating a complex intermittent distribution network. When hydraulic events of interest were captured by continuous monitoring, system operators were interviewed informally to better understand what had occurred. Also, when operators mentioned problems in the study zones, the relevant continuous monitoring data were reviewed. This back and forth between operators' observations and hydraulic monitoring data also permitted an assessment of whether and how such hydraulic monitoring might be useful to operators.

2.6. Pipe-Break Analysis

Pipe break repair records for the entire Arraiján network during 2012–2014 were analyzed to compare break rates in different parts of the network. Records were analyzed to identify areas of the network with particularly high break rates and assess whether there was an association between the frequency of pipe breaks and intermittent supply. Although these records represented repairs instead of breaks, for this analysis each repair is referred to as a break. Based on the location written on the form filled out by the repair crew, each break was assigned to a zone (a neighborhood or housing development). The length of the pipe in each zone was calculated using the utility's GIS database. Pipes with <2-inch or >12-inch diameter were excluded from the analysis. (The small-diameter pipes were not in the GIS database, so could not be included. The large-diameter pipes were normally transmission pipes and the pressure regime in those pipes was often not related to the supply regime in the zones they passed through.) Some zones of Arraiján for which pipe information was not available in the GIS database were excluded from the analysis.

To categorize supply continuity and the approximate age of the pipes in each zone, the utility's field supervisor, who had more than 30 years of experience working for the utility in Arraiján, was consulted. Pipe break data were analyzed in R [29]. Statistical tests for independence were done with permutation tests using the coin package [30] because these do not require assumptions regarding the distribution of the data. The threshold for statistical significance was $p < 0.05$.

3. Results and Discussion

3.1. Heterogeneous and Irregular Supply in Study Zones

Results from approximately one year of pressure monitoring revealed that supply continuity varied widely between the study zones and that supply in the intermittent study zones was often highly irregular.

3.1.1. Supply Schedule in Study Zones

Supply statistics for each study zone, based on continuous monitoring data, are summarized in Table 3. Zone 3 (valve controlled) was without water for the largest fraction of monitoring time (43%), as would be expected given the utility's plan for supply to be on for 3 days and then off for 3 days, followed by Zone 2 (17%), Zone 4 (13%), and Zone 1 (0.9%).

Table 3. Summary statistics of supply in each zone. Supply was considered to be off in Zones 1–3 when the pressure at the downstream station was <2 psi at ground level. Zone 4 was considered to be without supply when the pump station serving the zone was off. Average pressure for all zones is at the downstream monitoring station.

Study Zone	Zone 1 (Continuous)	Zone 2 (Tank-fed)	Zone 3 (Valve)	Zone 4 (Pumped)
Monitoring time (days)	350	318	317	349
Time without supply (days)	3.2	54.5	137	47
Fraction of monitoring time without supply	0.9%	17%	43%	13%
Average pressure when there was supply (psi)	22	38	36	47
Number of supply outages	11	107	114	336

The distributions of outage durations at each of the downstream monitoring stations are shown in Figure 4. The tank-fed zone (Zone 2) had 107 outages lasting up to 3 days. A typical outage began during the afternoon when the upstream storage tanks serving Zone 2 drained, and ended around midnight once the level in the tanks recovered. Longer interruptions occurred when the upstream storage tanks were without water for longer because of a supply deficit in the overall network caused by a pipe break or pump or treatment plant shutdown.

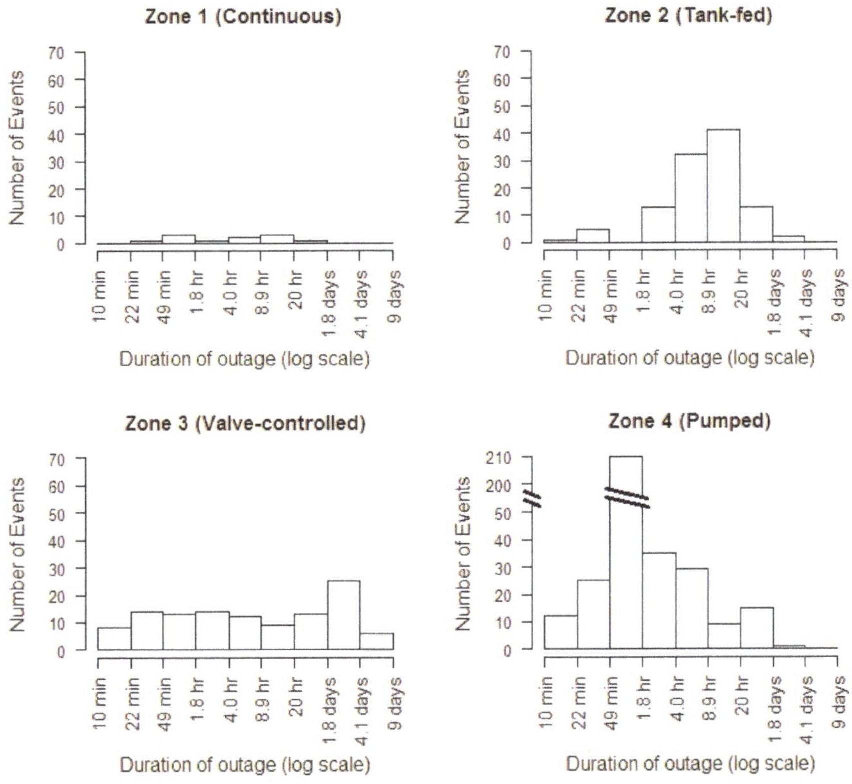

Figure 4. Distribution of outage durations for each study zone. Note that the x-axis is a log scale with nine evenly spaced bins between 10 min and 9 days; the y-axis for the Zone 4 graph is cut.

According to the operation plan for the valve-controlled zone (Zone 3), outages there should have lasted 3 days. While the most common outage length was about three days (22% of outages were between 1.8 and 4.1 days long), many measured interruptions were much longer or shorter. Eight interruptions of >4 days occurred. The three longest, lasting 6.3, 6.6, and 8.2 days, were associated with breaks in the single 4-inch pipe conveying water to Zone 3. Many shorter interruptions occurred when the Zone 3 pump station stopped temporarily during a supply period or the valve at the entrance was left partially open and supply at the downstream monitoring station fluctuated between off and on depending on demand elsewhere in Zone 3. (Operators sometimes left the valve at the entrance to Zone 3 partially open when it was scheduled to be closed so some flow entered Zone 3 and the rest went to a storage tank in an adjacent sector. This strategy was employed to supply the adjacent sector but avoid overflowing the storage tank there. During these times when the valve was partially open, the lower elevation portions of Zone 3 had supply and the higher portions did not, with the boundary between households with supply and without supply moving as demand fluctuated. Supply at the downstream monitoring station fluctuated between on and off during these times as the supply boundary moved back and forth across the monitoring location.)

Outage durations in the pumped zone (Zone 4) varied widely, since the pump station was not run according to a schedule, and stopped whenever the suction tank that it pumped from emptied or the electricity supply was interrupted. Seventy percent of the 336 outages, representing 24% of the time that the pump station was off, lasted between 30 and 120 min, likely because this was the approximate length of time it took for the suction tank to fill from the level at which the pump shut off to the level at which it turned back on. Ten outages lasted more than 24 h, with the longest one lasting 48 h. Between 29 April and 7 May 2015, the Zone 4 pump station stopped daily at 8:50 p.m. and started again at 4:52 a.m., causing the incident described in the introduction. During that time, one of the pumps was damaged without operators being aware of it and the other was programmed to stop during the night.

Data were also analyzed to determine whether outages were more common at certain times, days, or seasons. In Figure 5, the percentage of time during each hour of the day and each day of the week that each zone was without supply is shown. A high percentage means that the zone was without water more frequently during that hour of the day or day of the week. Service continuity did not vary noticeably by the hour of the day or day of the week in Zones 1 and 3. In Zone 1, the percentage of time that water was off was low at all times. In Zone 3, the valve was typically open or closed for several days at a time, which meant that there were no specific hours or days during which the zone was without service. On the other hand, supply in Zones 2 and 4 varied noticeably by the hour of the day and Zone 2 supply also appeared to vary by day of the week. In Zone 2, supply was off more often between 3 p.m. and 10 p.m., and on weekends, when it was more likely for the upstream storage tanks supplying the zone to be empty after being depleted by higher daytime demand. In Zone 4, supply was off more often between 8 a.m. and 5 p.m., probably because during those hours demand was higher in other parts of the network, which reduced system pressure and reduced supply to the Zone 4 pump station.

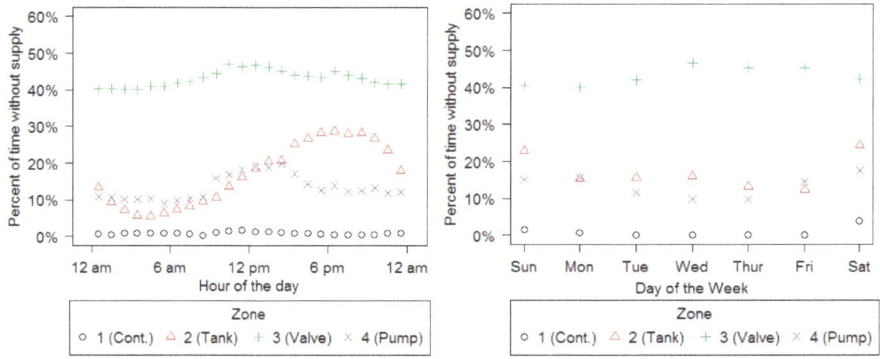

Figure 5. Percent of time each study zone was without supply by the hour of the day and day of the week.

In Figure 6, the variation in the percent of time that each zone was without supply throughout the year is shown. On some occasions, supply problems occurred in one zone and the other zones were unaffected. For example, at the end of October and beginning of November, Zone 3 was affected by two long outages (8.2 and 6.4 days) associated with breaks in the 4-inch pipe supplying that zone, but supply remained typical in the other zones. On other occasions, large-scale supply problems affected all four zones at once. For example, during the beginning of September and the end of January, all four zones were affected by breaks in the 24-inch transmission pipe from WTP A.

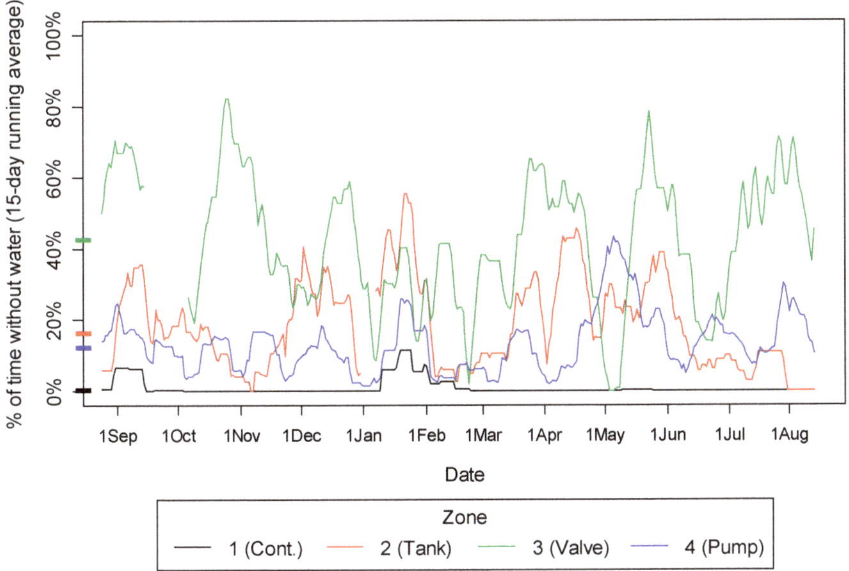

Figure 6. Fifteen-day running average of the percent of time that supply was off in each zone. For each day, the average percent of time water was off during the 15 nearest days is shown. Values are only shown if at least 7 days (out of 15) of data were available. Colored dashes on the y-axis mark the average percent of time water was off in each zone.

3.1.2. Variable Service Quality across the Arraiján Network

As can be seen from the data presented above, there were marked disparities in service quality across the four zones studied within the same distribution network. Supply ranged from nearly continuous in Zone 1 to very intermittent and prone to extended outages in Zone 3. It should be noted that the four study zones were not chosen to capture the full range of service quality found in the Arraiján network, but rather to focus on the different intermittent supply regimes found in the network. Much of the network had supply equally or more continuous than Zone 1, and, as mentioned in Section 2.1, portions of Arraiján received no piped supply at all and were supplied by tanker trucks. Thus, disparities across Arraiján as a whole were even wider than the disparities seen across the study zones.

While Section 3.1.1 describes approximate supply conditions in each zone based on conditions at the downstream monitoring station (Zones 1–3) or on whether the pump station serving the zone was on (Zone 4), supply also varied geographically within each zone based on elevation and distance from the zone's entrance. For instance, walking 50 yards up a hill in Zone 1 could take you from a house where supply rarely went out to a house where supply went out most afternoons. Spatial variation in supply continuity resulted in service quality being unequal between neighborhoods and even between neighbors.

Previous research has identified uneven service quality as a common problem with IWS [3,11,12] and provided evidence of such inequities in IWS networks [22]. Our continuous monitoring data from Arraiján supports those findings and provides a more detailed picture of uneven service quality.

3.1.3. Irregular and Unreliable Supply

Galaitsi et al. [15] proposed classifying IWS into Predictable Intermittency ("shut-offs that occur generally according to a predictable and anticipated schedule"), Irregular Intermittency ("supply arriving at unknown intervals within short time periods of no more than a few days"), and Unreliable

Intermittency ("uncertain delivery time and risk of insufficient water quantity, often exacerbated by limited storage and long periods of non-delivery").

Pressure data demonstrated that supply in all four Arraiján study zones had elements of predictability, irregularity, and unreliability, depending on the time scale considered. While supply in each zone cannot be strictly classified into one of the categories proposed by Galaitsi et al. [15], the categories are still useful for describing supply in each zone. Zone 1 supply was normally predictably continuous but did include a few unanticipated outages lasting up to 22 h. For a user accustomed to continuous supply and unlikely to have a large volume of water stored, an outage of 22 h may be perceived as unreliable. Supply in Zone 3 was intended to follow a predictable three-day-on, three-day-off schedule, but in practice, this zone often had the most unreliable supply, with unexpected extended outages. Zones 2 and 4 normally had irregular supply, with typical outages being short enough to be manageable for users. However, they also experienced bouts of unreliability when infrastructure failures caused more extended outages. As mentioned in Section 3.1.2, supply also varied within zones, and some users experienced a much more irregular and unreliable supply than that registered by our pressure monitoring.

Unreliable supply is an inconvenience and hardship for users. Burt and Ray [31] argue that customers' satisfaction with water supply is affected by quantity, quality, convenience, and reliability. Reliability is a particular concern in intermittent systems since supply often comes at irregular times and is unpredictable. In an intermittent system in Hubli-Dharwad, India, users who had never experienced continuous supply placed more value on punctual supply, increased frequency and duration of supply, and water quality than they did on receiving continuous supply [32]. Unpredictable supply can disproportionately affect lower-income households if they are less able to mitigate its effects. For example, in their study of intermittent supply in Hubli-Dharwad, India, Kumpel et al. [33] found that households that did not have rooftop tanks and used a public tap or a neighbor's tap shared with more than 2 households on average used only approximately 20 L per capita per day, a quantity likely insufficient for household uses such as bathing and laundry [34]. As will be discussed further in Section 3.4, irregular and unpredictable supply conditions also made the operation of the Arraiján network difficult for the utility.

3.2. Pipe-break Analysis

The average pipe break rate across the 142 zones analyzed was 1.42 breaks/km-year. Some breaks were fractioned between multiple zones when the recorded location was not specific enough to know in which of the zones the break occurred. Also, some breaks for which the diameter of the pipe was unknown (and may have been <2 inches and thus should have been excluded) were discounted by 39%, the portion of all breaks that occurred in pipes with diameter < 2 inches.

Break rates varied widely by zone (Figure 7a). In 54 zones there were no recorded pipe breaks during the 3 years. Twenty-three percent of all of the breaks occurred in just ten zones (see Supporting Information Table S1), even though they only had 2.9% of the pipe length. Some small zones (such as the zones ranked 1st and 7th) may have had artificially high pipe break rates either because some pipes in these zones were not registered in the GIS database (such that the total pipe length is actually longer than the value used for analysis) or because some breaks in nearby zones were classified as within these zones. The second-ranked zone included the 6-inch transmission pipe between the Zone 4 pump station and Zone 4. Thirty-nine of the 51 breaks in that zone were in the 6-inch pipe going to Zone 4. Zone 4 itself does not appear in the top-ten list but also had 20 breaks registered on the same 6-inch pipe. The frequent breaks in that pipe are likely due to high transient pressures from the intermittent pumping (see Supporting Information Figure S5). The zone with the third-highest break rate also had a pump station that stopped frequently and had a known problem with pressure surges. The fifth-ranked zone was Study Zone 3, controlled by intermittent valve operations. Study Zone 1 (continuous) ranked 45th, with a break rate of 1.08 breaks per km per year, and Study Zone 2 (tank-fed) ranked 29th, with a rate of 2.16.

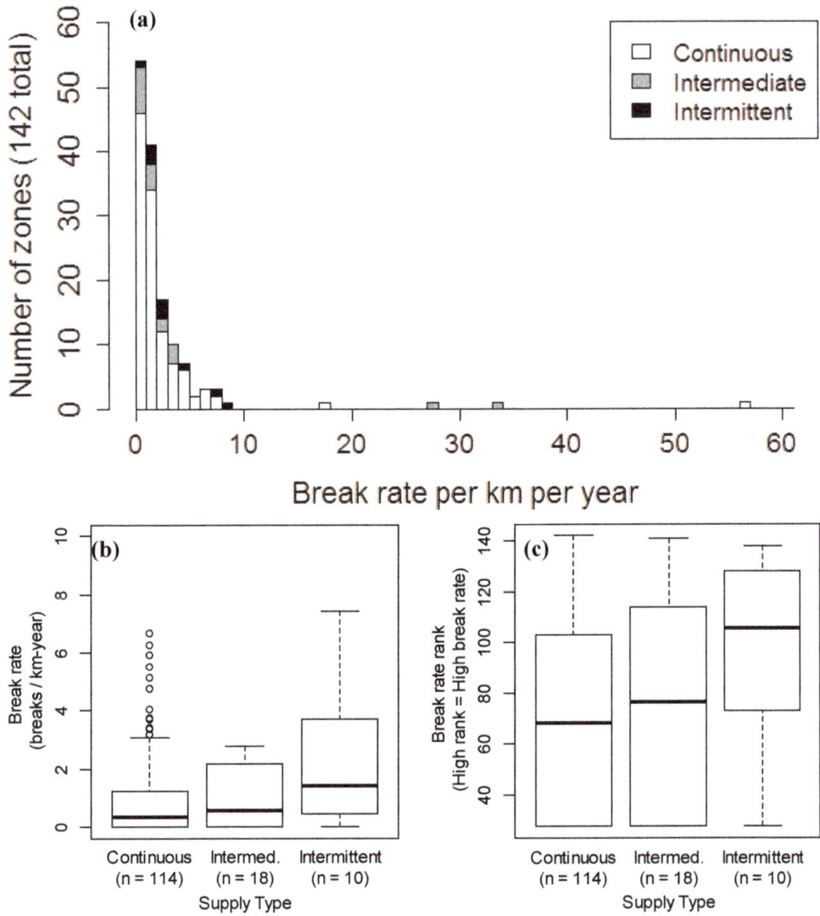

Figure 7. Analysis of 2012-14 pipe break data: (**a**) Distribution of break rates for each zone, (**b**) boxplots of break rate by supply type, and (**c**) and break-rate rank by supply type. In plot (**a**), the first bar (on the far left) represents the 54 zones with no breaks. Four zones with break rates > 10 breaks/km-year are excluded from the plot (**b**). Note that the minimum rank in (**c**) is 27.5 because 54 zones had no breaks and were each assigned an average rank.

The high fraction of breaks occurring in a small number of zones, some of them with known sources of transient pressures, suggests that by investigating and improving pressure conditions in a few key parts of the Arraiján network, the utility could drastically reduce the frequency of breaks and the associated supply interruptions and repair costs.

The potential relationship between the break rate and supply continuity was investigated by classifying supply into three categories. "Continuous" supply meant that the zone only lost supply when a large portion of Arraiján was without water; "Intermittent" supply meant that the zone regularly was without water, and; "Intermediate" supply meant that the zone normally had continuous supply but was vulnerable to losing supply when pressure was low in the main network. Break rates and break rate ranks are compared by supply type in Figure 7b,c. In one-way permutation tests for independence (two-tailed tests, classifying supply and age categories as ordered factors, and using break-rate ranks), higher break rates were significantly associated with more intermittent supply ($p = 0.042$) and higher pipe age ($p = 0.030$). In a permutation test for independence where both supply and pipe age categories were considered simultaneously, the association almost met the threshold for significance ($p = 0.058$).

While the permutation test showed a marginally significant association between more intermittent supply and high break rates, that association might have been driven mainly by a few zones affected by intermittent pumping and not indicate a general effect of IWS. As seen in Figure 7a, some intermittent zones had very low break rates, and some continuous zones had high break rates. Also, in some cases, other factors associated with IWS may be the actual cause of high break rates instead of the hydraulics of intermittent supply. For example, in Study Zone 3, many breaks occurred in a location were the pipe conveying water to the zone was suspended in the air to cross a stream, an installation constructed by the local residents. Intermittent areas, due to their often informal development, may tend to have more contributing factors that lead to pipe breaks.

Although the data did not show a clear-cut correlation between IWS and water main breaks, IWS could have been more strongly associated with service line breaks and leaks, which were not analyzed. A study in Cyprus [35] found that implementation of IWS caused a significant increase in the vulnerability to failure for household service lines but not for larger water mains.

3.3. Operational Challenges Contribute to Unreliable Supply

Examples from the study zones illustrate how a lack of information on the current hydraulic state of the distribution system delayed the detection of infrastructure failures and made supply less reliable. In one case, the capacity of a different pump station near the Zone 4 pump station was increased, which reduced flow to the Zone 4 station and reduced supply to Zone 4. The utility did not anticipate these effects, and changes were only made to resolve the situation after Zone 4 residents blocked a lane of Panama's largest highway to protest the decline in service quality (the first highway closure mentioned in the introduction). A second situation in Zone 4, where undetected pump station malfunctions led to poor service quality, user dissatisfaction, and eventually another highway lane closure in protest, was described in the introduction.

These problems with the Zone 4 pump station were all apparent when continuous pressure and flow data collected at the pump station discharge as part of this study were reviewed afterward. If operators had access to such data and monitored it routinely (or set up relevant alarms to alert them of problems), situations like these might be avoided.

In the valve-controlled zone (Zone 3), inconsistent operation and the inability to monitor supply conditions caused actual supply to deviate substantially from the utility's schedule of 3 days on and 3 days off. The control valve was sometimes not operated according to schedule, because operators were not available to open or close it due to another crisis or commitment, or because weekend operators were unaware of the intended schedule. Delay in detecting and repairing breaks in the 4-inch pipe supplying Zone 3 caused the three longest outages. Zone 3 valve operations and pipe breaks were also visible in continuous monitoring data. Such data could help the utility operate the valve more consistently and respond faster to pipe breaks. As one example of the latter, continuous pressure and flow data from the Zone 3 entrance during a pipe break are shown in Figure 8. Flow increased and pressure decreased to a negative value at the time of the break. Approximately 8 h later, the flow stopped and pressure increased to zero when an operator closed the control valve located just upstream of the entrance monitoring station.

The incidents described above from the study zones are indicative of operators' general inability to monitor the network and detect infrastructure failures, such as breaks in distribution pipes or pump malfunctions, before users had already been severely affected. Even though Arraiján's distribution network was quite complex, the utility operated it with little information about its current state. Some of the 27 pump stations frequently malfunctioned. Apart from the monitoring equipment installed for this project, only one of the pump stations could be monitored by telemetry. To monitor the others, operators had to do daily field inspections driving around in a truck.

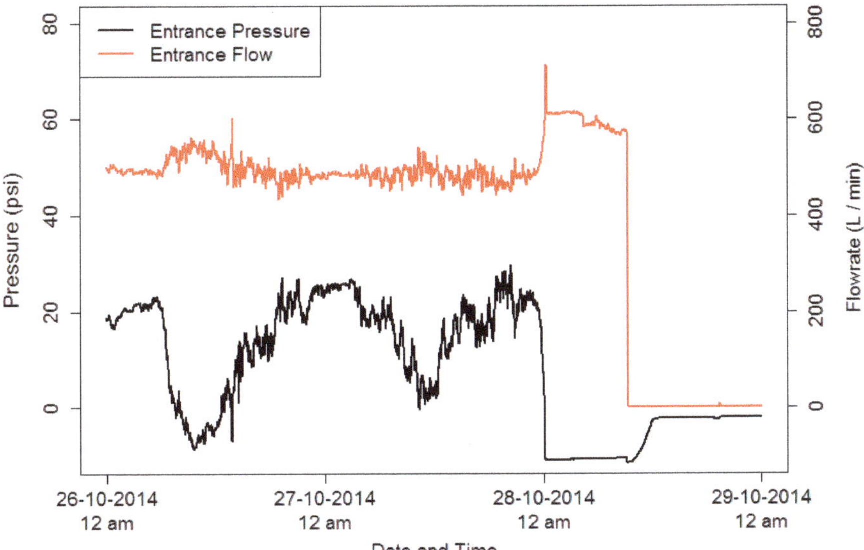

Figure 8. Continuous pressure and flow monitoring data from the Zone 3 (valve-controlled) entrance monitoring station during a break in the 4-inch pipe supplying Zone 3 (downstream of the entrance) at approximately 12 a.m. 28 October 2014.

While the localized failures discussed above caused many of the unplanned outages observed in the study zones, other outages observed in the study zones were the result of system-wide deficiencies. Although the Arraiján system had more than enough supply capacity, high rates of water loss, a lack distribution storage capacity, and limited ability to sectorize and control the system prevented the utility from being able to cope with temporary operational failures such as treatment plant stoppages, pump shutdowns, and pipe breaks. According to a log kept by utility managers, between August 2014 and July 2015, users in a large portion of Arraiján were without supply on 13 occasions (eight unexpected and five planned), due to such temporary operational problems.

Limited information on pipe connectivity and a lack of control valves often made these failures difficult to resolve and prolonged their effects. For example, incomplete maps of the pipe network often made it difficult for operators to determine the cause of an outage or to shut off a sector of the system to repair a pipe break. Repair crews lost time trying to determine how to depressurize the sector where a break was, and, due to a lack of control valves, sometimes had to cut off service to a large portion of the network in order to depressurize the area near a break.

3.4. *Opportunities to Improve Service Quality*

Observations from this study suggest that utilities could improve service in intermittent networks like Arraiján's by making both localized and system-wide operational improvements. Locally, intermittent supply could be made more consistent and reliable by using pressure and flow monitoring routinely and/or as a diagnostic tool. System-wide, reducing water losses, providing adequate storage capacity, and improving monitoring could make supply more continuous and reliable.

3.4.1. Local Monitoring for More Reliable and Transparent Intermittent Supply

Continuous hydraulic monitoring in sections of the distribution system with intermittent supply can be used as a diagnostic and tracking tool. As demonstrated in Section 3.1, such monitoring can identify areas where service is most deficient or unreliable. Operators, managers, and planners can then direct attention, resources, and capital improvements to those areas. Once corrective action has been taken, the same monitoring methods could be used to evaluate whether the measures taken resulted in the intended improvements, such as a more reliable supply schedule. Monitoring results could be made publically available to increase transparency and make users aware that the utility is working to identify and address supply problems.

In addition to being a tool for identifying chronic problems and prioritizing interventions over the medium- and long-terms, hydraulic monitoring can also help operators track and improve supply schedules on a daily basis and identify acute operational failures that require immediate operator attention (see Section 3.3). Equipped with monitoring tools, operators will be able to respond more rapidly to failures in the system and provide a more predictable and reliable service.

We were not the first to install online monitoring stations in Arraiján. As part of a 2010 project, the network was divided into eight District Metering Areas (DMAs), and 15 pressure and flowrate monitoring stations, not altogether different from the ones used in this study, were installed at the boundaries of each DMA [26]. At the time of our research, however, only some of the sensors were working, the original telemetry equipment to upload the data to the internet was not working, and water balances for the DMAs had not recently been calculated.

Several factors could explain why the previously installed sensors were not maintained. A specialized division of the utility, located 20 km away in Panama City, was in charge of maintaining and collecting data from the Arraiján monitoring stations and many others throughout the country but had few logistical and human resources to dedicate to the task. The stations were set up by a private contractor, and the capacity to train utility personnel on how to use and maintain them may have been inadequate. Apart from these resource limitations, utility staff may not have seen sufficient reason to prioritize maintenance and use of the sensors when their resources were already stretched thin by immediate operational problems. With limited personnel available to analyze and use the monitoring data, and local operators not involved in that process or able to use the data as an operational tool, the data's value, and thus the value of maintaining the monitoring stations, may have been seen as low. If new tasks such as information collection are seen as add-ons that are not integral to existing tasks, compliance from frontline workers may be low [36].

Thus, as seen from previous experience in Arraiján, the type of monitoring recommended herein will only be useful if the utility has the human and logistical resources required to maintain sensors, analyze and interpret the data they produce, and take corrective actions based on the data. Promptly detecting a pipe break will be of little value if a repair crew is not available to fix it.

As mentioned in the introduction, hydraulic conditions are often spatially heterogeneous in intermittent systems, and monitoring an entire system like Arraiján's in detail may be costly. Monitoring costs could perhaps be reduced by rotating monitoring equipment around to different problem areas or developing inexpensive monitoring systems that provide only the most critical information required by operators and managers, such as whether the supply is off or on. Also, while the cost of extensive monitoring may seem high when viewed in isolation, it may still be small in comparison to a utility's capital improvements budget or operators' and users' coping costs associated with intermittent supply.

3.4.2. Making Irregular Supply More Predictable

An irregular supply that does not follow a schedule is inherently more unpredictable for users. To mitigate irregularity, users can be notified of variations in the normal schedule. During the time of this study, IDAAN was providing some such notification on a large scale by alerting and updating its customers about outages and repairs via a national Twitter feed. However, that broad approach may not be efficient for notifying customers of schedule changes in small areas like the study zones considered here. Other more targeted notification approaches may have the potential to make supply more predictable even when it is not regular. NextDrop, a start-up company, has attempted to achieve that in intermittent systems in India by notifying customers via text message when supply is about to be turned on by a valve operator, however achieving consistent compliance with the program from valve operators has proved challenging [20,36].

3.4.3. System-wide Strategies for More Continuous and Reliable Supply

As discussed in Section 3.3, some unexpected outages in intermittent supply areas and in areas normally receiving continuous supply were the result of widespread loss of supply or reduction of pressure brought on by operational failures such as treatment plant outages and transmission main breaks. While it will be impossible to completely prevent such operational failures, measures can be taken to reduce their incidence and mitigate their effects when they do occur.

An analysis of system-wide pipe breaks in Arraiján (Section 3.2) indicated that the system's very high break rate is driven mostly by certain areas that make up a small portion of the system. While system operators were very much aware that certain areas and certain pipes were the most prone to breaks, these chronic problems often went unaddressed. Documentation and analysis of pipe break data would likely help operators to keep utility managers and decision-makers better informed, and could motivate engineering and optimization studies to control pressure transients in problematic areas. While we focused on pipe breaks, better tracking and documentation of other operational failures such as treatment plant failures and power outages might help the utility to better understand and address their root causes as well.

It is also important to mitigate the effects of operational failures when they do inevitably occur. A robust distribution system, with sufficient storage capacity and a reserve supply is better able to cope with operational failures. The Arraiján system had plenty of supply but was not robust due to a lack of storage capacity and water loss rates so high that even the existing storage capacity could not be filled reliably. Accurate information on pipe connectivity and the availability of operable control valves to isolate areas where problems occur would also help limit the effects of failures. Reducing water losses, increasing storage capacity, and improving operators' ability to control the system would go a long way in helping the utility to mitigate the effects of operational failures.

3.5. Applicability of Results to Improvement of IWS

Much has been written about the risks and challenges posed by IWS [3,6,12,15,16], and researchers have proposed innovative strategies to model [11,14,18,19,37,38] and optimize the planning and operation of IWS networks [11,12,38–43]. In the literature, these strategies are normally implemented theoretically for example distribution networks. To successfully implement such strategies in practice, it will be critical to consider not just the topology of the IWS networks where they are applied, but also the reality of how such networks are operated by humans and the supply that results under baseline conditions. Unfortunately, very few data have been published that characterize the supply conditions in the wide variety of IWS networks throughout the world.

This detailed account of operation and supply in four sectors of one distribution network in Arraiján, Panama is intended to help fill that void. By selecting four supply zones that each had a different supply situation, we aimed to capture a wide spectrum of IWS. We expect that there are commonalities between these supply zones and those in many other IWS networks, such that some of the findings and recommendations are transferable. Nevertheless, this is one study of one network, and more work is needed to document and better understand the wide variety of IWS situations throughout the world.

4. Conclusions

Continuous hydraulic monitoring in four distinct zones of the Arraiján distribution network revealed that supply continuity varied widely between the zones and also temporally within each zone. The supply schedule was often irregular and unreliable in the intermittent study zones and sometimes interrupted by infrastructure failures that were not corrected for several days. Such unreliability made intermittent supply, already an inconvenient situation, even worse for users, and exacerbated variability in service.

The direct and underlying causes of intermittent supply in each zone were described, and based on this understanding, opportunities are identified to improve supply reliability by addressing operational problems at the local and system-wide scale. Continuous pressure monitoring in intermittent supply sectors would alert operators to unexpected or prolonged supply interruptions, allow supervisors to monitor the supply schedule received by users, and provide data to help managers and planners prioritize infrastructure investments and optimization efforts. On a larger scale, reducing water losses and providing adequate distribution system storage would make the network more robust to operational failures that sometimes cannot be avoided.

The proposed strategies (with the exception of increasing distribution storage) are centered on monitoring, data analysis, and gradual operational improvements rather than large capital infrastructure investments. Such improvements will require investments in the utility's human capital–its staff, and the resources they need to do their jobs effectively.

Supplementary Materials: The following are available online at http://www.mdpi.com/2073-4441/12/8/2143/s1, Figure S1: Location of the study zones and upstream monitoring points, Figure S2: Schematic of Study Zone 1, Figure S3: Schematic of Study Zone 3, Figure S4: Schematic of Study Zone 4, Figure S5: Pressure transient at 3:30 a.m. 18 November 2014 at the discharge of the Zone 4 pump station caused by the startup of the second of two pumps, Table S1: The ten zones with the highest break rates.

Author Contributions: All three authors participated in the conceptualization, methodology design, and funding and resource acquisition for this research. J.J.E. conducted most of the investigation and analysis, with contributions from K.L.N. and Y.C.Q. J.J.E. led the writing of the manuscript with continual input from K.L.N. All authors reviewed and edited article. All authors have read and agreed to the published version of the manuscript.

Funding: This work was funded by the Inter-American Development Bank (IADB, RG-T2441), a USAID Research and Innovation Fellowship, an NSF Graduate Fellowship, the Blum Center for Developing Economies and Henry Wheeler Center for Emerging and Neglected Diseases at UC Berkeley.

Acknowledgments: Panama's National Institute of Aqueducts and Sewers (IDAAN) contributed valuable staff time and in-kind resources for the project. Pipe break data were compiled by Javier Agrazal. We are grateful for assistance from Alejandra Perroni, Stefan Buss and Gustavo Martinez at IADB; Mauro Romero and many others at IDAAN; Carlos I. González; Paul West; and Joshua Kennedy.

Conflicts of Interest: The authors declare no conflict of interest. IADB staff provided input into the methodology development and interpretation of results, but did not participate in the drafting of this article or the decision to publish it.

References

1. IWA Intermittent Water Supply. Available online: http://www.iwa-network.org/task/intermittent-water-supply-iws (accessed on 18 August 2019).
2. Shaheed, A.; Orgill, J.; Montgomery, M.A.; Jeuland, M.A.; Brown, J. Why "improved" water sources are not always safe. *Bull. World Health Organ.* **2014**, *92*, 283–289. [CrossRef]
3. Klingel, P. Technical causes and impacts of intermittent water distribution. *Water Sci. Technol. Water Supply* **2012**, *12*, 504. [CrossRef]
4. Kumpel, E.; Nelson, K.L. Intermittent Water Supply: Prevalence, Practice, and Microbial Water Quality. *Environ. Sci. Technol.* **2016**, *50*, 542–553. [CrossRef]
5. Rosenberg, D.E.; Talozi, S.; Lund, J.R. Intermittent water supplies: Challenges and opportunities for residential water users in Jordan. *Water Int.* **2008**, *33*, 488–504. [CrossRef]
6. Yepes, G.; Ringskog, K.; Sarkar, S. The High Cost of Intermittent Water Supplies. *J. Indian Water Works Assoc.* **2001**, *33*, 167–170.
7. Bivins, A.W.; Sumner, T.; Kumpel, E.; Howard, G.; Cumming, O.; Ross, I.; Nelson, K.; Brown, J. Estimating Infection Risks and the Global Burden of Diarrheal Disease Attributable to Intermittent Water Supply Using QMRA. *Environ. Sci. Technol.* **2017**, *51*, 7542–7551. [CrossRef]
8. McIntosh, A.C. *Asian Water Supplies: Reaching the Urban Poor: A Guide and Sourcebook on Urban Water Supplies in Asia for Governments, Utilities, Consultants, Development Agencies, and Nongovernment Organizations*; Asian Development Bank: Manila, Philippines; International Water Association: London, UK, 2003.
9. Lee, E.J.; Schwab, K.J. Deficiencies in drinking water distribution systems in developing countries. *J. Water Health* **2005**, *3*, 109–127. [CrossRef]

10. Cook, J.; Kimuyu, P.; Whittington, D. The costs of coping with poor water supply in rural Kenya. *Water Resour. Res.* **2016**, *52*, 841–859. [CrossRef]
11. Fontanazza, C.M.; Freni, G.; La Loggia, G. *Analysis of Intermittent Supply Systems in Water Scarcity Conditions and Evaluation of the Resource Distribution Equity Indices*; WIT Press: Southampton, UK, 2007; Volume I, pp. 635–644.
12. Vairavamoorthy, K.; Gorantiwar, S.; Mohan, S. Intermittent Water Supply under Water Scarcity Situations. *Water Int.* **2007**, *32*, 121–132. [CrossRef]
13. Christodoulou, S.; Agathokleous, A. A study on the effects of intermittent water supply on the vulnerability of urban water distribution networks. *Water Sci. Technol. Water Supply* **2012**, *12*, 523. [CrossRef]
14. Batish, R. *A New Approach to the Design of Intermittent Water Supply Networks*; ASCE, World Water Congress: Reston, VI, USA, 2003.
15. Galaitsi, S.; Russell, R.; Bishara, A.; Durant, J.; Bogle, J.; Huber-Lee, A. Intermittent Domestic Water Supply: A Critical Review and Analysis of Causal-Consequential Pathways. *Water* **2016**, *8*, 274. [CrossRef]
16. Coelho, S.T.; James, S.; Sunna, N.; Abu Jaish, A.; Chatila, J. Controlling water quality in intermittent supply systems. *Water Supply* **2003**, *3*, 119–125. [CrossRef]
17. Erickson, J.J.; Smith, C.D.; Goodridge, A.; Nelson, K.L. Water quality effects of intermittent water supply in Arraiján, Panama. *Water Res.* **2017**, *114*, 338–350.
18. Ingeduld, P.; Pradhan, A.; Svitak, Z.; Terrai, A. Modelling Intermittent Water Supply Systems with EPANET. In Proceedings of the 8th Annual Water Distribution Systems Analysis Symposium, Cincinnati, OH, USA, 27–30 August 2006.
19. Lieb, A.; Rycroft, C.; Wilkening, J. Optimizing intermittent water supply in urban pipe distribution networks. *SIAM J. Appl. Math.* **2016**, *76*, 1492–1514. [CrossRef]
20. Kumpel, E.; Sridharan, A.; Kote, T.; Olmos, A.; Parikh, T. NextDrop: Using Human Observations to Track Water Distribution. In Proceedings of the USENIX NSDR, Boston, MA, USA, 12–15 June 2012.
21. Kumpel, E.; Nelson, K.L. Comparing microbial water quality in an intermittent and continuous piped water supply. *Water Res.* **2013**, *47*, 5176–5188.
22. Guragai, B.; Takizawa, S.; Hashimoto, T.; Oguma, K. Effects of inequality of supply hours on consumers' coping strategies and perceptions of intermittent water supply in Kathmandu Valley, Nepal. *Sci. Total Environ.* **2017**, *599–600*, 431–441. [CrossRef]
23. Andey, S.P.; Kelkar, P.S. Performance of water distribution systems during intermittent versus continuous water supply. *J. AWWA* **2007**, *99*, 99–106. [CrossRef]
24. National Institute of Census and Statistics. *Panama Cuadro 11: Superficie, Población y Densidad de Población en la República, Según Provincia, Comarca Indígena, Distrito y Corregimiento: Censos de 1990 a 2010*; Contraloría General de la República de Panamá: Panama City, Panama, 2010.
25. National Institute of Census and Statistics. *Panama Cuadro 44: Estimación y proyección de la población del distrito de Arraiján, por corregimiento, según sexo y edad: Años 2010–2020*; Contraloría General de la República de Panamá: Panama City, Panama, 2010.
26. Louis Berger Group. *Fortalecimiento Institucional del IDAAN a Través de Acciones de Optimización Para la Ciudad de Panamá*; Louis Berger Group: Panama City, Panama, 2010.
27. Grupo Regional de Trabajo de Benchmarking de ADERASA. *Informe Anual-2013 (Datos año 2012)*; ADERASA: Asunción, Paraguay, 2013.
28. Grigg, N.S. Data and Analytics Combat Water Main Failures. *J. Am. Water Works Assoc.* **2019**, *111*, 35–41. [CrossRef]
29. R Core Team. *R: A Language and Environment for Statistical Computing*; R Foundation for Statistical Computing: Vienna, Austria, 2012.
30. Hothorn, T.; Hornik, K.; van de Wiel, M.A.; Winell, H.; Zeileis, A. *Package "Coin": Conditional Inference Procedures in a Permutation Test Framework*; R Foundation for Statistical Computing: Vienna, Austria, 2015.
31. Burt, Z.; Ray, I. Storage and Non-Payment: Persistent Informalities within the Formal Water Supply of Hubli-Dharwad, India. *Water Altern.* **2014**, *7*, 106–120.
32. Burt, Z.; VanGordon, M.; Vij, A. Continuous Piped Water or Improved Intermittency? Willingness to Pay for Improved Piped Water Services in Hubli-Dharwad, India. In Proceedings of the 11th Annual Meeting of the International Water Resource Economics Consortium (IWREC), World Bank Headquarters, Washington, DC, USA, 7–9 September 2014.

33. Kumpel, E.; Woelfle-Erskine, C.; Ray, I.; Nelson, K.L. Measuring household consumption and waste in unmetered, intermittent piped water systems. *Water Resour. Res.* **2017**, *53*, 302–315. [CrossRef]
34. Howard, G.; Bartram, J. *Domestic Water Quantity, Service Level and Health*; World Health Organization: Geneva, Switzerland, 2003.
35. Agathokleous, A.; Christodoulou, C.; Christodoulou, S.E. Influence of intermittent water supply operations on the vulnerability of water distribution networks. *J. Hydroinform.* **2017**, *19*, 838–852. [CrossRef]
36. Hyun, C.; Post, A.E.; Ray, I. Frontline worker compliance with transparency reforms: Barriers posed by family and financial responsibilities. *Governance* **2018**, *31*, 65–83. [CrossRef]
37. Taylor, D.D.J.; Slocum, A.H.; Whittle, A.J. Demand Satisfaction as a Framework for Understanding Intermittent Water Supply Systems. *Water Resour. Res.* **2019**, *55*, 5217–5237. [CrossRef]
38. Ilaya-Ayza, A.; Campbell, E.; Pérez-García, R.; Izquierdo, J. Network Capacity Assessment and Increase in Systems with Intermittent Water Supply. *Water* **2016**, *8*, 126. [CrossRef]
39. Ilaya-Ayza, A.E.; Benítez, J.; Izquierdo, J.; Pérez-García, R. Multi-criteria optimization of supply schedules in intermittent water supply systems. *J. Comput. Appl. Math.* **2017**, *309*, 695–703. [CrossRef]
40. Ilaya-Ayza, A.E.; Martins, C.; Campbell, E.; Izquierdo, J. Gradual transition from intermittent to continuous water supply based on multi-criteria optimization for network sector selection. *J. Comput. Appl. Math.* **2018**, *330*, 1016–1029. [CrossRef]
41. Ilaya-Ayza, A.; Martins, C.; Campbell, E.; Izquierdo, J. Implementation of DMAs in Intermittent Water Supply Networks Based on Equity Criteria. *Water* **2017**, *9*, 851. [CrossRef]
42. Zyoud, S.H.; Shaheen, H.; Samhan, S.; Rabi, A.; Al-Wadi, F.; Fuchs-Hanusch, D. Utilizing analytic hierarchy process (AHP) for decision making in water loss management of intermittent water supply systems. *J. Water Sanit. Hyg. Dev.* **2016**, *6*, 534–546. [CrossRef]
43. Nyahora, P.P.; Babel, M.S.; Ferras, D.; Emen, A. Multi-objective optimization for improving equity and reliability in intermittent water supply systems. *Water Supply* **2020**. [CrossRef]

Review

Investigation of the Current Situation and Prospects for the Development of Rainwater Harvesting as a Tool to Confront Water Scarcity Worldwide

Stavros Yannopoulos [1,*], Ioanna Giannopoulou [2] and Mina Kaiafa-Saropoulou [3]

1 Faculty of Engineering, School of Rural and Surveying Engineering, Aristotle University of Thessaloniki, GR-54124 Thessaloniki, Greece
2 Alamanas 29A, Kalamaria, GR-55132 Thessaloniki, Greece; igiannop@gmail.com
3 School of Architecture, Aristotle University, GR-54124 Thessaloniki, Greece; minakasar@gmail.com
* Correspondence: giann@vergina.eng.auth.gr; Tel.: +30-2310-996-114

Received: 15 September 2019; Accepted: 15 October 2019; Published: 18 October 2019

Abstract: Nowadays, available water resources face severe pressures due to demographic, economic, social causes, environmental degradation, climate change, and technological changes on a global scale. It is well known that rainwater harvesting, a simple and old method, has the potential to supplement surface and groundwater resources in areas that have inadequate water supply. In recent decades, many countries have supported the updated implementation of such a practice to confront the water demand increase and to reduce the frequency, peak, and volume of urban runoff. These considerations motivate interest in examining the current situation and the prospect of further development of this method worldwide. The present paper aims at the investigation of the current situation of rainwater harvesting (RWH) as an alternative water source to confront water scarcity in various countries around the world. In particular, the paper presents the following: (a) the causes of water shortage; (b) a concise historical overview of the temporal development of the RWH method; (c) the evolution of the concept of RWH; (d) the efforts to renew interest in RWH; and (e) incentives and perspectives for the spreading of the RWH method in various countries worldwide.

Keywords: water scarcity; alternative water source; rainwater harvesting; arid and semi-arid areas

1. Introduction

Water is a critical natural resource that plays an important role in the health, social, and economic development of a country, food production, and environment. Moreover, water is closely linked to fundamental human rights such as the right to life, to food, and to health and, consequently, access to safe water is a basic human right.

Globally, water availability has decreased more and more. Beyond the lack of rainfall and its erratic pattern in spatial and temporal scales, in many areas of the globe, available water resources face severe pressures due to demographic, economic, and social causes; environmental degradation; the impacts of climate change; and technological changes. Moreover, water demand is expected to increase in all sectors of production [1], while, by 2030, the world is projected to face a 40% global water deficit under the BAU (business as usual) climate scenario [2].

By 2025, water withdrawals are predicted to increase by 50% in developing countries and 18% in developed countries [3], while nearly 1.8 billion people will be living in areas under severe water stress, and the meeting of water requirements for different uses (agriculture, industry, domestic purposes, energy, and the environment, among others) will be on the threshold [4]. Furthermore, sources of pollution (non-point and diffuse) threaten the existing water resources, and they contribute to the degradation of the water quality of inland and coastal aquatic ecosystems.

The Organisation for Economic Co-operation and Development (OECD) recommends, to its members, research on alternative sources in terms of water supply and, specifically, it suggested rainwater harvesting (RWH), as well as grey and reclaimed water, as alternative water sources [5].

Regarding the physical alternatives to realize the sustainable management of freshwater, there are two solutions. The first one is to find alternative or additional water resources by using conventional centralized systems, while the second is the limited use of available water resources in a more efficient way. Until today, much attention has paid to the first case. Owing to the difficulty of developing new freshwater resources, rainwater harvesting, as well as water reclamation and reuse, are important additional water resources. Moreover, the collection, protection, and re-use of rainwater represent a viable process that can significantly increase available water resources and reduce flood risks. RWH is probably the most ancient practice in use in the world to confront water supply needs. Harvested rainwater is an alternative source of water in many parts around the world. In the last decades, many countries have supported a modern implementation of such a practice to address the increase in water demand pressures due to climatic, environmental, and societal changes. The overall aim of this paper is the investigation of the current situation on rainwater harvesting as a tool to confront water scarcity throughout the world, as well as prospects for spreading of the method.

2. A brief Outline of the History of Rainwater Harvesting

Archaeological findings in many parts of the world confirm that, since prehistory, people used to satisfy their domestic, irrigation, or livestock water needs by collecting and storing rainwater in simple cavities either in a low permeability soil or in rock. With the passage of the centuries, throughout the antiquity, rainwater was the main source of water for potable and non-potable use, thus rainwater harvesting was extremely important for their survival.

RWH was practiced about 4500 years ago by the people of the city Ur in the region of Sumer (in modern-day Iraq) and later by the Nabateans and other people of the Middle East [6]. The same process must have been applied in China 6000 years ago [7]. Moreover, archaeological evidence in southern Jordan indicates the existence of water collection systems, even 9000 years ago, for agricultural purposes, while similar water collection systems in the Negev desert of Israel were probably constructed about 4000 years ago (ca. 2000 BC), or even further back [8]. The collection and storage of rainwater in earthen tanks for domestic and agricultural uses has also been very common in India since the third millennium BC. In China, the history of RWH dates back 4000 years. Some of the early techniques that were used in the country included cisterns, roof open spaces, soil or rock pits, ditches, and micro-dams, among others [9]. Moreover, several archaeological findings suggest that RWH was common in many areas of the world, including Egypt, Thailand, Mexico, Pakistan, Ethiopia, Jordan, Korea, Sardinia, and so on. American Indians used similar systems 700 to 900 years ago in the southwestern United States [10]. Also, RWH has been practiced in rural Thailand for more than 4000 years [11].

Specific hydraulic structures for the harvesting and storage of rainwater were implemented in prehistoric times (ca. 3200–1100 BC) in different Minoan villages, cities, and palaces of Phaistos, Chamaizi, Myrtos-Pyrgos, Archanes, and Zarkos [12]. In addition, the Mycenaeans (1600 BC and ca. 1100 BC) built cisterns and several dams in the Peloponnese and Beotia that were essentially long, low dikes [13].

The cistern construction technology of the Minoans and Mycenaeans was improved by the ancient Greeks during the Archaic (c. 800–479 BC), Classical (478–323 BC), Hellenistic (323–30 BC), and Roman (30 BC–330 AD) periods.

In several Greek cities, rainwater was harvested through open spaces on the roofs, yards, and other open spaces into covered cisterns for storage and future use, in order to meet their daily water needs [14]. Numerous cisterns of the Classical and Hellenistic era have been found in private or public buildings all over Greece, and are quadrilateral or circular, for example, Delos, Akanthus, Santorini, Amorgos, Aiani (Kozani), and Pella. Bottle-shaped cisterns, with a narrow mouth, were quite common during the classical era in Athens, Piraeus, and Olynthus, and were still preferable during the Hellenistic period [15,16]. They were usually flat, pitched or vaulted roof, always coated with impervious material, and either built at ground level or carved deep into the earth, so as the harvested water could be kept cooler and thus more palatable, with its temperature constant. Many of them were multi-sectioned for water filtration. Their dimensions depended on their private or public use and the needs that had to be covered.

Afterwards, Roman private and public buildings included cisterns, very much similar to those in Minoan palaces or in Classical/Hellenistic constructions. They were usually located under paved courtyards, in order to collect rainwater and increase the water available from the city's aqueducts. Moreover, Roman rich houses and villas used to have shallow, oblong uncovered tanks, located on the ground, in the middle of the atrium, called an impluvium, in which rainwater was gathered from the ceiling, and particularly from a rectangular hole in the roof, known as a compluvium, so as to be available for household use or even for irrigating small gardens. Besides their decorative function, impluvia, especially in the early Roman era, were connected with underground storage cisterns, which is why they must be concerned as rainwater harvesting systems [16]. The majority of houses in Pompeii and Herculaneum possessed such hydraulic installations, while similar examples existed all over Greece, and in the eastern territories of the Roman Empire, as well as in North African cities [14].

3. Revival of Interest for Rainwater Harvesting

Rainwater Harvesting is both a simple and ancient method, which has been used for millennia in drier lands of the world, and particularly in regions where other water resources are scarce or difficult to access. It can vary from being small and basic, such as the attachment of a water butt to a rainwater downspout, to being large and complex, such as those that collect water from many hectares and serve large numbers of people [17]. However, the application of this method was reduced or even almost abandoned because of the following factors: (a) the available technical means during the industrial era, which made the water transfer from remote areas possible, through long and complex systems; (b) the ability to withdraw water from deep aquifers, so as to ensure the supply of large quantities of water for industry and urban water demands; and (c) the ability of the management of large quantities of water and supply, constantly and safely, via organized networks.

As Reddy [18] pointed out, the practice of RWH was then essentially abandoned until the early 1930s, except for collecting rainfall from rooftops in some areas. During the 20th century, and specifically before 1950, very few activities had taken place on the research and implementation of water collection techniques. In particular, farmers in Australia had already begun collecting water for domestic use and livestock after World War I. During World War II, there had been some water harvesting activities on islands with high rainfall, such as in Antiqua [19]. However, despite the fact that the revival of water harvesting techniques began in the early 1930s, the greatest activity in both construction and research began in the late 1950s [18]. Boers [20] noted that modern water-harvesting research started in the 1950s by H. J. Geddes, Professor of the University of Sydney in Australia.

According to Food and Agriculture Organisation (FAO) [21], renewed interest in the technology of water harvesting occurred in the 1950s in Australia, where "roaded catchments" based on the concept of compacted earth were constructed over more than 2000 hectares in order to collect water for agricultural purposes. These catchments were called "roaded catchments" because the soil was graded into a series of parallel roadways or gently sloping ridges that drained into the ditches separating them. These ditches carried the collected water to a storage reservoir by way of a collection ditch, which ran perpendicular to the roadways [18].

Prinz and Malik [19] pointed out that interest in water harvesting, both in the research and application level, was renewed partly because of the successful reconstruction of the water collection system for irrigation (1958 and 1959) by Evenari and his colleagues in the Negev desert of Israel. Pacey and Cullis [22] consider that the work of Evenari and his colleagues was of great significance because of the runoff farming models applied, the completeness of the research they conducted, and the historical sources of the models they used.

In the USA, water harvesting begun during the 1940s and was generalized in the early 1950s, when several small catchments were built from small sheets of steel and concrete to provide drinking water to animals and wildlife. In the 1950s, Lauritzen had pioneered an innovative technique of constructing catchments and reservoirs that required the evaluation and use of plastic and artificial rubber membranes [18].

In 1955, an important movement in research interest took place, when cooperative studies on the collection of water for the livestock between the U.S. Department of Agriculture and the Utah Agricultural Experiment Station started by using the soil itself as a catchment surface and by treating it with waterproofing and stabilizing materials. In these studies, various soil cover materials were evaluated, such as plastic vinyl films, polyethylene–butyl rubber sheets, chemical sealants, and so on [23]. Of these materials, plastic butyl films, when not under tension, exhibited excellent wear resistance from exposure to solar radiation, and their installation was relatively simple. However, many of these high cost structures, which were used only by public authorities on public lands, failed within 5–10 years mainly as a result of strong winds, which caused extensive damages. In the 1960s, systematic studies were initiated by various organizations (governmental, private, and universities) in the USA, as well as in other arid or semi-arid countries, that concerned both the development and the assessment of new methods and materials to be used for the construction of water collection systems at low installation costs and the improvement of system reliability [24].

Further incentives for investigating the possibilities of water collection to improve plant production were provided owing to the widespread droughts that occurred in the 1970s and 1980s in Africa and their effects on crops. Much of the experience with rainwater harvesting was gained in Israel, USA, and Australia. However, this experience has limited relevance to resource-poor areas in the semi-arid regions of Africa and Asia [25].

Moreover, interest in collecting and storing water for irrigation purposes was enhanced because of the improvements of the Earthmoving machinery and sealing soil materials, which reduced the cost and difficulty of preparing catchment for collecting water, and improved the efficiency of the collection system. In general, since the 1950s, a series of experiments has developed the variety and sophistication of water harvesting technology.

In recent years, researchers and policy-makers have shown a renewed interest in water use strategies owing to rising water demand, an increased interest in conservation of water and energy, and an increased regulatory emphasis on reducing stormwater runoff volumes and associated pollutant loads [26]. In addition, many countries have renewed their attention in water collection techniques, which are regarded as a viable decentralized water source. The renewal of interest is also related to the role that decentralized water collection systems can play to mitigate flood risks, among others, and because the decentralized multi-purpose rainwater harvesting systems constitute useful infrastructures to mitigate other water-related disasters, such as sudden water break and fire events, especially in highly developed urban areas. Nowadays, the art of collecting rainwater has received renewed attention and interest in many countries as a viable decentralized water source globally, like Germany, Italy, Spain, and France, among others, in Europe; India, China, Malaysia, Japan, and South Korea, among others, in Asia; Kenya, Ethiopia, and Syria, among others, in Africa; in several states of the USA (Nevada, Utah, and so on) and Canada; Brazil in South America; and Australia and New Zealand [14].

4. Rainwater Harvesting: Origin, Terminology, Concept

Koenig and Sperfeld [27] pointed out that the English term "rainwater harvesting" (RWH) is widely accepted internationally. However, there is not a unified definition about this term that is commonly accepted by the scientific community [14]. Researchers employ a wide variety of terms and definitions to describe the various methods aimed at the use of, collection, and storage of rain runoff in order to increase the availability of water for drinking, irrigation, and so on in arid and semi-arid areas [28]. In this way, their criterion is their own purposes and not a strict definition of the term "rainwater harvesting".

The harvesting of rainwater can be either directly from the atmosphere or via runoff. In the first case, rainwater quality is influenced by the atmospheric conditions, while in the second case, the quality is a major matter of concern and quality parameters as pH, COD, BOD_5, SS, nitrates, and so on are considered.

According to Pacey and Cullis [22], the term "rainwater harvesting" derives from the more general term "water harvesting". In the general sense, the term "water harvesting" describes a range of techniques for the collection and concentration of runoff. Usually, water harvesting is used as a generic term encompassing a whole range of methods for the collection and concentration of various forms of runoff (overland flow, rooftop runoff, ephemeral streams, and so on) from various sources (rain or dew) and for various purposes (agricultural, livestock, domestic water supply, environmental management) [14].

The origin of the term "water harvesting" is not well known. Probably, Professor H.J. Geddes of the Sydney University, Australia, is the first who used the term "water harvesting". He defined water harvesting as "the collection and storage of any farm of waters either runoff or creek flow, for irrigation use" [29]. In this definition, water harvesting includes stream flow in creeks and gullies, not just rainwater at the point where it falls [22]. Later, Currier [30] defined water harvesting as "the process of collecting natural precipitation from prepared watersheds for beneficial use".

In 1974, Geddes stated "The phrase 'water harvesting' was coined in the first instance to describe a project of the University of Sydney, which involved the collection and economic storage of farm runoff for irrigation, and to differentiate the work from normal farm water conservation to provide water for livestock or household purposes" and also he pointed out, "the phrase water harvesting has been adopted by others and given a wider connotation" [31]. Myers [10] generalized the definition of water harvesting as "the practice of collecting water from an area treated to increase runoff from rainfall and snowmelt".

In the past, a variety of expressions have been used in order to identify water harvesting techniques. The important basic factors in the various definitions of water harvesting are the surface runoff, the source of runoff, the form of runoff, and the harvesting techniques itself.

A brief review of the international literature shows that, in some definitions, the source of runoff is the key factor, such as "rainwater harvesting" [22,32], "rain harvesting" [33], "rainwater collection" [22,33], and "rainfall collection" [33]. Some definitions include dew and/or dew and mist and/or snow as a water source. However, the amounts of water that can be harvested from these sources are very small and of no particular interest.

In some cases, the term "water harvesting" includes only the surface runoff from the slopes and runoff from ephemeral streams. For example, Boers and Ben-Asher [32] are referred in "local surface runoff", MoALD [34] in "sheet runoff or ephemeral stream flows", Critchley [35] in "runoff before it reaches seasonal or permanent streams", Bruins et al. [36] in "runoff from whatever type of catchment or ephemeral stream", and Pacey and Cullis [22] in "water running off surfaces on which rain has directly fallen". In addition, there are definitions of "water harvesting" based on "stream flow", such as "harvesting stream flow" [37], "floodwater farming" [33], and "floodwater harvesting" [22,35].

Several specific terms have been used for the definition of "water harvesting" for agricultural purposes, such as "runoff agriculture" [38], "runoff farming" [10,22,39], "runoff culture" [40], "agricultural water harvesting" [39], and "floodwater farming" [33]. Bruins et al. [36] used the terms "runoff farming", "runoff agriculture", and "rainwater harvesting agriculture" alternatively to describe "farming in dry regions by means of the flow of rainwater from whatever type of catchment or ephemeral stream".

It is mentioned, however, that Bruins et al. [36] noticed that the term "water harvesting" is a purely hydrological sense, and it does not determine what kind of water is harvested nor the purposes for which the water will be used.

It is not the purpose of this paper to encompass an exhausting commentary on the semantics of the term "rainwater harvesting" and, consequently, this topic is not covered completely. In the framework of the present study, rainwater harvesting means the method by which rainfall that falls upon a surface catchment area (roof, sidewalks, parking lots, landscape areas, and so on) is collected and routed to a storage facility for direct or future use (domestic and agricultural use). It is indicated that RWH does not reduce the demand, but it can reduce the water abstraction needs. In the past, the rainwater harvesting practice was used in arid or semi-arid areas, while in the modern day, the use of rainwater harvesting also extended to sub-humid and humid regions.

5. Examples and Utilization of RWH Around the World

As aforementioned, in recent decades, the interest in rainwater harvesting for both developing and developed countries (including several EU Member States) is growing. The success of RWH systems depends to a great degree on their technical design and the identification of suitable sites, as the appropriate sites for the various RWH technologies in large areas represent a great challenge. For this reason, many researchers have developed and applied various methodologies and criteria to identify suitable sites and techniques for rainwater harvesting. The development in computer technology, hydrological modeling, multi-criteria analysis, geographical information systems (GIS), and remote sensing has made it possible to develop new procedures to identify suitable sites for RWH. In addition, researches and applications have been carried out at various levels on the following: (a) the use and management of RWH; (b) the quality of harvested rainwater; and (c) hydrological or economic data for RWH.

However, the degree of its modern implementation varies greatly across the world, frequently with systems that probably do not maximize potential benefits.

In several countries, both governments and local/regional authorities have promoted measures to install and use RWH systems, mostly under a legal framework, with financial incentives (subsidies, reductions or tax refunds, and so on). For example, some form of RWH is mandatory for buildings and houses in various cities and states of India (New Delhi, Indore, Chennai, Rajastan, and so on); in Catalonia of Spain; in Flanders of Belgium; in new buildings of some states of the USA (Tucson, Arizona, New Mexico, and so on); in many Caribbean islands; in Germany (Hessen, Baden-Württemberg, Saarland, Bremen, Thuringen, Hamburg, and so on); for newly constructed buildings in Seoul of South Korea; and in Malaysia only for large buildings like factories, schools, or bungalows, among others. The same is true in some Australian states, such as South Australia, New South Wales, and Queensland, where regulations stipulate a new rainwater collection system or alternative water source.

Meanwhile, manuals were developed about the design, construction, and management of rainwater harvesting systems, for example, in the United Kingdom, Malaysia, Japan, India, and Canada, among others. In the USA, the federal government does not regulate rainwater harvesting, but rather it is up to individual states to regulate the collection and use of rainwater. Some states, including Georgia, North Carolina, and Texas, among others, have published manuals that provide information on the types of processing systems and components needed for meeting specific water quality objectives. In addition, at the municipal level, several major cities, such as Los Angeles, San Francisco, Tucson, and Portland, have issued guidelines and/or policy documents on treatment and permitting requirements for rainwater collection systems [26].

To the best of our knowledge, there are neither European nor national regulations on the definition of quality standards for rainwater uses within the European Union. In several countries of the European Union, such as France (Décret du 2 Juillet 2008) and the United Kingdom (BS 815, 2009), some standards have been proposed, which are merely guidelines (directives) focusing on domestic uses of rainwater. In Spain, there is the Royal Decree 1620/2007, which establishes quality standards for possible uses of recycled water [41]. However, there is a lot of interest in rainwater harvesting in many European countries, including Germany, France, Spain, Italy, Cyprus, Malta, United Kingdom, Austria, Belgium, Denmark, Portugal, and so on.

RWH is not restricted to simple, small-scale roof collection systems, but is extended to the following: (a) larger systems usually used for providing water for schools, stadiums, airports, and so on; (b) collection systems for high-rise buildings in urbanized areas; and (c) land surface catchment systems and stormwater collection systems to prevent the pollution of water sources from roads, industrial sites, and agriculture. Large-scale RWH systems exist, as follows: (a) in Germany, such as in Berlin the Daimler Chrysler Potsdamer Platz and the building complex at Belss-Luedecke-Strasse; in Darmstadt, the Technical University; in Frankfurt, the Airport, among others; (b) in the United Kingdom, in London, the Millennium Dome, the Museum, and the Velodrome, among others; in Manchester, the Honda Dealership; in Bristol, the Imperial Tobacco Head; (c) in Singapore, the Changi Airport; (d) in Japan, in Sumida city, the Ryogoku Kokugikan Sumo-wrestling Arena and the Town Hall, among others; in Tokyo, the Rojison and the Sky Tower; and (e) in South Korea, in Seoul, the Star City Project in Kwangjin-Gu, among others.

In developed countries, including Japan, Singapore, Belgium, France, Germany, USA, Sweden, Canada, Spain, and so on, RWH is mostly used to supplement conventional systems for non-drinking water purposes such as irrigation, laundry, and toilet flushing, while in Australia, the collected water also has potable use. In developing countries, such as Bangladesh, Botswana, India, Kenya, Nepal, Namibia, Uganda, South Africa, and so on, RWH is mainly used to address water shortages for both potable (drinking, cooking, personnel hygiene, and so on) and non-potable purposes [42,43].

In several Latin American countries (Argentina, Brazil, Costa Rica, Chile, Mexico, and Peru), the RWH practice from roofs for domestic consumption is applied, while in the semi-arid areas of Argentina, Brazil, and Venezuela, runoff collection from roads with drainage ditches and street gutters is used, from which water is then transferred to cultivated areas for irrigation [44].

In Australian cities, RWH is popular. In urban areas, RWH systems are used to complement the main water system, whereas many rural and peri-urban communities completely rely on this. A total of 30% of rural Australians use RWH, while 7% use RWH in the capital cities. About 13% of all Australian households (2.6×10^6 people) use RWH systems as a primary source of drinking water [45]. Local authorities throughout Australia encourage the use of RWH systems in urban areas to supplement main water supplies and to manage urban stormwater runoff. For this reason, the Australian state and local governments adopted a wide range of policies, including subsidies and grants, to provide the installation of rainwater tanks in houses. These incentives vary from state to state, depending on the size of the water reservoir and the purpose of using the collected water. In South Australia, almost 50% of the population live in houses equipped with a rainwater tank. RWH is mandatory for new homes in Queensland [46].

In the USA, rainwater harvesting has become an increasingly common practice. Since 2004, it has been estimated that about 100,000 residential RWH systems were in use in the USA and its territories [7]. Some states and territories (Hawaii, Kentucky, New Mexico, North Carolina, Ohio, Oregon, Texas, Utah, and Washington, among others) consider RWH as a serious practice for protecting water resources, as well as for increasing the available volume of water for potable use. However, even though the major use of harvested rainwater is for landscape watering, flushing toilets, and so on, there are a number of systems that serve indoor uses as well.

In Bermuda, rooftop RWH is compulsory by law for all buildings and constitutes the primary source of water for domestic supply. The Public Health Act regulates the details for the maintenance and conservation of the catchments, tanks, gutters, pipes, vents, and screens in order for them to be maintained in a good situation [47].

In Canada, most of RWH systems are for residential use in rural areas, where there is no access to central public water supply systems. In cities, most cases relate to buildings that have been certified according to one of the green building rating systems, in which the reuse of rainwater and the reduction of runoff were taken into account [48]. Since 2010, the National Plumbing Code has been in force, which permits the use of rainwater for toilet and urinal flushing, as well as subsurface irrigation. In addition, it permits the use of rainwater, both indoor and outdoor, depending on the level of treatment. RWH is mandatory for new homes in Queensland [46]. In Ontario, several municipalities recognize RWH as an important tool for confronting problems arising from the management of water resources. The City of Toronto and the Regional Municipality of Waterloo have been active in promoting the technology through stormwater and green building policies [49].

In Mexico, RWH makes a significant contribution to reducing the water supply shortage that occurs in large areas of the country. Specifically, the National Water Plan to increase the percentage of the population with easy access to drinking water and improve the efficiencies of water services in the municipalities has adopted RWH as an alternative water supply. In addition, the National Water Commission (CONAGUA) has developed the National Program for Rainwater Harvesting and Eco-techniques in Rural Areas in order to provide water to rural populations. RWH practice is applied in various states of the country, such as State of Mexico, Guanajuato, Querétaro, Michoacán, Morelos, Zacatecas, San Luis Potosi, and others. In rural areas of the country, RWH use predominates, while in urban areas, it is mainly used for sanitation, watering, and cleaning. The potable use is accepted in some primary schools [50].

In Brazil, there is no legislation to cover RWH at the federal level. Since 2007, NBR-15227 has been in force, which has normative character and regulates the use of rainwater for non-potable purposes in urban areas. However, there are various cities and municipalities that have enacted guidelines that regulate the catchment and storage of rainwater for non-potable uses. Since 2002, the city of Sao Paulo has been pioneer as it implemented the first law regulating these aspects. Thence, other major cities including Rio de Janeiro, Curitiba, and Paraiba, among others, have implemented similar regulations. Owing to the large number of different laws and regulations in force in the different parts of the country, it is difficult to assess the extent to which Brazil is implementing RWH as an alternative to the municipal water supply systems [51].

In African countries, rainwater collection systems are being increasingly adopted. However, despite the rapid expansion of these systems, progress is slow because of the following factors: (a) the low rainfall and its seasonal nature, (b) the small number and size of impervious roofs, (c) the high cost of constructing catchment systems in relation to typical household incomes, (d) the lack of cement and pure graded sand in some parts of Africa, and (e) the lack of sufficient water for the construction industry, which burden the total cost. However, RWH systems are increasingly expanding in Africa with works in Botswana, Mali, Malawi, South Africa, Namibia, Zimbabwe, Tanzania, and so on [52].

The effort to develop rainwater collection systems in Africa is led by Kenya, which has a very long tradition in these systems through the centuries. Since the late 1970s, interest in RWH has rapidly grown. In different parts of the country, many RWH projects have been carried out, each one with their own designs and implementation strategies, in an effort to provide long-term solutions to water resource problems [52]. In the middle of the 20th century, the Government began to build rock catchment systems that served communities in the semi-arid area of Kitui district [47]. The variety of geographic and climatic conditions in the country has enabled the development of a very wide range of RWH technologies for water supply, agriculture, and livestock. In Nairobi, there are several manufacturers of water tanks from plastic, metal, and other materials. These tanks are sold everywhere in East Africa and beyond [47]. In many parts of Kenya, the United Nations Development Program and the World Bank consider rainwater storage tanks as an essential part of their program on water supply and sanitation [53]. It is noted that, although it is not mandatory for institutional buildings to dispose RWH facilities, many of them, especially in the rural areas, have those facilities. In 1994, the Kenya Rainwater Association was established, which is the first national RWH association in Africa. Since then, tens of thousands of rainwater collection systems have been built in Kenya by a wide range of organizations; as a result, millions of people are benefiting from these systems [47].

Japan is one of the developed countries in Asia that has a strong international exchange of experience in the use of RWH. Since the mid-1980s, Tokyo and other Japanese cities, as well as most municipalities and organizations of the country, have given particular importance to RWH so as to have safe water supplies to deal with emergencies, floods, rehabilitation of the natural hydrological cycle, and exploration of alternative water sources for non-potable use [54]. Moreover, the abnormal drought of 1994 and the Great Hanshin-Awaji Earthquake of 1995 highlighted the importance of securing water supplies from the viewpoint of disaster preparedness. A large number of municipalities re-evaluated the importance of RWH and tried to identify alternative water resources, as a means to prevent urban flooding and to secure emergency water for disaster responses. The issue was regulated by ordinance and guidelines according to the local conditions [55]. According to the survey of the Association for Rainwater Storage and Infiltration Technology held in April 2011, 208 municipalities are implementing subsidy programs for the establishment of facilities for storing or filtration systems of rainwater, and of these, 179 provide subsidies for installation of rainwater tanks. In April 2014, the Japanese Diet passed the Act to Advance the Utilization of Rainwater, which went into force in May 2015. Under this act, municipalities are obliged to make their best effort to define and work toward rainwater utilization targets, while the national government is required to grant financial support for subsidy programs. These arrangements are expected to provide a national mobilization to promote the technical rainwater use [56]. On 10 March 2015, the Japanese government, based on the above act, approved the wider usage of RWH systems in newly constructed buildings by the state government or incorporated administrative agencies, aiming for a 100% installation rate [57]. In Japan, the placement of small-sized RWH facilities with a storage capacity less than 1 m^3 is widespread in individual houses. It is noted that the Great Earthquake in Eastern Japan in March 2011 caused a sudden rise in the number of households that installed tanks to store rainwater for emergency [58]. In Japan, rainwater utilization is now flourishing at both the public and private levels [59]. There are approximately 2800 large-scale systems for water recycling or rainwater use in Japan [60].

In China, the growing interest in RWH was initiated in the 1980s owing to the widespread droughts of that decade, which was followed by serious shortages of drinking water and crop failures [61]. RWH practice and utilization are applied mainly in areas with the following types of water scarcity [53]: (a) in water deficient areas with a lack of water resources, such as Gansu and central Ningxia; (b) in areas with seasonal water deficit, such as Fujian, Guizhou, and other hilly areas; (c) in areas with water deficit, but also with difficulty in exploitation, such as the southwest mountainous areas of the country; (d) in water deficient areas with poor water quality, for example, brackish water, fluoride water, and high-arsenic water. In these areas, authorities have constructed water cellars, tanks, ponds, and other miniature water conservancy projects as an effective solution to the problem of water shortage. Unfortunately, there are still problems of water deficit. Seventeen provinces in the country have adopted the RWH practice by building 5,600,000 tanks with a total capacity of 1.8×10^9 m^3, supplying water to about 15×10^6 people, and supplemental irrigation of 1.2×10^6 ha of land [47]. RWH systems are also applied in the provinces of northwestern China (Ningxia Region, Shanxi Shaanxi, and the Inner Mongolia Region), as well as in the southwest and southeast provinces of the country (Guangxi Region and Guizhou Province). The implementation of RWH has a significant impact on the development of China's semi-arid rural areas and has practically solved the drinking water problems of populations living in semi-arid mountainous areas of the country [61].

In India, RWH was revived in the 1960s in response to declining groundwater availability caused by the rapid expansion of irrigation pumping. Many Indian cities have insufficient water supplies to meet their needs. Urban development makes it both difficult and expensive to build dams, pipelines, and canals commonly used nowadays in order to supply water to cities. RWH supported agriculture for many years in India, while there is a demand in urban areas for novel methods for decentralized water supply systems. Since 2000, the legislation on RWH has been changed in the various states and federal regions of the country, and it is compulsory for new buildings. Rooftop RWH systems are mandatory for new buildings in 18 of the 28 states and 4 of the 7 federal regions of the country [62].

In Malaysia, the promotion of RWH use began in 1999 with the introduction of the Guidelines for "Installing a Rainwater Collection and Utilization System" by the Ministry of Housing and Local Government. In 2006, the Ministry of Energy, Water, and Communication introduced the Water Services Industry Act 2006 and the Water Services Commission Act 2006, under which the

Ministry encouraged the citizens to implement RWH systems. In 2009, guidelines entitled "Rainwater Harvesting: Guidebook for Planning and Design" were introduced by the Department of Irrigation and Drainage (DID), followed by "Guideline on Eco-Efficiency in Water Infrastructure for Public Buildings in Malaysia" by the National Hydraulic Research Institute of Malaysia in 2011 and "Urban Stormwater Management Manual for Malaysia" (2nd Ed.) by the DID in 2012. In 2011, the Malaysian government imposed the installation of RWH systems in several cases of buildings, such as all types of buildings with a roof area equal or more than 100 m^2 and residential buildings (bungalow and semi-detached) of the country. In addition, in 2013, the "Guidelines for Installing a Rainwater Collection and Utilization System" of 1999 were revised [63]. As Mohd.-Shawahid et al. [64] stated, the most encouraging development to make RWH mandatory in Malaysia was the two guidelines of 2006, although this policy was applied only to large buildings like factories, schools, or bungalows. In addition, they considered that it is certainly a step in the right direction to make RWH mandatory, despite the fact that there is still a lack of robust policy to promote the installation of RWH systems in Malaysia. According to Lee et al. [63], rainwater is collected for general washing and gardening purposes, but the common use of rainwater in a building is for toilet flushing, whereby the rainwater cistern is connected to the water closet fittings to minimize the use of treated water for non-potable use.

In 2009, the government of Taiwan introduced new water regulations, which included wastewater reuse, seawater desalination, and rainwater harvesting as alternative water resources for domestic water supply. In 2003, a green building policy was developed. This policy requires new buildings with a total floor area greater than 10,000 m^2 to have installed domestic rainwater harvesting equipment to supply at least 5% of the total water required by the building [65].

Germany has developed new and sophisticated RWH systems and techniques and is considered as one of the leading countries in the world in this field. In particular, Germany has more than 1.5×10^6 integrated rainwater systems not only in homes for toilet use, but also for car washes and garden irrigation, as well as in service water demanding industries [66]. According to the Environmental Agency [48], 35% of new buildings in the country are equipped with a RWH system, and from 50,000 [67] to 80,000 [68] of such new systems are installed every year. As Partzsch [68] pointed out, in 2005, every third new building in Germany was supplied with a rainwater storage tank. In Germany, the promotion of RWH in households has become widespread since the 1980s [67]. Schuetze [69] presented a documented description on rainwater harvesting and management in Germany.

In the United Kingdom, modern RWH systems have been introduced relatively recently [48], as the interest in RWH research, technology, development, and utilization has yet to mature, although several initiatives are in place to promote RWH [70]. The Code for Sustainable Homes, which is in force in England, Northern Ireland, and Wales, supports and encourages the promotion of installation of RWH systems in new houses. In particular, owners of new homes are encouraged to save money and water resources by installing RWH systems for toilets, washing clothes, garden watering, and car washing. According to UKRHA (United Kingdom Rainwater Harvesting Association) [71], approximately 100,000 RWH systems already exist in the United Kingdom and approximately 4000 systems per year are installed, which are commonly internally plumbed to supply toilet flushing as well as garden irrigation.

In France, the interest in RWH for indoor and outdoor uses has increased, and constitutes a serious issue even in urban areas. As Gerolin et al. [72] pointed out, "rainwater harvesting (RWH) has known a revival of interest since the establishment in 2006 of a national tax credit for households implementing a rainwater collection system". Since 2008, a decree (French Government Order of 21 August 2008) concerning RWH has been in force. In reality, it is about regulations, which define better management of the use of rainwater and the precise technical requirements to be met by the components of the collection systems supplying both outdoor and indoor uses. Specifically, the regulations prohibit the use of harvested rainwater for drinking, showering, or bathing, but they allow its use for toilet flushing; cleaning the ground; and, under certain conditions, washing clothes [73]. In France, according to a survey conducted in 2009, 15% of the population have an RWH system in urban areas [74].

Belgium has national legislation that supports RWH, which stipulates that all new constructions must have a rainwater collection system, the water of which can be used for washing the toilet and for external water uses. In Flanders, it is estimated that 10% of current household water consumption comes from RWH, which could be increased to 25% by 2025. It is estimated that households account for 72% of the total rainwater use in Flanders [75].

In Portugal, the ERSAR (Water and Waste Services Regulation) guidelines allow the use of RWH only for non-potable use and, in particular, for irrigation purposes. In 2012, ANQIP (National Association for the Quality of Building Installations), a nonprofit organization promoting water sustainability at the building level, published a technical document (ETA 0701, 2012) describing the procedures to be taken into account regarding the installation of rainwater collection systems in Portuguese buildings [76].

In Malta, a significant proportion of 35.4% of households are currently using RWH, of which 33.6% collect it in underground cisterns and a small percentage of 1.8% in plastic containers. Since 2004, the exploitation of RWH in new constructions has been regulated by the plan of the MEPA (Malta Environment and Planning Authority), which regulates the creation of water collecting surfaces on roofs, the possible size and capacity of the tanks, and so on [77].

6. Conclusions

Worldwide, many countries are facing water shortages more and more growing as a result of the continued increasing demand for water from various competing users like domestic, agriculture, industry, and environment use, as well as because of urbanization, climate change, water pollution, and so on. All of these factors exert pressures on the existing water resources. One of the biggest challenges of the 21st century is to overcome the growing water shortage.

In general, the flawed strategy was the construction of large-scale projects (dams, long-distance pipelines, pumping stations, and so on), as their construction has not proven sufficient to meet the water needs of the different users, while at the same time, they had significant social, economic, and environmental impacts, and required significant investment. Therefore, searching for alternative water sources (grey water, desalination, and RWH) has attracted worldwide interest. RWH is seen as a more promising alternative or supplementary water resource owing to its minimal environmental impact, the low treatment needs in comparison with other alternative water sources, the benefits from flood mitigation, and many other reasons.

RWH is a very old traditional and sustainable practice that has been adopted in many ancient sites as a water supply method for both potable and non-potable purposes. Nevertheless, RWH, which was a worldwide technology, has been neglected over the past 150 years because of new technologies, which enable us to store, pump from deep groundwater, and transport huge volumes of water (ground and surface) via dams, pump stations, and long length pipelines.

The literature reveals that the interest and use of RWH systems, on a global basis, has been continually increasing from about the beginning of the second half of the previous century onwards. There is a trend for the revival of traditional technologies, blending them with modern methods to meet the required needs of water, in the present and future. As Yannopoulos et al. [14] stated, "Worldwide, rainwater harvesting has retrieved its importance as a valuable water resource, alternative or supplementary, in conjunction with more conventional water supply technologies. If rainwater harvesting is practiced more widely, many water shortages, actual or potential, can be alleviated".

Nowadays, a large number of countries all over the world consider RWH as a viable decentralized water source and have started considering and practicing rainwater harvesting as a sustainable development strategy. However, a significant push to extend this technique is required. Specifically, significant efforts are still needed in research, investments, information, education of the public on the importance of rainwater harvesting, economic incentives (subsidies and tax exemptions), suitable legislation, and regulations.

Until today, two approaches have been applied concerning the extension of RWH, namely either voluntary via incentive-based programs or via mandated regulations. In several countries, government subsidies and rebate programs can be particularly effective in promoting RWH implementation. Contrary to regulations that require compliance, subsidies target individuals with an appreciation for RWH and provide an incentive for them to pursue adoption of this practice.

It is noted that the main variable of interest for the design of an RWH system in a particular locality is rainfall (intensity, frequency, and temporal variability). Therefore, the reliability of rainfall in terms of both spatial and temporal distribution constitutes a critical factor in the performance of an RWH system. Moreover, alteration of the rainfall pattern and its variability in the future is likely to take place as a result of climate change, which can introduce uncertainty in the design of RWH systems (tank storage). RWH systems have many benefits, for example, (a) they provide important economic advantages for consumers because they reduce the amount of water purchased from public systems; (b) they give the possibility of an alternative water supply and, consequently, reduce pressure on surface water sources and aquifers; (c) they are simple and inexpensive technologies that are easy to install, maintain, expand, reconfigure, or relocate to meet the needs of each household. In this way, RWH systems are an effective tool to minimize the use of treated water for non-potable uses and supply drinking water in places where the existing sources cannot meet the water needs; furthermore, they are an effective adaptive strategy to climate change against the reduction of water availability and for the control of runoff. In addition, RWH systems decrease (a) the flow of stormwater drains; (b) nonpoint source pollution; (c) onsite erosion and flooding; and (d) the costs of managing runoff.

Despite the aforementioned benefits, RWH systems cannot supply water for all domestic uses and it is almost impossible for them to make the households independent of conventional water supply systems. However, RWH systems can serve as the main water source in rural areas in which the availability of water resources is a critical issue; furthermore, they can constitute a complementary water supply in urbanized areas, provided that they will connect with existing conventional water supply systems.

Author Contributions: S.Y. had the original idea, supervised the research, prepared the manuscript, and mainly carried out the data collection; I.G. and M.K.-S. conducted the review of it, made English corrections, and contributed to manuscript preparation.

Funding: This research received no external funding.

Acknowledgments: Part of this paper was presented at the XIV International Conference Protection and Restoration of the Environment, 3–6 July 2018, Thessaloniki, Greece, and published in its Proceedings: (Theodosiou, N., Christodoulatos, C., Koutsospyros, A., Karpouzos, D., Mallios, Z., Eds.): Yannopoulos, S.; Giannopoulou, I.; Kaiafa-Saropoulou, M. Rainwater Harvesting as an Alternative Source to Confront Water Scarcity Worldwide—Current Situation and Perspectives, pp. 64–75.

Conflicts of Interest: The authors declare no conflict of interest.

References

1. WWAP (World Water Assessment Program). *The United Nations World Water Development Report 4: Managing Water under Uncertainty and Risk*; UNESCO: Paris, France, 2012; p. 391. ISBN 978-92-3-104235-5.
2. 2030 WRG (2030 Water Resources Group). *Charting Our Water Future: Economic Frameworks to Inform Decision-Making*; 2030 WRG: Washington, DC, USA, 2009; p. 185.
3. WWAP (World Water Assessment Program). *The State of the Resource, World Water Development Report 2, Chapter 4*; U.N. Educational, Scientific and Cultural Organization: Paris, France, 2006.
4. UN Water. Coping with Water Scarcity: Challenge of the Twenty-First Century. Prepared for World Water Day 2007. Available online: http://www.unwater.org/wwd07/downloads/documents/escarcity.pdf (accessed on 23 March 2007).
5. OECD (Organization for Economic Co-Operation and Development). *Alternative Ways of Providing Water: Emerging Options and Their Policy Implications*; Advance Copy for 5th World Water Forum; OECD: Paris, France, 2009; p. 33. Available online: https://www.oecd.org/env/resources/42349741.pdf (accessed on 10 March 2018).

6. Sivanappan, R.K. Rain water harvesting, conservation and management strategies for urban and rural sectors. In Proceedings of the 22nd National Convention of Environmental Engineers and National Seminar on Rainwater Harvesting and Water Management, Nagpur, India, 11–12 November 2006; Institution of Engineers: Nagpur, India, 2006; pp. 1–5.
7. TWDB (Texas Water Development Board). *The Texas Manual on Rainwater Harvesting*, 3rd ed.; Texas Water Development Board: Austin, TX, USA, 2005.
8. Evenari, M.L.; Shanan, L.; Tadmor, N.; Aharoni, Y. Ancient agriculture in the Negev. *Science* **1961**, *133*, 979–996. [CrossRef]
9. Akpinar-Ferrand, E.; Cecunjanin, F. Potential of rainwater harvesting in a thirsty world: A survey of ancient and traditional rainwater harvesting applications. *Geogr. Compass* **2014**, *8*, 395–413. [CrossRef]
10. Myers, L.E. Water harvesting 2000 BC to 1974 AD. In Proceedings of the Water Harvesting Symposium, Phoenix, AZ, USA, 26–28 March 1974; USDA, February 1975, ARS W-22. pp. 1–7.
11. Visvanathan, C.; Vigneswaran, S.; Kandasamy, J. Rainwater Collection and Storage in Thailand: Design, Practices and Operation. *J. Water Sustain.* **2015**, *5*, 129–139.
12. Angelakis, A.N. Evolution of rainwater harvesting and use in Crete, Hellas, through the millennia. *Water Sci. Technol. Water Supply* **2016**, *16*, 1624–1638. [CrossRef]
13. Mays, L.W. A Brief History of Water Technology during Antiquity: Before the Romans. In *Ancient Water Technologies*; Mays, L.W., Ed.; Springer Science and Business Media: Dordrecht, The Netherlands, 2010; pp. 1–28.
14. Yannopoulos, S.; Antoniou, G.; Kaiafa-Saropoulou, M.; Angelakis, A.N. Historical development of rainwater harvesting and use in Hellas: A preliminary review. *Water Sci. Technol. Water Supply* **2017**, *17*, 1022–1034. [CrossRef]
15. Robinson, D. *Excavations at Olynthus, Part VIII, The Hellenic House, a study of the houses found at Olynthus with a detailed account of those excavated in 1931–1934*; Oxford University Press: London, UK, 1938.
16. Hodge, T. *Roman Aqueducts and Water Supply*; Duckworth & Co.: London, UK, 1992; p. 504.
17. Leggett, D.J.; Brown, R.; Brewer, D.; Stanfield, G.; Holliday, E. *Rainwater and Greywater Use in Buildings: Best Practice Guidance*; CIRIA Report C539; CIRIA: London, UK, 2001; p. 134.
18. Reddy, Y.A. Water Harvesting: Limitations in Implementation. In Proceedings of the 22nd National Convention of Environmental Engineers and National Seminar on Rainwater Harvesting and Water Management, Nagpur, India, 11–12 November 2006; Institution of Engineers: Nagpur, India, 2006; pp. 70–77.
19. Prinz, D.; Malik, A.H. *Runoff Farming*; WCA InfoNET: Rome, Italy, 2002; p. 39.
20. Boers, T.M. *Rainwater Harvesting in Arid and Semi-Arid Zones*; Publication No. 55; International Institute for Land Reclamation and Improvement (ILRI): Wageningen, The Netherlands, 1994; p. 127.
21. FAO (Food and Agriculture Organization). Role of forestry in combating desertification. In Proceedings of the FAO Expert Consultation on the Role of Forestry in Combating Desertification, Saltillo, Mexico, 24–28 June 1985; FAO Conservation Guide No. 21. FAO: Rome, Italy, 1989; p. 333.
22. Pacey, A.; Cullis, A. *Rainwater Harvesting: The Collection of Rainfall and Runoff in Rural Areas*; Intermediate Technology Publications: London, UK, 1986; p. 216.
23. Lauritzen, C.W. Ground Covers for Collecting Precipitation. *Farm Home Sci. Utah Sci.* **1960**, *21*, 66–67, 87. Available online: http://digitalcommons.usu.edu/utscience/vol21/iss3/1 (accessed on 20 January 2018).
24. Frasier, G.W.; Myers, L.E. *Handbook of Water Harvesting*; Agriculture Handbook No. 600; U.S. Department of Agriculture: Washington, DC, USA, 1983; p. 45.
25. Critchley, W.; Siegert, W.K. *Water Harvesting: A Manual for the Design and Construction of Water Harvesting Schemes for Plant Production*; AGL/MICS/17/91; FAO: Rome, Italy, 1991; p. 154.
26. USEPA (U.S. Environmental Protection Agency). *Rainwater Harvesting: Conservation, Credit, Codes, and Cost Literature Review and Case Studies*; U.S. EPA, EPA-841-R-13-002; USEPA: Washington, DC, USA, 2013; p. 36.
27. Koenig, K.W.; Sperfeld, D. *Rainwater Harvesting—A Global Issue Matures*; Fachvereinigung Betriebs- und Regenwassernutzung e.V; Association for Rainwater Harvesting and Water Utilisation: Darmstadt, Germany, 2006.
28. Haut, B.; Zheng, X.-Y.; Mays, L.; Han, M.; Passchier, C.; Angelakis, A.N. Evolution of rainwater harvesting and heritage in urban areas through the millennia: A sustainable technology for increasing water availability. In *Water and Heritage: Material, Conceptual, and Spiritual Connections*; Willems, W.J.H., van Schaik, H.P.J., Eds.; Sidestone Press: Leiden, The Netherlands, 2015; pp. 37–56.

29. Geddes, H.J. Water Harvesting. In Proceedings of the National Symposium of Water Resources, Use and Management, Canberra, Australia, 9–13 September 1963; Australian Academy of Sciences: Melbourne, Australia; University Press: Melbourne, Australia, 1964.
30. Currier, W.F. Water Harvesting by Trick Tanks, Rain Traps and Guzzlers. In Proceedings of the Water-Animal Relations Symposium, Twin Falls, ID, USA, 1–8 June 1973; Mayland, H.F., Ed.; USDA-ARS Snake River Conservation Research Center: Kimberly, ID, USA, 1973.
31. Geddes, H.J. Achievements in Water Harvesting Technology in Australia. In Proceedings of the International Workshop on Farming Systems, 18–21 November 1974; ICRISAT: Begumpet, India, 1974; pp. 165–176.
32. Boers, T.M.; Ben-Asher, J. A review of rainwater harvesting. *Agric. Water Manag.* **1982**, *5*, 145–158. [CrossRef]
33. Matlock, W.G.; Dutt, G.R. *A Primer on Water Harvesting and Runoff Farming*, 2nd ed.; Agricultural Engineering Department, College of Agriculture, University of Arizona: Tucson, Arizona, USA, 1986; p. 126.
34. MoALD. *Runoff Harvesting for Crop, Range and Tree Production in the BPSAAP-Area*; Baringo Pilot Semi-Arid Area Project (BPSAAP) Interim Report Baringo Pilot Semi-Arid Area Project; Ministry of Agriculture and Livestock Development, Government of Kenya: Nairobi, Kenya, 1984.
35. Critchley, W.R.S.; Finkel, M.M. *Some Lessons from Water Harvesting in Sub-Saharan Africa*; Report from a Worlshop held in Baringo, Kenya, 13–17 October 1986; The World Bank: Washington, DC, USA, 1987; p. 58.
36. Bruins, H.J.; Evenari, M.; Nessler, U. Rainwater-harvesting agriculture for food production in arid zones: The challenge of the African famine. *Appl. Geogr.* **1986**, *6*, 13–32. [CrossRef]
37. Das, D.C. Surface water development in arid zones. In *Sand Dune Stabilisation, Shelterbelts and Afforestation in Dry Zones*; FAO, Conservation Guide No. 10; FAO: Rome, Italy, 1985; pp. 203–219.
38. Evenari, M.; Shanan, L.; Tadmor, N. *The Negev: The Challenge of a Desert*; Harvard University Press: Cambridge, MA, USA, 1971; p. 437.
39. Huibers, F.P. Rainfed Agriculture in a Semi-Arid Tropical Climate: Aspects of Land and Water Management for Red Soils in India. Ph.D. Thesis, Agricultural University of Wageningen, Wageningen, The Netherlands, 1985.
40. Kutsch, H. Currently used techniques in rainfed water concentrating culture: The example of the Anti-Atlas. *Appl. Geogr. Dev.* **1983**, *21*, 108–117.
41. Llopart-Mascaró, A.; Ruiz, R.; Martínez, M.; Malgrat, P.; Rusiñol, M.; Gil, A.; Suárez, J.; Puertas, J.; Rio, H.; Paraira, M.; et al. Analysis of Rainwater Quality: Towards Sustainable Rainwater Management in Urban Environments—Sostaqua Project. In Proceedings of the 7th International Conference on Sustainable Techniques and Strategies in Urban Water Management (NOVATECH 2010), Lyon, France, 27 June–1 July 2010.
42. Lade, O.; Oloke, D. Modelling Rainwater System Harvesting in Ibadan, Nigeria: Application to a Residential Apartment. *Am. J. Civ. Eng. Architect.* **2015**, *3*, 86–100.
43. Kisakye, V.; Van der Bruggen, B. Effects of climate change on water savings and water security from rainwater harvesting systems. *Resour. Conserv. Recycl.* **2018**, *138*, 49–63. [CrossRef]
44. Ringler, C.; Rosegrant, M.W.; Paisner, M.S. *Irrigation and Water Resources in Latin America and the Caribbean: Challenges and Strategies*; EPTD Discussion Paper No. 64; International Food Policy Research Institute (IFPRI): Washington, DC, USA, 2000; p. 92.
45. Coombes, P. Guidance on the Use of Rainwater Harvesting Systems for Rain Harvesting. Available online: http://rainharvesting.com.au/wp-content/uploads/media_Research_Assoc_Prof_Peter_Coomber_Expert_Report_for_Guidance_of_Rainwater_Harvesting_Systems_2006.pdf (accessed on 20 January 2017).
46. CMHC (Canada Mortgage and Housing Corporation). *Collecting and Using Rainwater at Home: A Guide for Homeowners*; CMHC: Ottawa, ON, Canada, 2013; p. 61.
47. Lo, A.G.; Gould, J. Rainwater Harvesting: Global Overview. In *Rainwater Harvesting for Agriculture and Water Supply*; Zhu, Q., Gould, J., Li, Y.-H., Ma, C.-X., Eds.; Springer: Singapore, 2015; pp. 213–233.
48. E.A. (Environmental Agency). *Harvesting Rainwater for Domestic Uses: An Information Guide*; Environment Agency: Bristol, UK, 2010; p. 30.
49. TRCA (Toronto and Region Conservation Authority). *Performance Evaluation of Rainwater Harvesting Systems*; Final Report 2010 (revised June 2011); Sustainable Technologies Evaluation Program (STEP): Toronto, ON, Canada, 2010.
50. Fuentes-Galván, M.L.; Ortiz Medel, J.; Arias Hernández, L.A. Roof Rainwater Harvesting in Central Mexico: Uses, Benefits, and Factors of Adoption. *Water* **2018**, *10*, 116. [CrossRef]
51. Da Costa Pacheco, P.R.; Dumit Gómez, Y.; Ferreira de Oliveira, I.; Girard Teixeira, L.C. A view of the legislative scenario for rainwater harvesting in Brazil. *J. Clean. Prod.* **2017**, *141*, 290–294. [CrossRef]

52. UN-HABITAT. *Rainwater Harvesting and Utilisation*; Blue Drop Series, Book 2: Beneficiaries & Capacity Builders; UN-HABITAT: Mtwapa, Kenya, 2005; p. 77.
53. Liu, L.-S.; Liu, L.-G.; Wu, L.-X.; Wu, J.L.-P.; Huo, W.J. Problems and countermeasures on the safety of rainwater harvesting for drinking in China. *MATEC Web Conf.* **2016**, *68*, 14006. [CrossRef]
54. Koenig, K. *The Rainwater Technology Handbook*, 1st ed.; Wilo-Brain: Dortmund, Germany, 2001; p. 143.
55. Furumai, H.; Kim, J.; Imbe, M.; Okui, H. Recent application of rainwater storage and harvesting in Japan. In Proceedings of the 3rd IWA International Rainwater Harvesting and Management Workshop as a part of IWA-Viena World Water Congress & Exhibition, Vienna, Austria, 11 September 2008; p. 7.
56. JFS (Japan for Sustainability). Let's Use Rainwater! Recent Trends in Rainwater Use in Japan. Available online: http://www.japanfs.org/en/news/archives/news_id035023.html (accessed on 20 January 2017).
57. JFS (Japan for Sustainability). Cabinet of Japan Decide to Make Full Use of Rainwater in Newly Constructed Buildings. Available online: http://www.japanfs.org/en/news/archives/news_id035257.html (accessed on 20 January 2017).
58. Campisano, A.; Butler, D.; Ward, S.; Burns, M.J.; Friedler, E.; DeBusk, K.; Fisher-Jeffes, L.N.; Ghisi, E.; Rahman, A.; Furumai, H.; et al. Urban rainwater harvesting systems: Research, implementation and future perspectives. *Water Res.* **2017**, *115*, 195–209. [CrossRef]
59. Lo, K.F.A. Multi-Faceted Use of Rainwater Harvesting to Combat Water Problems. In Proceedings of the 11th IRCS Conference Towards a New Green Revolution and Sustainable Development Through an Efficient Use of Rainwater, Texcoco, Mexico, 25–29 August 2003.
60. De Graaf, R.E. Innovations in Urban Water Management to Reduce the Vulnerability of Cities. Ph.D. Thesis, Delft University of Technology, Delft, The Netherlands, 2009.
61. Li, X.-Y. Rainwater harvesting for agricultural production in the semiarid loess region of China. *J. Food Agric. Environ.* **2003**, *1*, 282–285.
62. MARWAS, A.G. *Water & Water Treatment in India. Market Opportunities for Swiss Companies*; OSEC: Zurich, Switzerland, 2010; p. 28.
63. Lee, K.E.; Mokhtar, M.; Mohd-Hanafiah, M.; Halim, A.A.; Badusah, J. Rainwater harvesting as an alternative water resource in Malaysia: Potential, policies and development. *J. Clean. Prod.* **2016**, *126*, 218–222. [CrossRef]
64. Mohd Shawahid, H.O.; Suhaimi, A.R.; Rasyikah, M.K.; Jamaluddin, S.A.; Huang, Y.F.; Farah, M.S. Policies and incentives for rainwater harvesting in Malaysia. In Proceedings of the Colloquium on Rainwater Utilization, Putrajaya, Malaysia, 19–20 April 2007; Ministry of Natural Resources and National Hydraulic Research Institute of Malaysia (NAHRIM): Putrajaya, Malaysia, 2007; pp. 1–15.
65. Liaw, C.-H.; Chiang, Y.-C. Framework for Assessing the Rainwater Harvesting Potential of Residential Buildings at a National Level as an Alternative Water Resource for Domestic Water Supply in Taiwan. *Water* **2014**, *6*, 3224–3246. [CrossRef]
66. Herrmann, T.; Schmida, U. Rainwater Utilisation in Germany: Efficiency, dimensioning, hydraulic and environmental aspects. *Urban Water* **1999**, *1*, 307–316. [CrossRef]
67. Nolde, E. Possibilities of rainwater utilisation in densely populated areas including precipitation runoffs from traffic surfaces. *Desalination* **2007**, *215*, 1–11. [CrossRef]
68. Partzsch, L. Smart regulation for water innovation-the case of decentralized rainwater technology. *J. Clean. Prod.* **2009**, *17*, 985–991. [CrossRef]
69. Schuetze, T. Rainwater harvesting and management—Policy and regulations in Germany. *Water Sci. Technol. Water Supply* **2013**, *13*, 376–385. [CrossRef]
70. Ward, S. Rainwater Harvesting in the UK - Current Practice and Future Trends. Available online: http://www.harvesth2o.com/rainwater_harvesting_UK.shtml (accessed on 15 January 2018).
71. UKRHA (United Kingdom Rainwater Harvesting Association). *Enhanced Capital Allowance Scheme as it Applies to Rainwater Harvesting Systems*; Industry Fact Sheet No 3; UKRHA: Nottinghamshire, UK, 2006.
72. Gerolin, A.; Le Nouveau, N.; de Gouvello, B. Rainwater harvesting for stormwater management: Example-based typology and current approaches for evaluation to question French practices. In Proceedings of the 8th International Conference, Novatech, Lyon, France, 23–27 June 2013; p. 10.
73. Vialle, C.; Busset, G.; Tanfin, L.; Montrejaud-Vignoles, M.; Huau, M.-C.; Sablayrolles, C. Environmental analysis of a domestic rainwater harvesting system: A case study in France. *Resour. Conserv. Recycl.* **2015**, *102*, 178–184. [CrossRef]

74. Belmeziti, A.; Coutard, O.; de Gouvello, B. A New Methodology for Evaluating Potential for Potable Water Savings (PPWS) by Using Rainwater Harvesting at the Urban Level: The Case of the Municipality of Colombes (Paris Region). *Water* **2013**, *5*, 312–326. [CrossRef]
75. Campling, P.; De Nocker, L.; Schiettecatte, W.; Iacovides, A.I.; Dworak, T.; Arenas, M.A.; Pozo, C.C.; Le Mat, O.; Mattheiß, V.; Kervarec, F. *Assessment of the Risks and Impacts of Four Alternative Water Supply Options*; European Commission—DG Environment: Brussels, Belgium, 2008.
76. Silva, C.M.; Sousa, V.; Carvalho, N.V. Evaluation of rainwater harvesting in Portugal: Application to single-family residences. *Resour. Conserv. Recycl.* **2015**, *94*, 21–34. [CrossRef]
77. Reitano, R. Water Harvesting and Water Collection Systems in Mediterranean Area. The Case of Malta. *Procedia Eng.* **2011**, *21*, 81–88. [CrossRef]

Article

Assessment of A Rainwater Harvesting System in A Multi-Storey Residential Building in Brazil

Jéssica Kuntz Maykot * and Enedir Ghisi

Laboratory of Energy Efficiency in Buildings, Department of Civil Engineering, Federal University of Santa Catarina, Florianópolis-SC 88040-900, Brazil; enedir.ghisi@ufsc.br

* Correspondence: jessica.k.maykot@gmail.com; Tel.: +55-48-3721-2115

Received: 6 January 2020; Accepted: 13 February 2020; Published: 15 February 2020

Abstract: This article aims to present an economic feasibility and user satisfaction analysis of a rainwater harvesting system in a multi-storey residential building (where there is rainwater to supply toilets) located in Florianópolis, southern Brazil. This research used detailed methods and also considered the opinion and habits of users regarding the use of a rainwater harvesting system. The water end-uses were estimated through questionnaire survey in each flat. The potential for potable water savings was estimated using computer simulations. Simulations were performed using the computer programme Netuno, version 4 and economic feasibility analyses were performed considering different rainwater demands. Analyses associated with the habits of the residents, the satisfaction of users and the importance of saving potable water were also carried out. Showers were responsible for the highest share (54.2%) of water consumption in the flats, followed by the other end-uses: washing machine (21.3%), kitchen tap (9.3%), toilet flush (9.2%) and washbasins (2.6%). The most economically feasible system, which presented lower payback and higher internal rate of return, corresponds to the system sized to supply rainwater only to toilet flushing. Such a system would need a rainwater tank with a capacity smaller than the capacity of the one currently in use. In general, residents expressed satisfaction regarding the rainwater harvesting system installed in the building. The study is important because, besides obtaining water end-uses in the flats, it also investigates the perception of residents related to rainwater harvesting, which has been little explored in the scientific literature.

Keywords: rainwater harvesting system; multi-storey residential building; end-uses; economic feasibility; satisfaction survey

---

1. Introduction

Multiple factors contribute to water scarcity, including its heterogeneous distribution, population growth and increased water consumption in the agricultural, industrial and energy production sectors [1–3]. One of the solutions that can be used in buildings in order to save potable water is the rainwater harvesting system.

In general, Brazil has a high potential for saving potable water [4–7] through the use of rainwater, since the annual rainfall average ranges from 1146 mm to 2182 mm [5], depending on the region of the country. In residential buildings located in Santa Catarina, it is estimated that the potential for saving potable water varies between 23% and 100% [8]. According to Lopes et al. [9], it is possible to save between 75 and 471 L/household/day when using rainwater in houses in Santa Catarina. Therefore, it is important that the management of water consists of reusing, recycling and recovering the resource [3].

In Florianópolis, for example, there is a law [10] that requires the use of alternative sources of water supply in buildings. The city has a high potential for potable water savings (73%) by using rainwater, primarily in February [8]. Results from research conducted in a multi-storey residential building in the city show that the largest potential for saving potable water corresponds to 17% [11]. These variations in the potential for potable water savings through the use of rainwater occur due to different demands for potable water and different rainwater catchment areas. Research conducted in public buildings in Florianópolis showed that between 56% and 86% of the potable water consumption could be replaced with rainwater [12,13].

Regarding the water end-uses in residential buildings in different countries, it is noticeable that, in general, showers contribute to the highest water consumption [14–19]. Beyond that, the percentage of water consumed for toilet flushes is significant. In Rathnayaka et al. [17] and Matos et al. [20] research, for example, the toilet flush occupies the third place among the end-uses with a higher percentage of consumption of potable water in the residential sector. In a study conducted by Jordán-Cuebas et al. [19] the toilet flush was the second largest end-use in two multi-storey residential buildings. In schools and office buildings, toilet flushes contribute with a percentage of water consumption even larger when compared to the percentage obtained in residential buildings [13,21].

Even though Florianópolis has a high potential for potable water savings, before implementing a rainwater harvesting system in a building it is necessary to measure the feasibility of implementing this type of solution using financial indicators, i.e., to perform an economic feasibility analysis. In Athayde Júnior et al. [22] research, for example, it is noted that the implementation of this type of system is only feasible in high-standard buildings, where the tariffs and consumption of water are higher than the tariffs and consumption of water in medium and low-standard buildings. According to Dòmenech and Saurí [23], for both single-family and multi-storey residential buildings, there is only economic feasibility on the implementation of rainwater harvesting systems when the water consumption is high. According to Matos et al. [24], there is economic feasibility on the implementation of rainwater harvesting systems in commercial buildings. In wetter years or in intermediate humidity conditions in India, rainwater harvesting systems were economically feasible as payback period ranged from 2 to 10 years [25]. In China, Jing et al. [26] corroborate this information by stating that there is economic feasibility of the system only in humid and semi-humid regions of the country. Severis et al. [27] and Abas and Mahlia [28] state that rainwater harvesting systems that demand larger volumes of water tend to be more economically feasible. According to Amos et al. [29] the economic feasibility of the implementation of a rainwater harvesting system is affected by the regional cost of potable water, which is also supported by Abdulla [30].

User acceptance and satisfaction with domestic rainwater harvesting systems should also be taken into consideration before adopting this solution in buildings. In Bangladesh, Islam et al. [31] interviewed low-income residents and found that there is generally good acceptance for the use of rainwater as an alternative source for community water supply. Domènech and Saurí [23] conducted a survey in Spain to verify satisfaction of residents with rainwater harvesting systems in their homes. The research was conducted in high-standard single-family buildings and in a high-standard multi-storey residential building. According to the authors, the residents were satisfied with this type of system. However, there is still a gap to be filled in the scientific literature on this subject. Few studies have investigated user acceptance or satisfaction with this type of solution.

The use of rainwater harvesting systems in buildings is an alternative in order to enable the rational use of potable water. In addition of that, rainwater harvesting systems do not always provide water with adequate visual characteristics [32], which can be a barrier for acceptance of this type of solution. Thus, the purpose of this article is to perform economic analysis resizing a rainwater tank taking into account the end-uses of a high-standard multi-storey residential building in Florianópolis, southern Brazil. Furthermore, a user satisfaction assessment was conducted related to the rainwater harvesting system installed in the building.

2. Method

In order to assess economic and user satisfaction aspects related to the rainwater harvesting system in a multi-storey residential building, some procedures were conducted. Water consumption data were collected from the flats; water consumption monitoring questionnaires were applied to the residents;

the water end-uses were estimated and two scenarios considering different water consumption for non potable purposes were simulated to verify the ideal capacity of rainwater tank. Finally, the costs of equipment used in the current rainwater harvesting system were surveyed. Labour and electricity costs were also included in the feasibility analysis. The satisfaction survey was conducted by applying questionnaires to the residents. Figure 1 shows the steps performed during the research period.

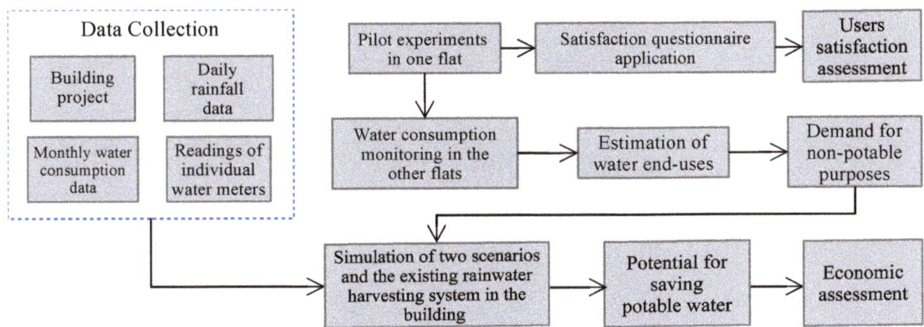

Figure 1. Performed for economic and user satisfaction assessments regarding rainwater harvesting system.

2.1. Characteristics of the Study Area

Florianópolis is located between the parallels 27°10′ and 27°50′ of south latitude and between the meridians 48°25′ and 48°35′ of west longitude [33]. The lowest rainfalls occur during winter, while the maximum rainfall in three consecutive months occurs in January, February and March (over summer) [34]. Figure 2 shows average rainfall over 2000 to 2015, obtained from EPAGRI/CIRAM (Santa Catarina Agricultural Research and Rural Extension Enterprise/Santa Catarina Environmental Resources and Hydrometeorology Information Center) weather station (latitude: 27°38′50″ S, longitude: 48°30′ W and altitude equal to two meters).

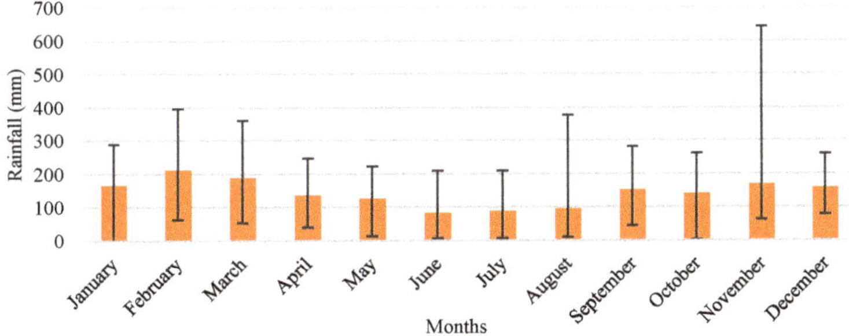

Figure 2. Monthly average, maximum and minimum rainfall in Florianópolis from 2000 to 2015.

2.2. The Multi-Storey Residential Building

This research was conducted in a multi-storey residential building which holds a rainwater harvesting system that supplies the flush of toilets. In Brazil the building is considered as a high-standard one. It consists of fourteen floors, and each floor contains four flats. The residents who live in the building are between the middle and upper social classes. Figure 3 shows the building facade and Table 1 shows the floor-plan area of the rooms in the flats.

Figure 3. Building facade.

Table 1. Areas of flat rooms.

Rooms	Areas (m^2)	
	Flats Type 1	Flats Type 2
Living and dining room	30.72	29.98
Kitchen	13.00	11.55
Bathroom in the service area	2.25	2.25
Laundry room	4.85	5.45
Small bedroom	7.02	-
Main balcony	9.45	8.60
Corridor	5.10	4.05
Toilet	1.82	-
Suite 1	11.15	10.45
Bathroom 1	3.20	3.20
Master suite	15.55	15.40
Bathroom 2	5.60	4.76
Second balcony	2.35	-
Suite 2	10.20	-
Bathroom 3	3.50	-
Third balcony	1.85	-
Bathroom for visitors	-	3.55
Bedroom	-	9.25
Total	127.61	108.49

2.3. Data Collection

Daily rainfall data recorded from 2000 to 2015 were used in this research. Analysing the architectural project of the building, the available roof area for rainwater catchment was obtained, i.e., 561.60 m^2.

Regarding water consumption in the flats, two types of data were collected. First, daily readings were taken on the water meters for each flat. These readings were performed over 21 days, between February and March (summer), at 10 pm. (due to lower water consumption). The difference between two readings performed on consecutive days resulted in the daily water consumption per flat. In addition, readings were recorded on the same date and start time of a questionnaire which was completed after 24 h in the flats—the procedure is better detailed in Section 2.4. Monthly water consumption data for each flat, from 2011 to 2018, were also collected from the water bills.

2.4. Water End-Uses

The water end-uses in the flats were estimated using a questionnaire with information about frequency, time of use and water flows at the water fixtures.

2.4.1. Pilot Experiment

Two different questionnaires were developed and two pilot experiments were conducted in order to define the questionnaire that minimised the error associated with the responses of the residents to water consumption. The most accurate questionnaire was applied to as many flats as possible.

Water Consumption

The water consumption on taps and showers where the pilot experiment was conducted was estimated through the product between the water flow of the water fixture and the time of use of the water fixture. The consumption of washing machines and dishwashers was obtained from the manufacturers with brand and model information of the equipment, provided by the residents according to Table 2. Cells marked with "-" indicate that the equipment was not used during monitoring of water consumption.

Table 2. Water consumption in washing machines and dishwashers which were used during monitoring in flats.

Flat Number	Brand and Washing Machine Model	Brand and Dishwasher Model	Washing Machine Water Consumption (m^3/Cycle)	Dishwasher Water Consumption (m^3/Cycle)
104	Brastemp 11 kg	-	0.136	-
202	Consul Facilite	-	0.130	-
204	Brastempadvantechwash 6 kg	Brastemp solution–8 services	0.156	0.020
402	LG inverter direct drive 8,5 kg	-	0.061	-
503	Consul 7,5 kg–Tide	-	0.097	-
601	BrastempAtive! 11 kg	-	0.139	-
602	LG Tromm	-	0.056	-
603	Electrolux ecoturbo wash and dry 10.5 kg	-	0.090	-
703	Eletrolux wash and dry 8.5 kg	-	0.072	-
801	Eletrolux 15.2 kg	-	0.126	-
901	LG Direct drive 10.2/6 kg	BrastempAtive 12 services	0.089	0.016
903	Brastemp 9 kg	-	0.097	-

There are dual-flush bowl-and-tank toilets in all flats; 3 litres of water are used in a half flush and 6 litres of water in a full flush.

The drinking water consumption was estimated considering the number of glasses (200 mL) of water the residents consume in a day. Consumption of bottled water was not included. The potable water consumption in the flat was calculated summing the daily consumption of drinking water, washing machine, the dishwasher, taps and showers.

Although potable water is also consumed for flushing the toilets, it was not possible to quantify separately the volumes of potable water and rainwater used in these appliances. This is because there is no measurement of the rainwater used nor measurement of the potable water used to flush the toilets. Therefore, the water consumption for toilet flushing was calculated separately from the water consumption in the flats.

Flow Rates Calculation

The water flow rates of taps and showers were calculated through the ratio of the recipient capacity (500 mL) to the time required to fill it. The taps were opened to half of their maximum aperture, in order to obtain an average flow of water when compared to their maximum flow rates. The valve of the shower was completely open. The final flow rates were calculated as the average of three flow rate measurements performed at each water fixture.

The First Questionnaire Developed

The first questionnaire was developed to allow residents to answer questions about their daily water consumption at each water fixture and questions related to their satisfaction with the rainwater harvesting system. Participants also expressed their opinion on the importance of water savings. This questionnaire was applied to a family of five people. After their responses, the average water consumption was calculated, excluding water consumption for toilet flushing. For verification purposes, the water consumption recorded in the water meter during the day of the experiment was used as a reference value.

The Second Questionnaire Developed

In order to improve the precision of the residents' answers, another questionnaire was elaborated, consisting of two parts. The first part, filled out by each person, questioned the participants about their particular opinions and habits. The second part aimed to monitor the water consumption in the flats over 24 h. The questionnaire was prepared in order that residents could register the usage time of each water fixture when they would use it. An area was also reserved for registering the amount of full and half toilet flushes that were set during the monitoring period, as explained in Section 2.4.2. In the questionnaires of the kitchen and laundry room, it was requested that the characteristics (brand and model) of the dishwasher and washing machine be registered. An area was also reserved to ask residents to mark the number of glasses of water consumed in the day. Residents were asked to write their names on the questionnaires in order to allow their identification in case of non comprehension of some data. This questionnaire was applied to the same family.

The second questionnaire provided more accurate results than the first questionnaire. Therefore, it was used to give continuity to the experiments in the other flats.

2.4.2. Experimental Procedure

Both parts of the second questionnaire were applied to the residents from February to April 2018. First, the residents were asked about the possibility of completing and delivering the questionnaire. On the delivery date of the questionnaire, the experimental procedure was explained in detail to the residents. They were asked about the possibility of differences in their water consumption during the weekdays and on weekends. In the flats whose residents answered that this possibility was true, the questionnaire was applied during a working day, as well as during a day on a weekend. Next, the day of the experiment was scheduled. It was clarified that the experiment should be concluded 24 h after its beginning. Besides, residents were instructed to keep water consumption as close as possible to their daily consumption and to send to the researcher photographs of showers and taps they used during monitoring.

Water consumption was calculated according to Section 2.4.1. In order to obtain tap and shower flow rates, some measurements were performed on taps in the common areas of the building due to the non consent by the residents to measure the flow rates in some flats. Thus, the flow rates used for the kitchen and laundry room taps correspond to the average of the flow rate of the kitchen tap obtained in the pilot experiment and the flow rate measured in the kitchen of the building common area. The flow rate in the washbasins was estimated as the average between the washbasin flows of the pilot experiment and the flow rates measured in the common areas. For the gas showers, it was considered the shower flow rate obtained in the pilot experiment. For electric showers, the flow rates specified by the manufacturer were used.

2.4.3. Differences Between Measured and Estimated Consumption

The potable water transported through the rainwater harvesting system is measured in the building general water meter, and it is monthly prorated between flats. Therefore, the water consumptions obtained using the questionnaire should approach the maximum of the consumption registered in the individual water meters. In this way, it will be possible to admit that there were no errors of registration in the questionnaire associated with the water consumption in the toilet flushes. The difference between the water consumption resulting from the questionnaire and the water consumption recorded in the water meter will be considered as an error.

2.4.4. Percentage of Consumption at the Points of Use

The water end-uses were estimated through the proportion between the daily water consumption in the water fixture and the total daily water consumption in the flat resulting from monitoring.

The average daily water consumption in the building was estimated through the sum of the average daily water consumption in the flats. These water consumptions were estimated by the ratio between the sum of the daily water consumptions of flats and the number of water meters read at 10 pm., according to Section 2.3.

2.5. Potential for Saving Potable Water

2.5.1. Percentage of Non Potable Water that can be Replaced with Rainwater and Simulated Scenarios

In order to estimate the potential for saving potable water in the building by using rainwater, the percentage of water used for non potable purposes was calculated. It corresponds to the sum of percentages of water used in toilet flushes, tanks in the laundry room and the washing machines of the flats. Thus, the first scenario was simulated in order that the rainwater harvesting system would provide rainwater for use in toilet flushes, laundry troughs and washing machines. The second scenario considered that the rainwater harvesting system would provide rainwater only for use in toilets (which is the current real situation of the existing rainwater harvesting system in the building).

2.5.2. Rainwater Tank Capacity

Although there is already a rainwater tank in the building, the method used to size such a tank is not known. In addition, it is sought to verify the feasibility of the rainwater harvesting system not only to supply the toilets flushes but also to supply the washing machines and the tap of the laundry tub.

The sizing of the rainwater tank was performed by using the "Netuno" programme [35], a computer programme developed in order to simulate rainwater harvesting systems. The programme simulates a rainwater catchment system for a set of known variables, supplied as input: daily rainfall data; discharge of the initial rainfall (equal to 2.0 mm [36]); catchment area; daily consumption of potable water per capita; total number of residents in the building; runoff coefficient of 0.85 in fibre cement tiles [37]; percentage of potable water to be replaced with rainwater; capacity of lower and upper rainwater tanks. The daily potable water consumption per capita was established through the proportion between the average monthly water consumption of the building, from historical series of water consumption data and the product between the residents of the building and the number of days in the month.

Several rainwater tank capacities were simulated at 1.0 m^3 intervals for each of the two different percentages of potable water to be replaced with rainwater. For each simulation, the "Netuno" programme calculated the potential for potable water savings through the use of rainwater. Thus, a graph was created where the x-axis represents the simulated lower rainwater tank capacities and the y-axis indicates the corresponding potential for potable water savings.

First, the total percentage of potable water that could be replaced with rainwater was considered. Then, only the percentage of water regarding the use in toilet flushes was considered. Ideal rainwater tank capacities were those whose last volume variation resulted in an increase in the potential of water savings equal to or less than 0.21% considering the use of rainwater in toilet flushes, washing machines and tubs in the laundry rooms and 1.1% when the rainwater harvesting system was sized to use rainwater only in the toilet flushes. The maximum capacities of the lower rainwater tanks were approximately 10 times higher than the daily rainwater demands considered in this study.

The upper rainwater tank capacities were considered equal to the daily rainwater demands: firstly the daily water consumption in the toilets, tubs in the laundry rooms and washing machines were considered and, finally, only the daily water consumption in the toilet flushes of the flats was taken into account.

2.6. Satisfaction of Users

Participants were individually asked about the use of full and half toilet flushes and the appearance of water in the toilets. In addition, they were asked about their satisfaction with the rainwater harvesting system, their knowledge related to the existence of this system, their satisfaction regarding the water pressure at the water fixtures and personal questions on water saving. Name, age, number of residents in the flat and gender were also registered.

Several answers were obtained since the questions were open. In order to facilitate data interpretation, each response was framed in a category defined by the researcher. Regarding the question about water saving importance, the answers associated with the environmental preservation argument were included in the category "resource preservation". Responses on water scarcity were included in the category "ensuring water supply to the population", and responses in which water savings were associated with a reduction of financial expenses were included in the "financial savings" category.

After data collection, the answers were transferred to a spreadsheet. Data were summarised, mainly through tables and frequency charts. The analyses verified both satisfaction of residents with the rainwater harvesting system and awareness of residents about water saving.

2.7. Economic Analyses

The economic feasibility analyses of the systems were performed using the "Netuno" programme [35], which estimates the internal rate of return, payback and the net present value. First, the feasibility of the system currently installed in the building was analysed. Subsequently, the economic feasibility analyses of two other systems were performed: one considering the total percentage of potable water that may be replaced with rainwater and the other considering only the percentage of water consumed to flush the toilets. In the first simulation, the capacities of the upper and lower tanks were considered according to the hydro-sanitary building project. In the second and third simulations, the rainwater tank capacities were modified to the ideal capacities found through the analyses explained in Section 2.5.2.

Since many costs could not be obtained for the time when the building was constructed, economic analyses were performed based on costs available in 2018. Equation (1), proposed by Tomaz in 2010 [38], was used to calculate the costs of the tanks constructed with reinforced concrete. The value (from 2010) was corrected for actuality considering Brazilian inflation from 2011 to 2018.

$$C = 336 \times V^{0.85} \quad (1)$$

where: C is the cost of the rainwater tank (USD) and V is the capacity of the tank (m^3).

Costs were considered for concreting as well as manufacturing, installation and removal of forms of the tanks [39]. For the system components installation, labour costs were also included [40].

Costs associated with the cleanliness of the tanks were estimated based on previous expenses of the building over the rainwater harvesting system operation (USD 125.00 per tank). Cleanliness of the tank was considered every six months. Pipe costs were estimated considering that they represent 15% of the contingency obtained when including costs of tanks and accessories of rainwater harvesting system [41].

For inflation, it was considered a monthly rate whose annual inflation corresponds to 4.0%—inflation target for 2020, according to resolution 4582 of the Central Bank of Brazil [42]. The minimum attractiveness rate was obtained through the monthly averages of interest rates (SELIC) from 2011 to May 2018, according to data provided by Brazilian Federal Revenue [43]. The analysis period considered was 20 years, the lifetime of the system [24,44].

Specifications in the sanitary project regarding motor pumps were also considered, as well as the tariffs of energy from the local energy company. In addition, the daily operation time of the pump was estimated according to the established rainwater demand (Equations (2) and (3)). The monthly cost of electric energy for the pump operation is given by Equation (4).

$$T_{op.npu} = \frac{C_{Mmonth} \times P_{npu}}{30 \times Q_{pump}} \quad (2)$$

$$T_{.toilet} = \frac{C_{Mmonth} \times P_{P.toilet}}{30 \times Q_{pump}} \quad (3)$$

$$C_{monthly} = P_{mp} \times T_{.} \times N_{d/month} \times V_{CELESC} \quad (4)$$

where: $T_{op.npu}$ is the time of daily pump operation when the rainwater harvesting system supplies all nonpotable uses (h/day); C_{Mmonth} is the monthly average consumption of water in the building, including the water volumes spent on toilets (m^3); P_{npu} is the percentage of water regarding nonpotable uses (%); Q_{pump} is the flow rate of water holding pressure of the motor pump assembly, specified in project and equal to 2.18 m^3/h; $T_{func.toilet}$ is the daily operating time of the pump when the rainwater harvesting system only supplies toilets (h/day); $P_{p.toilet}$ is the potable water percentage which may be replaced with rainwater regarding to the consumption in the toilets (%); $C_{monthly}$ is the monthly energy cost demanded by motor pumps (USD); P_{mp} is the power of the motor pumps (kW); $T_{func.}$ is the daily operation time of the motor pumps (h/day); $N_{d/month}$ is the number of days in the month when the pump is operating, and V_{CELESC} is the electric energy cost charged by the local energy company (USD/kWh).

Water and sewage tariffs were obtained from CASAN (Santa Catarina Water Sanitation Company), the local water company. In the case of the studied building the tariff corresponds to USD 1.9328 per m^2 consumed. The sewage rate charged corresponds to 100% of the water tariff. Two Brazilian taxes were also considered: PIS (1.65%)—Social Integration Programme and COFINS (7.60%)—Contribution to Social Security Financing, obtained from the water bills over 2018. A complete economic analysis should include assessment of the benefits from rainwater use, such as: reducing the costs of water bills, prevention of possible shortages of potable water and environmental preservation. The economic analyses performed in this article, however, did not quantify the last two benefits, since to perform this evaluation it would be necessary to perform a more complex analysis, which deviates from the main objectives of this study.

3. Results

3.1. Historical Data on Water Consumption

The building studied was inaugurated in 2009. Flats' water consumption data were available from 2011 to March 2018. Then, averages of monthly water consumption were calculated for each year and each flat. Subsequently, average daily consumptions were calculated for each flat using average monthly consumption data. Finally, the average water consumption per capita for the entire building was estimated. In addition, monthly water consumption averages were calculated as well as the average of the total monthly water consumption, regarding total monthly data available (which was 12.63 m^3). Figure 4 shows the maximum, minimum and monthly average water consumptions. It is noted that in February the average monthly water consumption of the flats is lower. This probably happens because in that month some residents tend to travel for summer holidays, thus they are not consuming water in the flats.

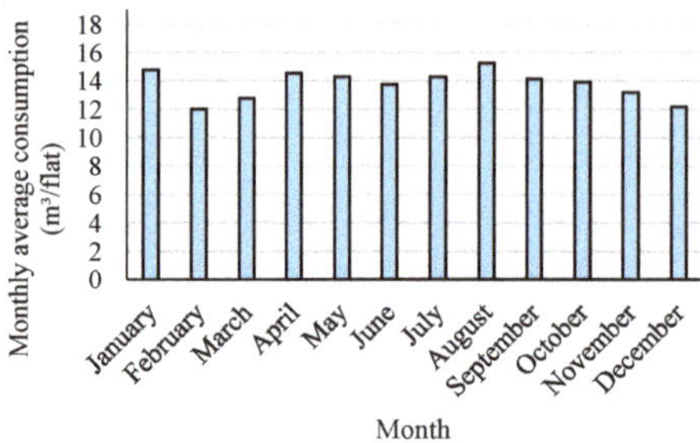

Figure 4. Average, maximum and minimum monthly consumption of water per flat (data from 2011 to 2018, excluding consumption of water from toilet flushes).

Figure 5 shows the distribution of frequencies for average monthly water consumption from 2011 to 2018. In Figure 5, it is noted that the consumption frequency between 0 and 3 m^3 is higher in 2011. In 2018 it was also observed that the number of flats whose consumption of water was between 0 and 3 m^3 was relatively higher than in 2012 and 2017. This may be explained because residents travel more during January and March. It is notable that in 2011, 2012 and 2017 one of the highest frequencies of water consumption in the flats corresponds to the range 9–12 m^3. In 2013, 2014, 2016 and in the first quarter of 2018 the highest frequency of average monthly water consumption was from 12 to 15 m^3. In 2015 and 2017, one of the highest frequencies of average monthly water consumption was from 6 to 9 m^3.

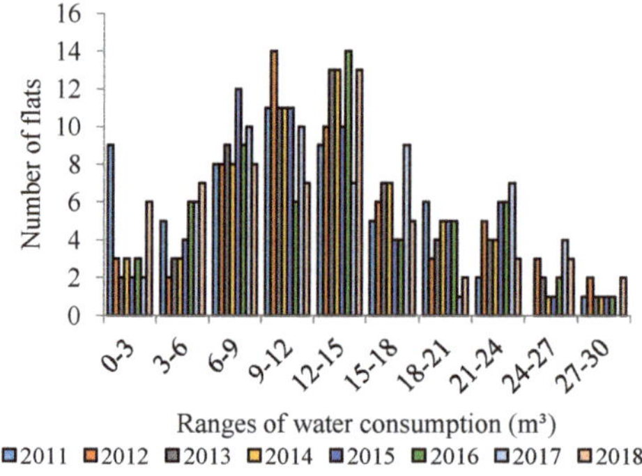

Figure 5. Histogram, separated per class of water consumption and per year, relative to monthly averages of water consumption in the flats (it does not include water consumption in toilets).

There are 150 residents living in the building (three residents per flat, on average). Thus, average consumption per capita was obtained in the building. Considering consumption data from January 2011 to March 2018, it is estimated that the average corresponds to 0.157 m^3/person/day. Figure 6 shows average daily water consumption per capita for each year of the analysis period. It should be noted that the lowest average water consumption per capita occurred in 2011 (0.136 m^3/person/day) while the highest water consumption per capita occurred in 2012 (0.167 m^3/person/day). The lowest consumption, in 2011, may be related to the fact that the number of residents in the building this year was lower than that considered for all years analysed (150 residents). The same number of inhabitants as 2018 was considered for the previous years since the number of inhabitants of these years is unknown.

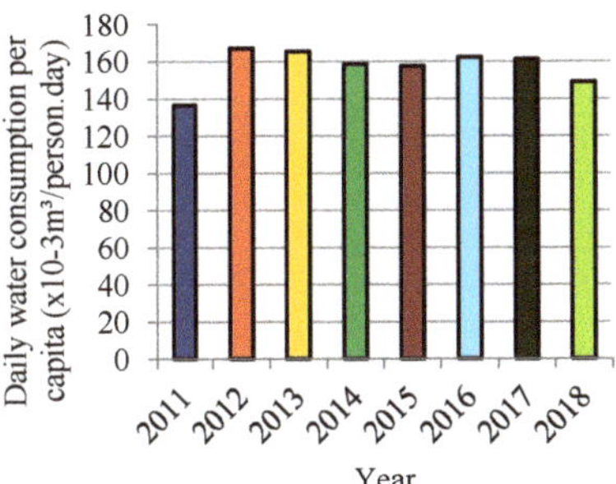

Figure 6. Average daily consumption per capita, based on annual consumption data (it does not include water consumption in toilets).

Water consumption data were also separated by the seasons of the year, and they were summarised as shown in Figure 7. Average daily water consumption per capita were also calculated for each season, including all consumption data (from 2011 to 2018)—Table 3. Thus, it is noticed that, in general, residents consume less water over the summer. However, in such a season, water consumption per capita data show a greater dispersion around the average consumption when compared to data from other seasons of the year. The lower consumption during summer may be associated with the period in which some residents travel or go on vacations, for example. In general, the highest consumption per capita occurs during autumn and winter. Some hypotheses may justify the higher consumption of water during the winter, such as long hot showers and a greater amount of time spent in the flats.

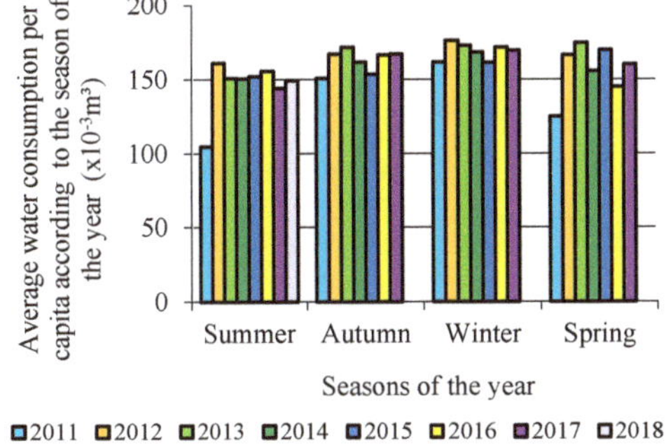

Figure 7. Average water consumption per capita separated by season (it does not include water consumption in toilets).

Table 3. Standard deviation, maximum and minimum water consumption per capita, including all water consumption data, separated by season of the year (it does not include water consumption in toilets).

Seasons	Average (m^3)	Standard Deviation (m^3)	Maximum (m^3)	Minimum (m^3)
Summer	0.146	0.017	0.161	0.105
Autumn	0.162	0.008	0.171	0.151
Winter	0.169	0.006	0.176	0.161
Spring	0.157	0.017	0.175	0.125

3.2. Daily Readings

The daily readings of the 56 water meters in the building indicate a trend towards lower water consumption on weekends. The average daily water consumption in the flats was 0.427 m^3, and the average daily consumption per capita was 0.154 m^3. The maximum daily consumption was 1.788 m^3 and, in some flats, there was no water consumption (these flats were empty during the reading period). It is important to emphasise that these water consumptions do not include the portion of water consumed in the toilets.

3.3. Pilot Experiment

After the application of the questionnaire in the first pilot experiment, it was verified that the difference between the water consumptions estimated and measured was 39.3%. This difference is due to mistaken estimates by residents on their average water consumption. When applying the second questionnaire, such a difference between consumptions was reduced to approximately 5.6%. Table 4 shows the results achieved with the application of these questionnaires.

Table 4. Results obtained in the first two experiments (pilot experiments).

Experiments	Consumption Registered in the Questionnaire (m^3/day)							Consumption Registered by Water Meter (m^3/day)	Difference of Consumption to Consumption Registered in the Questionnaire (%)
	Taps	Toilets	Showers	Washing Machine	Dishwasher	Filter	Total		
First Experiment	0.141	0.084	0.963	0.089	0.000	0.003	1.280	0.859	39.3
Second experiment	0.067	0.075	0.687	0.000	0.016	0.007	0.853	0.824	5.6

3.4. Water Flow Rates

In the showers of the flats (except for flats whose showers have an electric heating system), the flow rates measured in flat 1001 (0.00942 m^3/min) were considered. Table 5 shows the flow rates used in this study.

Table 5. Flow rates used to calculate the water consumption in the flats.

Water Fixture	Taps of Kitchen and Laundry Tub (m^3/s)	Taps of Washbasins (m^3/s)	Tap of the Bathroom in the Laundry (m^3/s)	Tap of the washbasin of the suite (m^3/s)	Tap of the barbecue area (m^3/s)	Showers with gas heating (m^3/min)
Flow rates used for all flats	0.000077	0.000063	0.000063	0.000063	0.000079	0.00942
Flow rates used in the pilot experiment	0.000069	0.000073	0.000076	0.000048	0.000079	0.00942

3.5. Water End-Uses

Among the 56 flats in the building, 23 accepted to participate in the monitoring of water consumption. In general, residents completed the questionnaire just once. Most of them reported being absent more frequently on weekends when compared to working days. As only two flats monitored the water consumption also on a weekend, such data was not considered separately from consumptions obtained on working days.

After calculating the daily water consumption in the flats (which excludes the water used for toilet flushing) and comparing it with the water consumption recorded using the water meters, a difference between real and estimated water consumption of less than 25% was verified for 15 flats. This difference was assumed once the number of households that accepted monitoring water consumption in the flats was restricted. Differences below 25% would greatly reduce the sample collected, impairing the representativeness of the sample. Among the other participating flats, three presented data with extremely high errors—for which reason they were not considered—(298.0%; 223.0% and 373.5%). In these three cases, the actual water consumption registered in the water meter was lower than the water consumption estimated using the questionnaire. These errors may be the result of mistaken estimates of appliance usage time (especially showers), inaccuracies in flow rates, high water consumption in the washing machine (since manufacturers usually provide its maximum water consumption). Figure 8 shows the percentage of water used at each water fixture to the total water consumption average in a flat.

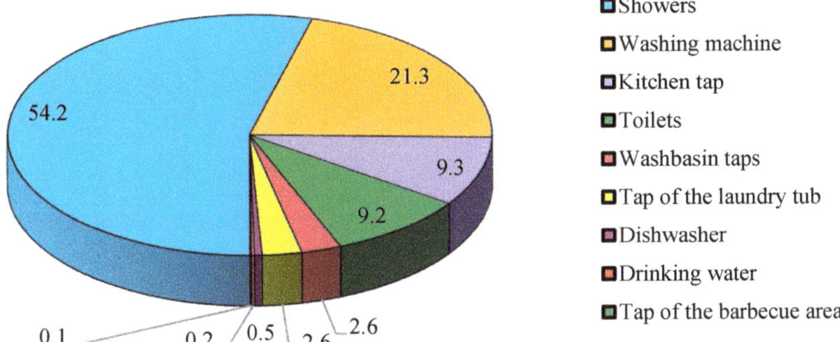

Figure 8. Water end-uses in the flats.

There are differences between actual and estimated water consumption (0.533 m^3 and 0.563 m^3, respectively). However, this difference is relatively low (5.4%). It is noted that the water consumption in showers exceeds half of the total water consumption in flats. The second largest consumption of water occurs with the use of washing machines, followed by the kitchen tap and toilets.

The monitoring of water consumption was performed for 24 h at each flat, since participants could not perform it for a full week. Thus, the day when participants performed the monitoring may not have been the most representative of their water consumption routine. Residents may have performed it when they usually wash their clothes, for example, which may raise the percentage of water used in washing machines. In addition, most washing machine manufacturers provide maximum water consumption values per cycle, but the programme used in the washing machine will not always be the one whose water consumption is the maximum. There are also flats where it is usual to use the washing machine more than once in a day. In one of the flats, for example, the use of the washing machine was observed four times during a day in the economic cycle. The manufacturer, however, only provided maximum consumption data, which makes it difficult to estimate the water consumption in the washing machines with more accuracy.

3.6. Potential for Rainwater Harvesting

Considering water consumption for laundry tubs, washing machines and toilets, it is possible to replace 33.1% of potable water with rainwater. The second scenario aims to verify if the rainwater tank was adequately sized to supply the demand for toilets, i.e., 9.2%. Table 6 shows the input data considered.

The upper tank capacity was assumed to be equal to the daily rainwater demand. Total water demand was calculated considering that the daily average of water consumed per person (0.157 m^3) corresponds to 90.81% of the total water demand. It was assumed that the remaining percentage is the portion of water consumption per capita used in the toilet. Therefore, the total water consumption of 0.173 m^3/person/day was considered.

Table 6. Data used in the "Netuno" programme for sizing the lower rainwater tank under two conditions of demand.

Input Variables	Different Rainwater Uses	
	Use of Rainwater in Toilets, Taps of Laundry Tubs and Washing Machines	Use of Rainwater only in Toilets
Percentage of total demand to be replaced with rainwater (%)	33.1	9.2
Upper rainwater tank capacity (m^3)	9.09	2.53
Lower tank maximum capacity (m^3)	90.00	30.00
Difference between potentials of potable water savings by using rainwater (%/m^3)	0.21	1.10
Interval between simulated capacities (m^3)	1.0	
First-flush diversion (mm)	2.0	
Catchment area (m^2)	561.60	
Total water demand (m^3/person/day)	0.173	
Number of residents	159	
Runoff coefficient	0.85	

It was verified that 150 residents were living in the building at the time. However, three flats were empty. Therefore, to perform the simulations in "Netuno", three residents were considered in each of these flats, adding up to 159 residents.

The simulation which took into account the water demand required to supply all the non potable uses indicated that the ideal lower tank capacity should be 20 m^3, with a potential of water savings equal to 6.5%. The tank capacity when rainwater was used only for toilet flushing was 16.0 m^3, with potable water savings equal to 5.0%. Figure 9 shows the simulated scenarios.

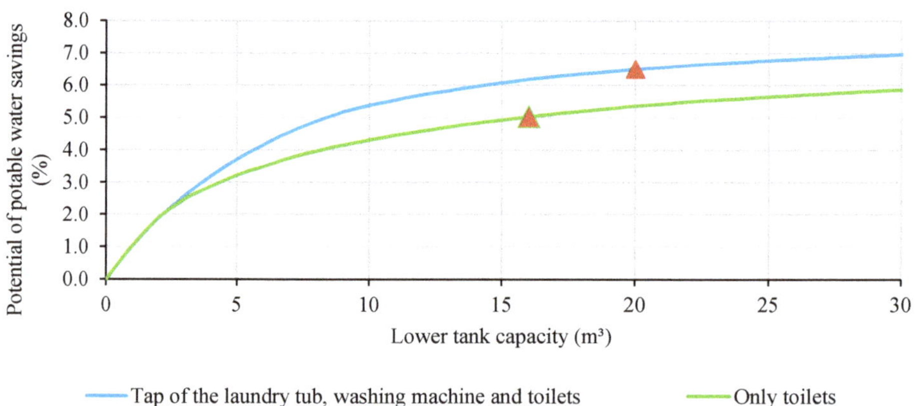

Figure 9. Water tank capacities versus the potential of potable water savings according to simulated scenarios.

The capacity of the lower rainwater tank which is currently in use in the building is 30.9 m^3. Such a capacity is much larger than the ideal capacities obtained for both scenarios. In addition, the capacity for the upper rainwater tank currently in use is 28.59 m^3. Figure 10 shows the results from Netuno when considering capacities from the rainwater tank currently in use in the building. The potential for potable water savings is approximately 6.15%. Therefore, the rainwater tanks that were constructed in the building were oversized.

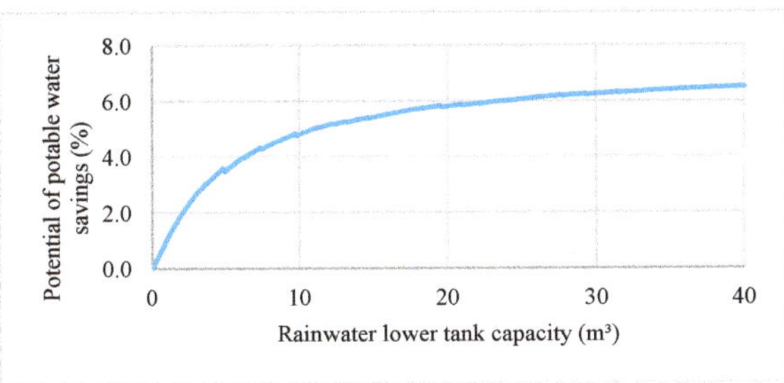

Figure 10. Potential for potable water savings as a function of the current rainwater tank capacities used in the building.

3.7. Economic Analysis

The average water consumption—based on the historical series of water consumption in the building—corresponds to 13.9 m^3/month/flat (including the percentage used in toilets). Table 7 shows other data used in the economic analysis in each of the scenarios. In 2010, the year in which the work of Tomaz [38] showed an equation for calculating costs of rainwater tanks constructed with reinforced concrete was published, one dollar was equivalent (on average) to 1.76 reais.

Table 7. Used in the "Netuno" programme for economic analysis of rainwater harvesting systems according to different tank capacities.

Rainwater Use	Toilets, Washing Machines and Taps of the Laundry Tubs		Toilets			
Potential for saving potable water (%)	6.5		5.0		6.1 (Existent system)	
Tanks costs (USD)	Lower: 2928.95	Upper: 1498.36	Lower: 2422.92	Upper: 505.23	Lower: 4239.37	Upper 3968.43
Labour costs (USD)	1402.74		1075.82		2008.59	
Pipes costs (USD)	946.50		721.62		1513.57	
Motor pump run time (h/day)	2.10		0.59			
Monthly cost with electric energy in the motor pump (USD/month)	16.08		4.47			
Motor pump cost (USD)	1169.75					
Cost of vortex WFF 150 Filter (USD)	507.50					
Two-way solenoid valve cost (USD)	180.00					
Cost of float switch (USD)	25.75					
Tank installation month	July					
Motor pump flow (m^3/h)	2.180					
Motor pump power (cv)	3.0					
Motor pump efficiency (%)	32					
Operating days per month (days)	30					
Cost of electric energy (USD/kW)	0.11496					
ICMS (%)—Tax on the circulation of goods and services	25					
Monthly inflation (%)	0.33					
Minimum rate of attractiveness per month (%)	0.82					
Adjustment of water and electricity tariffs (months)	12					

Table 8 shows the results found after performing the economic analyses.

Table 8. Analyses conducted for different potentials for saving potable water.

Indicators	Washing Machines, Toilets and Taps of Laundry Tubs	Only Toilets (Potential of Saving Potable Water: 5.0%)	Only Toilets (Potential of Saving Potable Water: 6.1%)
Present net value (USD)	12,540.01	11,930.77	6137.91
Discounted payback (months)	67	57	131
Internal rate of return (% per month)	2.07	2.34	1.25

According to the economic analysis, all the scenarios are economically feasible. The highest net present value was obtained for the scenario designed to supply all non potable uses in the flats, i.e., laundry tubs, washing machines and toilet flushing. However, payback was higher than the one for the system designed to supply rainwater to toilet flushing only. Among the two scenarios and the current rainwater harvesting system of the building, this last one is the least economically feasible, and this is due to the large rainwater tank capacities used in the building. The best economic feasibility was the one for the scenario that supplies rainwater to toilet flushing only, in which the rainwater tank capacities are smaller than those installed in the building.

3.8. User Satisfaction

Questionnaires were completed in 27 flats but not all the residents of the flats answered them. From all answers collected, 31 were from females (55.4%) and 25 (44.6%) from males. The average age of the sample was 49 years, ranging from 14 to 94 years. On average there were three residents per flat but this ranged from one to five.

The first four questions were related to their habits on full and half toilet flushes. Table 9 summarises the results associated with such questions. Most of the residents use the half flush and many of them admit that using a single half flush is enough to transport the waste (42.9%). Thus, a higher percentage of not using total flush was expected. Although many participants agreed with this, 82% of the participants stated that they use the full flush. This can occur due to the lack of attention when flushing the toilet, a simple habit of using full flush without concern about the real need for its use or a disregard by users since the water used in the toilets can be rainwater.

The residents were satisfied with the appearance of rainwater used in the toilets. Twelve residents reported having noticed turbidity in the toilet rainwater; most of them stated that this had been caused by incorrect cleaning of the rainwater tank. Regarding the water pressure at the water fixtures, almost all residents were satisfied; only two of them expressed dissatisfaction associated with water pressure. Table 9 also shows the results related to water appearance and water pressure satisfaction.

Regarding the questions about the rational use of water, 21.4% of participants were unaware of the existence of a rainwater harvesting system in the building and 7.1% of them did not answer this question. Figures 11 and 12 show the percentage of answers regarding the function of the rainwater harvesting system and the importance of saving water, respectively.

Almost half of the residents who answered the questionnaire know that the rainwater harvesting system supplies only toilets. In general, residents associate the importance of saving water with the possible scarcity of the resource, making it insufficient to supply the population. However, 20% of residents stated that water savings also represents economic savings in the monthly water bill. All participants understood the importance of saving water, but 6% admitted that they could save more, or that they do not care about saving water. Some residents did not answer if they usually save water (16%). Most participants said they were trying to save it (78%), however, many answers resembled the following statement: "whenever I can, I save water" or "I save water whenever it is possible". These statements, however, are very vague to define whether the residents are concerned or not about saving water.

Table 9. Questions related to toilet usage habits, toilet flush performance, satisfaction with the appearance of water and the water pressure used in taps and showers.

Question	Categories	Number of People	Percentage to the Total (50 answers) (%)
Do you usually use the partial flush in the toilets of your flat?	Yes	47	83.9
	No	9	16.1
Is a partial flush sufficient to transport the waste (solids)?	Without answer	2	3.6
	Yes	24	42.9
	Sometimes	12	21.4
	No	18	32.1
Do you usually use the full flush in the toilets of your flat?	Yes	47	83.9
	No	9	16.1
Is a full flush sufficient to transport the waste (solids)?	Yes	54	96.4
	Sometimes	1	1.8
	No	1	1.8
Are you satisfied with the appearance of the toilet water used in your flat?	Satisfied	54	96.4
	Dissatisfied	1	1.8
	Without answer	1	1.8
Have you noticed signs of turbidity in the water used in the toilets of your flat?	Yes	10	17.9
	Rarely	5	8.9
	No	41	73.2
Are you satisfied with the water pressure in the taps and showers in your flat?	Satisfied	54	96.4
	Dissatisfied	2	3.6

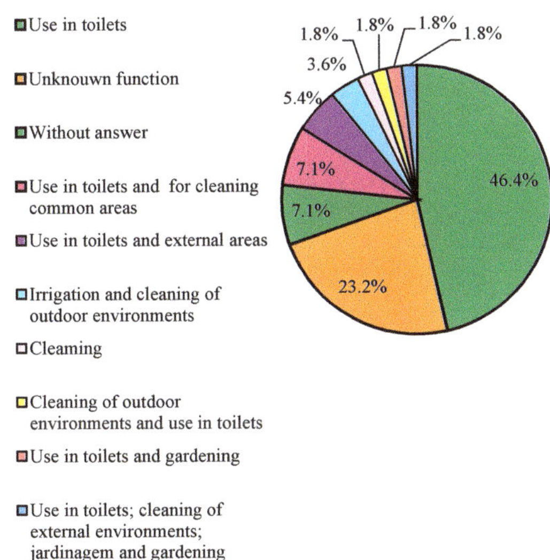

Figure 11. Responses about the function of the rainwater harvesting system used in the building.

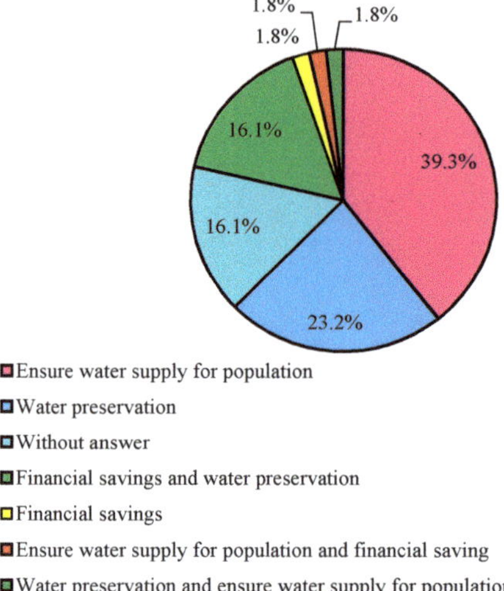

Figure 12. Responses about the purposes of using rainwater harvesting system in the building.

4. Discussion

It was observed that daily water consumption per capita (including consumption in toilets) is similar to the consumption of water found in surveys conducted by Willis et al. [15] in Australian residential buildings and by Loh and Coghlan [16], who also analysed water consumption in Australian households. In addition, the increasing order of water consumption found in the two surveys is very similar to that found in this work (showers, washing machines, kitchen taps, toilets and washbasin taps). However, in both studies, the percentage of water consumed in the shower is lower (33% in both surveys) than that found in our study (more than 50%). The percentage of water used in the washing machines is slightly higher (21.3%) than the percentage observed in the Australian residential buildings of the research of Willis et al. [15] (19%) and higher than the percentage of water used in the same type of equipment in the Loh and Coghlan study [16]. There was also a difference in the use of water in toilets. The research of Willis et al. [15] showed that water consumption in toilets equals 13% of the total water consumption in residential buildings. The research of Loh and Coghlan [16] estimated that water consumption used in toilets is 17% of the total water consumption in residential buildings, i.e., higher values than those obtained in our research. The report about rational water use and energy efficiency in social housing [45] also found that in popular houses, the shower represents the largest share in water consumption.

Regarding the estimation of potable water savings by using rainwater harvesting system, it was verified that the savings obtained in our work are considerably lower than several surveys conducted not only in Florianópolis but also in other regions of Brazil [5–8,11]. One of the reasons why this discrepancy was observed may be associated with the small size of the catchment area in relation to the demand required to meet the non potable uses of flats.

Some authors [22,23] agree that the implementation of a rainwater harvesting system is only economically feasible when there is high daily water consumption per capita. However, the economic analyses performed in this research, for different potentials of potable water savings, confirmed the economic feasibility of the system implementation even if the average daily consumption per capita is not that high.

The building where this research was conducted is considered to be a high standard one, since the construction company values the quality of buildings using high quality materials, optimized construction processes and continuous improvement (certified by ISO 9001 (International Organization for Standardization)). Even so, it is important that research be performed to investigate the behaviour of users in these buildings in order to verify whether a certain constructive solution, especially in the area of sustainability, is well used. The analysis about the answers on the habits related to the use of the toilet flushes revealed that the majority of residents use the half flush and most of them answered that a single half flush is sufficient to transport the solid waste. In general, the residents of the building are satisfied with the pressure and appearance of water used in toilets, which demonstrates satisfaction with the rainwater harvesting system performance. Most residents did not notice signs of turbidity. On issues related to water savings, all residents are aware of this importance and many of them affirm that they save water whenever it is possible.

During the research some limitations were assumed. One of them corresponds to the number of flats participating in the survey. Although the building consists of 56 flats, only 26 households agreed to participate in the experiment. Even with 26 data resulting from water consumption monitoring in the flats, some were excluded from the analysis due to their high consumption differences. Other limitations were: the impossibility of making measurements of water end-uses, being necessary the monitoring questionnaire application; impossibility to conduct monitoring for one week in each flat, which could produce more stringent results regarding the water consumption routine in the flats (the probability of using equipment with higher water consumption, such as washing machine and dishwasher would increase); use of average flow rate values for taps and showers of flats, rather than real values, measured on site; the water consumption in the toilets was considered to be real (without comparing it with the measured water consumption, since there was no measurement of such water consumption); no life-cycle analyses were conducted associated with the components of the rainwater harvesting system.

Although the study was applied to a multi-storey residential building in Brazil, the results described can be applied to multi-storey residential buildings with similar characteristics, in addition to residents with similar socioeconomic status.

5. Conclusions

The current rainwater tanks used in the building were oversized, which resulted in unnecessary costs and greater payback period. The upper rainwater tank could have a capacity close to the daily average rainwater demand. The capacity of the lower rainwater tank could be closer to 20.0 m^3 (in the case of using rainwater to supply toilets, washing machines and laundry tubs) or 16.0 m^3 (in the case of using rainwater to supply only toilets). The highest potential for potable water savings among the scenarios simulated was 6.50% for a lower tank capacity equal to 20.0 m^3.

Data from satisfaction analysis for the water used in toilets showed good acceptance by residents. Results may indicate that visual water quality parameters are not essential for water to be used for nonpotable purposes.

The results show that the implementation of rainwater harvesting systems in multi-storey residential buildings is an economically feasible alternative that satisfies users between middle and upper socioeconomic classes in Florianópolis. The most feasible scenario is the one which is adequately sized to supply only toilets. It should be noted that the most economically feasible alternative will not always be the one that uses rainwater to supply larger amounts of water appliances. The implementation of rainwater harvesting systems may be a marketing strategy for construction companies to expand the sale of residential buildings that promote more rational use of water. On the other hand, it is suggested to construction companies to provide rainwater meters for each flat so that residents can be more aware about the amount of rainwater being used and the amount of potable water required to supply the water appliances when rainwater is not available. This can make residents value the water resources more, avoiding waste.

Some factors, however, do not support the decision to implement the rainwater harvesting system in buildings located in Florianópolis. One of them is the system of charging the water tariff in the municipality. The local water company establishes a fixed (minimum) rate of water for monthly consumption up to 10 m^3. Buildings whose monthly water consumption is less than 10 m^3 pay the fixed rate and, therefore, the implementation of rainwater harvesting systems in these buildings would not provide economic savings. In addition, it is not very attractive for construction companies to install this type of system, since it raises the cost of the enterprise and does not provide direct benefit to the companies. In order for rainwater harvesting systems to be more widespread, there is a need to change the mentality of many consumers, change the method of water tariff charging (especially in buildings where consumption is low) and create public policies that encourage the use and installation of rainwater harvesting systems.

Author Contributions: Conceptualization: J.K.M. and E.G.; methodology: J.K.M. and E.G.; software: E.G.; validation: J.K.M. and E.G.; formal analysis: J.K.M.; investigation: J.K.M.; resources: J.K.M.; data curation: J.K.M.; writing: J.K.M.; writing—review and editing: J.K.M. and E.G.; visualization: J.K.M. and E.G.; supervision: E.G.; project administration: J.K.M. and E.G. All authors have read and agreed to the published version of the manuscript.

Funding: This research received no external funding.

Acknowledgments: The authors would like to thank CAPES and CNPq (agencies of the Brazilian Government for research) for the financial support. The authors would also like to thank all the subjects involved in the field study.

Conflicts of Interest: The authors declare no conflict of interest.

References

1. WWAP (United Nations World Water Assessment Programme); UNESCO. *The United Nations World Water Development Report 2015: Water for a Sustainable World*; UNESCO: Paris, France, 2015; Available online: http://unesdoc.unesco.org/images/0023/002318/231823E.pdf (accessed on 5 September 2017).
2. United Nations. *World Population Prospects. The 2017 Revision: Key Findings and Advance Tables*; United Nations: New York, NY, USA, 2017.
3. WWAP (United Nations World Water Assessment Programme); UNESCO. *The United Nations World Water Development Report 2017. Wastewater: The Untapped Resource*; UNESCO: Paris, France, 2017; Available online: http://unesdoc.unesco.org/images/0024/002471/247153e.pdf (accessed on 5 September 2017).
4. Ghisi, E.; Bressan, D.L.; Martini, M. Rainwater tank capacity and potential for potable water savings by using rainwater in the residential sector of southeastern Brazil. *Build. Environ.* **2007**, *42*, 1654–1666. [CrossRef]
5. De Lima, J.A.; Dambros, M.V.R.; De Antonio, M.A.P.M.; Jansen, J.G.; Maechetto, M. Potencial da economia de água potável pelo uso de água pluvial: Análise de 40 cidades da Amazônia. [Potential of saving potable water by means of rainwater use: Analysis of 40 cities in the Amazon]. *Eng. Sanit. Ambient* **2011**, *16*, 291–298. (In Portuguese) [CrossRef]
6. Maia, A.G.; Dos Santos, A.L.; De Oliveira Filho, P.C. Avaliação da economia de água potável com a implantação de um sistema de aproveitamento de água de chuva: Estudo de caso no município de Irati, Paraná [Assessment of saving potable water with the implantation of a Rainwater harvesting system: A case of study in the city of Irati, Paraná]. *Ambiência* **2011**, *7*, 51–63. (In Portuguese)
7. Dos Santos, S.M.; De Farias, M.M.M.W.E.C. Potential for rainwater harvesting in a dry climate: Assessments in a semiarid region in northeast Brazil. *J. Clean. Prod.* **2017**, *164*, 1007–1015. [CrossRef]
8. Ghisi, E.; Montibeller, A.; Schmidt, R.W. Potential for potable water savings by using rainwater: An analysis over 62 cities in southern Brazil. *Build. Environ.* **2006**, *41*, 204–210. [CrossRef]
9. Lopes, A.C.; Rupp, R.F.; Ghisi, E. Assessment of the potential for potable water savings by using rainwater in houses in Southern Brazil. *Water Sci Tech-W Sup* **2016**, *16*, 533–541. [CrossRef]
10. FLORIANÓPOLIS. Lei ordinária n° 8080, de 09 de novembro de 2009. [Ordinary law n° 8080 of 9th of November, 2009]. Institui o programa municipal de conservação, uso racional e reuso da água em edificações e dá outras providências. [Establishes a municipal programme for the conservation, rational use and reuse of water in buildings and other measures]. Sistema leis municipais. [System of municipal laws]. Available online: https://leismunicipais.com.br/a1/sc/f/florianopolis/lei-ordinaria/2009/808/8080/lei-ordinaria-n-8080-2009-institui-programa-municipal-de-conservacao-uso-racional-e-reuso-da-agua-em-edificacoes-e-da-outras-providencias?q=8080 (accessed on 26 June 2018). (In Portuguese).

11. Ghisi, E.; Ferreira, D.F. Potential for potable water savings by using rainwater and greywater in a multi-storey residential building in southern Brazil. *Build. Environ.* **2007**, *42*, 2512–2522. [CrossRef]
12. Kammers, P.C.; Ghisi, E. Usos finais de água em edifícios públicos localizados em Florianópolis, SC. [Water end-uses in public buildings located in Florianópolis, SC]. *Ambient Constr.* **2006**, *6*, 75–90. (In Portuguese)
13. Proença, L.C.; Ghisi, E. Water end-uses in Brazilian office buildings. *Resour. Conserv. Recycl.* **2010**, *54*, 489–500. [CrossRef]
14. Marinoski, A.K.; Vieira, A.S.; Silva, A.S.; Ghisi, E. Water End-Uses in Low-Income Houses in Southern Brazil. *Water* **2014**, *6*, 1985–1999. [CrossRef]
15. Willis, R.; Stewart, R.A.; Panuwatwanich, K.; Capati, B.; Giurco, D. Gold Coast domestic water end use study. *Water: J Austr Water Assoc* **2009**, *36*, 3679–3685.
16. Loh, M.; Coghlan, P. Domestic Water Use Study: In Perth, Western Australia 1998–2001. *Water Corp.* **2003**. Available online: https://www.water.wa.gov.au/__data/assets/pdf_file/0016/5029/42338.pdf (accessed on 19 September 2017).
17. Rathnayaka, K.; Malano, H.; Maheepala, S.; George, B.; Nawarathna, B.; Aora, M.; Roberts, P. Seasonal Demand Dynamics of Residential Water End-Uses. *Water* **2015**, *7*, 202–2016. [CrossRef]
18. Barreto, D. Perfil do consumo residencial e usos finais da água [Profile of residential consumption and water end-uses]. *Ambient Constr.* **2008**, *8*, 23–40. (In Portuguese)
19. Jordán-Cueba, F.; Masce, U.K.; Andrews, C.J.; Senick, J.A.; Hewitt, E.L.; Wener, R.E.; Allaci, M.S.; Plotnik, D. Understanding Apartment End-Use Water Consumption in Two Green Residential Multistory Buildings. *J. Water Resour. Plan. Manag.* **2018**, *144*, 1–20. [CrossRef]
20. Matos, C.; Teixeira, C.A.; Bento, R.; Varajão, J.; Bentes, I. An exploratory study on the influence of socio-demographic characteristics on water end uses inside buildings. *Sci. Total Environ.* **2014**, *466–467*, 467–474. [CrossRef]
21. Fasola, G.B.; Ghisi, E.; Marinoski, A.K.; Borinelli, J.B. Potencial de economia de água em duas escolas em Florianópolis, SC. [Potential of water savings in two schools in Florianópolis]. *Ambient Constr.* **2011**, *11*, 65–78. (In Portuguese) [CrossRef]
22. Athayde Júnior, G.B.; Dias, I.C.S.; Gadelha, C.L.M. Viabilidade econômica e aceitação social do aproveitamento de águas pluviais em residências na cidade de João Pessoa. [Economic feasibility and social acceptance of rainwater use in homes in the city of João Pessoa]. *Ambient Constr.* **2008**, *8*, 85–98. (In Portuguese)
23. Domènech, L.; Saurí, D. A comparative appraisal of the use of rainwater harvesting in single and multi-family buildings on the Metropolitan Area of Barcelona (Spain): Social experience, drinking water savings and economic costs. *J. Clean. Prod.* **2011**, *19*, 598–608. [CrossRef]
24. Matos, C.; Bentes, I.; Santos, C.; Imteaz, M.; Pereira, S. Economic analysis of a rainwater harvesting system in a commercial building. *Water Resour. Manag.* **2015**, *29*, 3971–3986. [CrossRef]
25. Bashar, M.Z.I.; Karim, M.R.; Imteaz, M.A. Reliability and economic analysis of urban Rainwater harvesting: A comparative study within six major cities of Bangladesh. *Resour. Concerv. Recycl.* **2018**, *133*, 146–154. [CrossRef]
26. Jing, X.; Zhang, S.; Zhang, J.; Wang, Y.; Wang, Y. Assessing efficiency and economic viability of rainwater harvesting systems for meeting non-potable water demands in four climatic zones of China. *Resour. Conserv. Recycl.* **2017**, *126*, 74–85. [CrossRef]
27. Severis, R.M.; Silva, F.A.; Wahrlich, J.; Skoronski, E.; Simioni, F.J. Economic analysis and risk-based assessment of the financial losses of domestic rainwater harvesting systems. *Resour. Conserv. Recycl.* **2019**, *146*, 206–217. [CrossRef]
28. Abas, P.E.; Mahlia, T.M.I. Techno-Economic and Sensitivity Analysis of Rainwater Harvesting System as Alternative Water Resource. *Sustainability* **2019**, *11*, 2365. [CrossRef]
29. Amos, C.C.; Rahman, A.; Gathenya, J.M. Economic Analysis of Rainwater Harvesting Systems Comparing Developing and Developed Countries: A Case Study of Australia and Kenya. *J. Clean. Prod.* **2018**, *172*, 196–207. [CrossRef]
30. Abdulla, F. Rainwater harvesting in Jordan: Potential water saving, optimal tank sizing and economic analysis. *Urban Water J.* **2019**. [CrossRef]
31. Islam, M.M.; Chou, F.N.-F.; Kabir, M.R. Feasibility and acceptability study of rainwater use to the acute shortage areas in Dhaka city, Bangladesh. *Nat. Hazards* **2011**, *56*, 93–111. [CrossRef]

32. Mendez, C.B.; Klenzendorf, J.B.; Afshar, B.R.; Simmons, M.T.; Barrett, M.E.; Kinney, K.A.; Kirisits, M.J. The effect of roofing material on the quality of harvested rainwater. *Water Res.* **2011**, *45*, 2049–2059. [CrossRef]
33. Papst, A.L. Uso de Inércia Térmica no Clima Subtropical: Estudo de Caso em Florianópolis—SC. [Use of Thermal Inertia in the Subtropical Climate: A Case of Study in Florianópolis—SC]. Master's Thesis, Federal University of Santa Catarina, Florianópolis, Brazil, 1999. (In Portuguese).
34. Nimer, E. *Climatologia do Brasil, [Climatology of Brazil]*, 2nd ed.; IBGE, Department of Natural Resources and Environmental Studies: Rio de Janeiro, Brazil, 1989. (In Portuguese)
35. Ghisi, E.; Cordova, M.M. Netuno 4 2014. Computer programme. Federal University of Santa Catarina, Department of Civil Engineering. Available online: http://www.labeee.ufsc.br/ (accessed on 5 September 2017). (In Portuguese).
36. Associação Brasileira de Normas Técnicas. *[Brazilian Association of Technical Standards] NBR 15527: Água de Chuva—Aproveitamento de Coberturas em áreas Urbanas Para Fins não Potáveis—Requisitos, [NBR 15527: Rainwater—Use of Roofs in Urban Areas for Non-Potable Purposes—Requirements]*; ABNT: Rio de Janeiro, Brazil, 2007. (In Portuguese)
37. DTU Development Technology Unit. Very-low-cost domestic roof water harvesting in the humid tropics: Existing practice. School of Engineering, University of Warwick: Coventry, UK, 2002. Available online: http://www.eng.warwick.ac.uk (accessed on 2 December 2017).
38. Tomaz, P. Aproveitamento de água de chuva em áreas urbanas para fins não potáveis 2010. [Rainwater harvesting in urban areas for non-potable purposes]. Available online: http://www.pliniotomaz.com.br/downloads/livros/Livro_aprov._aguadechuva/LivroAproveitamentodeáguadechuva5dez2015.pdf (accessed on 24 October 2017). (In Portuguese).
39. TCPO. Tabelas de composições de preços para orçamentos [Price Composition Tables for Budgets]. In *Civil Engineering, Construction and Architecture*, 13th ed.; PINI: São Paulo, Brazil, 2010. (In Portuguese)
40. SINAPI—Índices da Construção Civil. [Civil Construction Indexes]. Available online: http://www.caixa.gov.br/poder-publico/apoio-poder-publico/sinapi/Paginas/default.aspx (accessed on 30 May 2018). (In Portuguese)
41. Ferreira, M.I.P.; Silva, J.A.F.D.; Pinheiro, M.R.D.C. Recursos hídricos: Água no mundo, no Brasil e no Estado do Rio de Janeiro. [Water resources: Water in the world, in Brazil and the state of Rio de Janeiro]. *Bol. Obs. Ambient. Alberto Ribeiro Lamego* **2008**, *2*, 29–36. (In Portuguese) [CrossRef]
42. Brazil. Brazilian Central Bank. Resolução n° 4.582, fixa a meta para a inflação e seu intervalo de tolerância para os anos de 2019 e 2020. [Resolution n°. 4,582, sets the target for inflation and its tolerance interval for the years 2019 and 2020]. Resolução N° 4.582, de 29 de Junho de 2017. [Resolution n°. 4,582 of 29th of June, 2017]. Available online: http://www.bcb.gov.br/pre/normativos/busca/downloadNormativo.asp?arquivo=/Lists/Normativos/Attachments/50402/Res_4582_v1_O.pdf (accessed on 30 May 2018).
43. Brazil. Under-secretariat of Collection and Attendance. Federal Revenue. Taxa de Juros Selic [Selic interest rate]. 2018. Available online: http://idg.receita.fazenda.gov.br/orientacao/tributaria/pagamentos-e-parcelamentos/taxa-de-juros-selic (accessed on 30 May 2018).
44. Antunes, L.N.; Ghisi, E. Water and energy consumption in schools: Case studies in Brazil. *Environ. Dev. Sustain.* **2019**. [CrossRef]
45. Ghisi, E.; Vieira, A.S.; Da Rosa, A.S.; Marinoski, A.K.; Silva, A.S.; Balvedi, B.F.; Almeida, L.S.S. *Uso Racional de água e Eficiência Energética em Habitações de Interesse social: Volume 1—Hábitos e Indicadores de Consumo de água e Energia, [Rational Use of Water and Energy Efficiency in Social Housing: Volume 1—Habits and Indicators of Water and Energy Consumption]*; Laboratory of Energy Efficiency in Buildings: Florianópolis, Brazil, 2015; Available online: http://www.labeee.ufsc.br/publicacoes/relatorios_pesquisa/RelatorioFINEP-VOL01.pdf (accessed on 1 June 2018). (In Portuguese)

Correction

Correction: Maykot, J.K. and Ghisi, E. Assessment of A Rainwater Harvesting System in A Multi-Storey Residential Building in Brazil. *Water* 2020, *12*, 546

Jéssica Kuntz Maykot *and Enedir Ghisi

Laboratory of Energy Efficiency in Buildings, Department of Civil Engineering, Federal University of Santa Catarina, Florianópolis-SC 88040-900, Brazil; enedir.ghisi@ufsc.br

* Correspondence: jessica.k.maykot@gmail.com; Tel.: +55-48-3721-2115

Received: 8 May 2020; Accepted: 11 May 2020; Published: 22 May 2020

The authors wish to make the following correction to this paper [1]. About Figure 2, due to mislabeling, we replace:

with

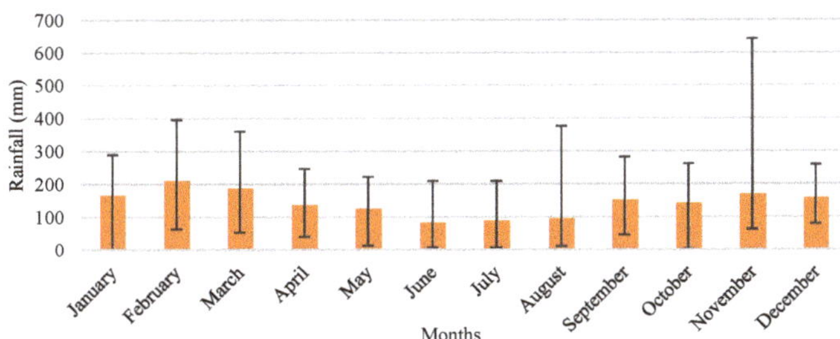

The authors would like to apologize for any inconvenience caused to the readers by these changes.

Reference

1. Maykot, J.K.; Ghisi, E. Assessment of A Rainwater Harvesting System in A Multi-Storey Residential Building in Brazil. *Water* **2020**, *12*, 546. [CrossRef]

Communication

Assessment of Water Quality in Indo-Gangetic Plain of South-Eastern Asia under Organic vs. Conventional Rice Farming

Debjani Sihi [1,*,†], Biswanath Dari [2,*,†], Zhengjuan Yan [3], Dinesh Kumar Sharma [4], Himanshu Pathak [4], Om Prakash Sharma [5] and Lata Nain [6]

1. Climate Change Science Institute and Environmental Sciences Division, Oak Ridge National Laboratory, 1 Bethel Valley Rd, Oak Ridge, TN 37830, USA
2. Department of Crop and Soil Science, Klamath Basin Research and Extension Center, Oregon State University, 6923 Washburn Way, Klamath Falls, OR 97603, USA
3. School of Chemical Engineering, Sichuan University, No. 24 South Section 1, Yihuan Road, Chengdu 610065, China; zjyan@scu.edu.cn
4. Division of Environmental Sciences, Indian Agricultural Research Institute, New Delhi 110012, India; dks_env@rediffmail.com (D.K.S.); hpathak.iari@gmail.com (H.P.)
5. National Centre of Integrated Pest Management, New Delhi 110012, India; opsharmadelhi@gmail.com
6. Division of Microbiology, Indian Agricultural Research Institute, New Delhi 110012, India; latarajat@yahoo.co.in

* Correspondence: sihid@ornl.gov (D.S.); b.dari@oregonstate.edu (B.D.); Tel.: +1-352-222-5655 (D.S.); +1-352-222-9565 (B.D.)

† These two authors have contributed equally in every aspect of this article. Therefore they both are leading/first authors.

Received: 1 March 2020; Accepted: 24 March 2020; Published: 28 March 2020

Abstract: Water contamination is often reported in agriculturally intensive areas such as the Indo-Gangetic Plain (IGP) in south-eastern Asia. We evaluated the impact of the organic and conventional farming of basmati rice on water quality during the rainy season (July to October) of 2011 and 2016 at Kaithal, Haryana, India. The study area comprised seven organic and seven conventional fields where organic farming has been practiced for more than two decades. Water quality parameters used for drinking (nitrate, NO_3; total dissolved solids (TDS); electrical conductivity (EC) pH) and irrigation (sodium adsorption ratio (SAR) and residual sodium carbonate (RSC)) purposes were below permissible limits for all samples collected from organic fields and those from conventional fields over the long-term (~15 and ~20 years). Importantly, the magnitude of water NO_3 contamination in conventional fields was approximately double that of organic fields, which is quite alarming and needs attention in future for farming practices in the IGP in south-eastern Asia.

Keywords: water quality; conventional farming; organic farming; nitrate; residual sodium carbonate; sodium adsorption ratio; total dissolved solids

1. Introduction

The enormous rate of agricultural production required to feed the burgeoning population in India (and other parts of south-eastern Asia) needs a constant supply of irrigation water. One such reliable source of irrigation is well water (~borewell water) in the Indo-Gangetic Plain (IGP), India, where deep wells extract water from the underlying aquifer [1]. The global importance of well water for both irrigation and drinking purposes in south-eastern Asia is obvious, but its deteriorating quality has become a serious decadal concern in IGP for intensive agriculture operations [2,3]. Basmati rice is an important commodity for the Indian economy related to the export business, and reports are limited to the effects of basmati rice cultivation on groundwater quality.

Nitrate (NO_3) concentrations exceeding the permissible limits in well water due to agricultural management practices (e.g., fertilizer/manure application, irrigation etc.) have been reported throughout the world including the United States [4], Europe [5], Australia [6], China [7], Germany [8] and France [9]. Other than well water nitrates, pH, electrical conductivity (EC), and total dissolved solids (TDS) (i.e., factors directly related to dissolved mineral matter) are other useful indicators of well water quality for drinking purposes due to their close association with human health problems [1]. The assessment of well water quality for irrigation purposes is usually achieved by using indices including the sodium adsorption ratio (SAR) and residual sodium carbonate (RSC), which are calculated using the concentrations of various nutrients such as carbonate (CO_3^{2-}), bicarbonate (HCO_3^-), calcium (Ca^{2+}), magnesium (Mg^{2+}), and sodium (Na^+) in well water [1].

The simultaneous evaluation of the long-term (~15–20 years) effects of farming practices (organic vs. conventional) on well water quality parameters has still seldom been practiced. To that end, we focused on assessing water quality parameters, for both drinking (NO_3^-, pH, EC, TDS) and irrigation (SAR, RSC) purposes, under long-term organic vs. conventional practices of basmati rice in the Indo-Gangetic Plain (IGP)—the bread basket of south-eastern Asia. We have assessed these well water quality parameters with the hypothesis that well water pollution will be higher in conventional farming systems than in organic farming systems under long-term (~15 and ~20 years) cultivation practices.

2. Materials and Methods

2.1. Site Description and Well Water Samples Collection

The study area is situated in Kaithal, Haryana, India (Table S1 and Figure S1). Seven farmers' fields, from both organic and conventional basmati rice systems, were chosen for comparative analysis with farmers' participation during the Kharif (or rainy) season (July to October) of 2011 [10,11] and 2016 (unpublished data) (see Appendix A for details). The well water samples were collected at the end of the growing season after 15 (2011) and 20 (2016) years of rice cultivation in both systems (see Appendix B for details). All samples were collected and preserved following the operating procedure for well water sampling [12]. Well water samples were collected at each site from the wells or bore well (or tube wells) with depths ranging from approximately 28.7 to 89.3 m. Details of study sites, farming practices and sample collection are described in the Appendix section (as Supplementary Materials).

2.2. Analysis of Well Water Samples

Well water quality parameters for drinking purposes were measured following the standard methods used by the American Public Health Association and American Water Works Association [13]. Well water pH and EC were measured using a pH and EC meter following the standard method of water sample analysis [14]. TDS was measured following the procedure in [15]. Nitrate was measured using salicylic acid nitration methods [16]. RSC and SAR were calculated following the method described in [14].

2.3. Statistical Analysis

The effects of farming practices on well water quality parameters were determined based on the completely randomized design using an analysis of variance (ANOVA) in JMP Pro 14.0 [17]. The year was treated as a random factor. Tukey's test was used to perform post-hoc multiple comparisons. All statistical analyses were conducted at a 5% level of significance.

3. Results and Discussion

3.1. Evaluation of Well Water Quality Parameters for Drinking Purposes

We evaluated the influence of farming practices on well water quality parameters after ~15 (2011) and ~20 (2016) years of the onset of organic farming practices. In general, the farming practice significantly affected all well water quality parameters (Figure 1, Figure 2 and Table 1). In the biplot, two groups of well water samples collected from organic vs. conventional fields could be easily identified, and the first two principal components explained 95.2% of the total variance (Figure 2A,B). Similar results were obtained from the cluster analysis (Figure S2). The dendrogram consisted of two distinct clusters, where the first cluster included well water samples collected from the organic fields and the second cluster included well water samples collected solely from conventional farming systems. All well water quality indices were better in the organic farming system than in the conventional farming system, indicating the potential reduction of well water pollution risks from organic farming systems [14]. For instance, the values of all the parameters, such as well water pH, EC, TDS, NO_3, RSC, and SAR, were lower in organic fields than conventional fields for both years, where the corresponding values fall within the standard range [3] (Figure 1).

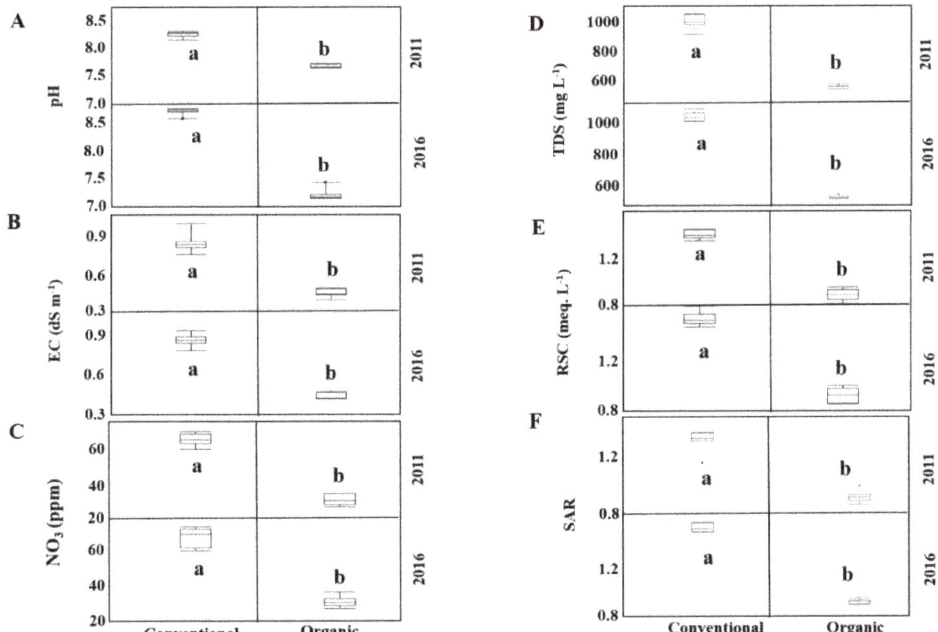

Figure 1. Comparison between the environmental impacts of organic vs. conventional rice cultivation on groundwater quality parameters ((**A**). pH, (**B**). electrical conductivity (EC), (**C**). NO_3, (**D**). total dissolved solids (TDS), (**E**). residual sodium carbonate (RSC), and (**F**). sodium absorption ratio (SAR)) (n = 21 for each year) in Kaithal, Haryana, India. Different letters indicate significant difference at a 5% level of significance.

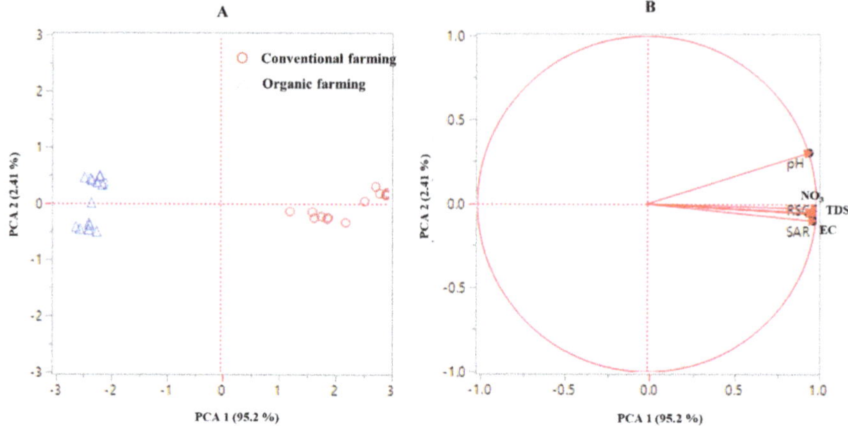

Figure 2. Biplot (**A**) and loading plot (**B**) from the principal component analysis of groundwater quality parameters under organic and conventional rice cultivation systems (n = 42 by combining both years) in Kaithal, Haryana, India.

Table 1. Effect of farming practice on groundwater quality parameters under organic and conventional rice cultivation systems (n = 42 by combining both years) in Kaithal, Haryana, India.

Parameters	Farming Practices	
	F Ratio	p-Value
Groundwater pH	623	<0.0001 *
Groundwater EC	204	<0.0001 *
Groundwater total dissolved solids (TDS)	619	<0.0001 *
Groundwater nitrate (NO_3)	529	<0.0001 *
Residual sodium carbonate (RSC)	633	<0.0001 *
Sodium adsorption ratio (SAR)	156	<0.0001 *

* The level of significance for each parameter was determined at $p < 0.05$.

Over the longer term, the well water pH was 0.54 (for 2011) and 1.41 (for 2016) units higher in conventional farming systems than in the organic farming system. Well water pH values from both farming systems fell within the safe limit (6.5 to 8.5) as per the drinking water guidelines [18] for 2011, but not for 2016, when the well water pH values were approximately 0.1 unit higher than the threshold value (see corresponding values for samples collected before transplanting in Figure 1A). The pH of well water depends on the acidity or basicity of the leachate from the soil accessing and/or recharging water table; the high pH leachate is possibly responsible for raising the well water pH in conventional fields. Thus, the basic nature of chemical fertilizers applied to the conventional fields resulted in leachate with a high pH. Conversely, the application of organic manures in organic fields conceivably enhanced the formation of acids, leading to a decrease in the pH of the leachate as well as well water pH.

Soil EC—similar to soil pH—can also control the EC of the leachate reaching the well water. Thus, higher values of EC in the well water of conventional fields (see corresponding values for samples collected before transplanting Figure 1B) may be attributed to the higher amounts of dissolved ions in soil solution leaching down due to the excess of the available form of nutrients (i.e., surplus amount) from applied fertilizers. The EC in well water usually depends on the presence of dissolved mineral matter (solid content; TDS). The TDS values of well water samples collected from conventional fields were 1.97 (for 2011) and 2.09 (for 2016) times higher than those collected from organic fields (see corresponding values for samples collected before transplanting in Figure 1D). Moreover, the TDS values of well water samples from our conventional fields were always above the drinking water threshold limit of 500 mg L^{-1}; however, corresponding values from organic fields indicated that water samples from organic fields may be suitable for drinking purposes [18].

Similar to TDS, the well water NO_3 levels in conventional fields were also beyond threshold level for drinking water (50 mg L^{-1}), as specified in [19], and may pose a health risk to humans or other animals. The lower NO_3 content in well water under organic cultivation (see corresponding values for samples collected before transplanting in Figure 1C) may be due to the N mineralization rate, which matches with the plant uptake and a reduced risk for N leaching and well water pollution. In contrast, the residual NO_3 accumulated from the excessive application of inorganic fertilizers in conventional farming [20] is subject to leaching and becomes part of the sub-surface or well water [7] and continuously moves downward even after the season of application, particularly in semi-arid regions [21]. It has been further reported that NO_3 leaching could be 50% greater in conventionally managed fields compared to organically managed fields [22].

3.2. Evaluation of Well Water Quality Parameters for Irrigation Purposes

The well water quality indices used for irrigation purposes (RSC and SAR) also showed a similar trend to that of well water quality parameters used for drinking purposes (see Figure 1E,F). Higher RSC and SAR values in conventional farming indicated an appreciable hazard in soils in the long-term since the corresponding values passed the safe limits for irrigation water (1.24 and 1.0 for RSC and SAR, respectively), but the water could be used for irrigation purposes after proper management (e.g., gypsum application). On the contrary, both RSC and SAR values in the organic farming system were within safe limits and would not pose any hazard when applied for irrigation purposes (see Table S2 and standard permissible limits [1]). The higher organic matter applications in organic fields may have helped to ameliorate the soil and water sodicity by mobilizing Ca (and Mg) from soil minerals [23].

4. Conclusions

Long-term organic rice farming practices are characterized by favorable pH, EC, TDS, NO_3, RSC, and SAR values of water compared to the conventional systems. Our findings indicated that well water NO_3 may lead to the water having poor drinking quality when adjacent to an agricultural field with conventional cultivation practices. The novelty of this study is its characterization and comparison of water quality indices under long-term organic vs. conventional systems and the sustainability of farming practices. To the best of our knowledge, our approach has resulted in some unique findings as we evaluated the long-term (~15 and ~20 years) effects of farming practices on water quality parameters in the Indo-Gangetic Plain of south-eastern Asia for the very first time. Additionally, we have documented the comparison of water quality parameters conducted with farmers' participation in certified organic and conventional fields. This finding will be useful for decision-makers to identify farming practices that enhance well water qualities in agriculture-intensive areas such as the IGP, south-eastern Asia. However, a long-term research approach involving both the economic and environmental implications of well water and associated farming practices would provide insights into the subject matter in a broader context.

Supplementary Materials: The following are available online at http://www.mdpi.com/2073-4441/12/4/960/s1, Table S1: An overview of the study area and fertilizer application under organic and conventional "Taraori Basmati" rice cultivation systems in Kaithal, Haryana, India, Table S2: Permissible limits (or threshold values) of groundwater quality parameters for drinking and irrigation purposes, Figure S1: Study locations in Kaithal district of Haryana, India, and geographical positions of the organic and conventional fields used for groundwater sample collection, Figure S2: Dendrogram of the clusters identified from the analysis of answers related to the groundwater quality parameters of the samples collected from organic and conventional "Taraori Basmati" rice cultivation systems in Kaithal, Haryana, India. Cluster I and II explain the groundwater samples collected from organic and conventional fields, respectively. EC; electrical conductivity, EC; electrical conductivity, TDS; total dissolved solids, NO_3; nitrate, RSC; residual sodium carbonate, SAR; sodium adsorption ratio.

Author Contributions: The contributions of individual authors are as follows: Conceptualization, D.S. and B.D.; methodology, D.S., D.K.S., H.P., O.P.S., L.N.; software, B.D., Z.Y.; validation, D.S., B.D., Z.Y., D.K.S.; formal analysis, D.S., B.D., Z.Y., H.P.; investigation, D.S., D.K.S, O.P.S.; resources, D.K.S., H.P., O.P.S., L.N.; data curation, D.S.;

writing—original draft preparation, D.S., B.D., Z.Y.; writing—review and editing, D.S., B.D., Z.Y., D.K.S., H.P., O.P.S., L.N.; visualization, D.S., B.D., Z.Y., supervision, D.K.S, H.P.; project administration, D.K.S., H.P., funding acquisition, D.S., D.K.S. All authors have read and agreed to the published version of the manuscript.

Funding: This research received no external funding.

Acknowledgments: The author wish to thank the Junior Research Fellowship (JRF) program of the Indian Council of Agricultural Research (ICAR) and Sushmita Munda, Totan Adak, and all fourteen farmers for the support of this study.

Conflicts of Interest: The authors declare no conflict of interest. The funders had no role in the design of the study; in the collection, analyses, or interpretation of data; in the writing of the manuscript, or in the decision to publish the results.

Appendix A

Site Description and Details of Organic Farming Practices and Related Management Practices

Organic farming practices have been conducted in partnership with Agrocel Pvt. Ltd., a private company certified by SKAL (also known as the Control Union certification, which is accredited to APEDA, Agricultural and Processed Food Products Export Development Authority) since late 1990's in Kaithal (Haryana, India). This company is the second largest organic rice exporter of India and mainly focused on Taraori Basmati Rice (i.e., CSR-30 or Yamini). The certified farming practices under contract organic farming have been conducted since last 15 years with the same partnership in Kaithal area and was recognized by FLO and Fair-Trade in 2007. Originally, the activities were developed in parallel strategies with IFOAM (International Federation of Organic Agriculture Movements) under farmers' participatory mode. Currently, 5750 ha land has been included in this contract organic farming with a public-private partnership.

A tropical steppe, semi-arid and hot climate usually prevails in the Kaithal district in Haryana, India (study area). The average annual precipitation and temperature of the region are 568 mm and 24.6 °C, respectively. The soils of study region can generally be categorized as sandy to sandy loam, marginally fertile, and named as Sierozem soil (under major zonal soil classification system in India). The lands were prepared by plowing the plots using a tractor draw mold board plow followed by cross-disking and leveling and pressing with a tractor drawn leveler. In organic fields, recommended dose of certified organic fertilizers in terms of farm yard manure i.e., FYM (0.5% N, 0.2% P_2O_5 and 0.5% K_2O) and decorticated neem cake (a bio-fertilizer-cum-organic soil amendment made of neem [*Azadirachta indica* L.] seed kernels after removing the husk, 2.0 to 5.0% N, 0.5 to 1% P and 1.0 to 2.0% K) were applied @ 5 t ha^{-1} and 125 kg ha^{-1}, respectively. In conventional fields, the inorganic fertilizers in terms of urea, diammonium phosphate (DAP) or single super phosphate (SSP), and muriate of potash (MOP) were applied as recommended doses (150 kg N ha^{-1}, 40 kg P_2O_5 ha^{-1}, and 40 kg K_2O ha^{-1}, respectively). A recommended practices i.e., wet method of nursery raising was followed to raise the nursery of rice followed by transplanting using 30 d old seedlings @2–3 seedlings per hill in rows and 20 cm apart. A 2–3 cm water level was maintained during the initial stage followed by 4–5 cm to maintain a standard water level in the rice field up to milk or dough stage. No synthetic chemical or plant growth regulator has been applied in organic fields. The hexaconazole @300 mL ha^{-1}, tricyclazole @120 g ha^{-1}, carbendazim @1.35 kg ai ha^{-1} were applied for controlling bacterial leaf blight (BLB), sheath blight, stem borer and leaf folder diseases, respectively in the conventional fields. The harvesting was performed with sickle at 110 d after transplanting when about 90% of the grains in panicle had ripened.

Appendix B

Well Water Sample Collection

The depth of water table in the entire study area ranged between 31 to 37 m below ground level, which closely matches with that a standard method used in the area [24]. The information on the water table depth was collected from individual farmers who estimated the depth while digging the tube wells in <2 years before our first field campaign. Thus, we carefully chose farmers' fields such that there was potentially negligible gravity-driven lateral flow of groundwater as adjacent fields which were of the similar hydraulic head (or, groundwater table depth) [2]. Additionally, a considerable distance (a minimum of 10 km radius) was always maintained between an organic and an adjacent conventional field used in this study to avoid any contamination between these two farming systems. Given the geology of the region (unconsolidated alluvial deposits of Quaternary age) [25], climate (hot semi-arid climate as per Köppen-Geiger climate classification system; Geiger 1954), and soil type (sandy loam) [11] are similar in both conventional and organic fields, the management practices associated with the type of fertilizers/manures played an instrumental role in controlling the groundwater quality parameters.

Appendix C

Collection of Groundwater Samples

Field campaigns were conducted before transplanting and after harvest of the growing basmati rice crop in both the systems. All samples were collected and preserved following the operating procedure for groundwater sampling [12]. Three field replicates of groundwater samples were collected at each site from the boreholes (or tube wells) with depth ranging from approximately 120 to 180 feet (~37–55 m). Thus, the number of our collected samples equate to a total of twenty-one samples (3 field replicates × 7 farmer's field) from both organic and conventional farming systems. We used boreholes (or tube wells) because of its increased use in extracting groundwater for both drinking and irrigation water purposes in the Indian subcontinent [24]. All samples were poured in high-density polyethylene (HDPE) round plastic bottles with a threaded cap. After collection, all samples were transported to the laboratory facility of the Division of Environmental Sciences, Indian Agricultural Research Institute (IARI) and stored at 4 °C until processed.

Analysis of Groundwater Samples

Groundwater quality parameters were measured following the standard protocol for water quality parameters analysis [13]. To measure total dissolved solids (TDS), the water samples were filtered with glass microfiber filter, and then 50 mL of filtrate was added to a pre-weighed ceramic dish and placed in a drying oven to evaporate at 105 °C. Quantification of TDS was done by subtracting the initial weight of the empty ceramic dish from the weight of the ceramic dish with the dried residue. Carbonate (CO_3^-) and bicarbonate (HCO_3^-) in groundwater sample were determined by titrating 25 mL of water sample against standardized 0.1 N H_2SO_4 using phenolphthalein and methyl red as indicators, respectively. The concentration of calcium (Ca^{++}) and magnesium (Mg^{++}) was determined by versenate titration method by titrating against 0.01 (N) EDTA-disodium salt solution using Erichrome black T dye as an indicator and ammonium chloride-ammonium hydroxide buffer. Sodium (Na^+) concentration was measured using a flame photometer. The pH of groundwater samples was measured by shaking 50 mL of water sample in a 100-mL beaker using a glass electrode in a pH meter. The electrical conductivity (EC) (dS m^{-1} at 25 °C) was measured in 50 mL of the water sample using conductivity meter [14].

The NO_3^- concentration in groundwater was derived on the basis of nitration of salicylic acid [16]. One mL standard or sample was transferred into a 50 mL Erlenmeyer flask and 0.5 mL TRI solution (1 g sodium salicylate + 0.2 g NaCl + 0.1 g $NH_4SO_3NH_2$) and swirled thoroughly. Evaporated to dryness for 5 min at 100–120 °C) and cooled thereafter. The residue was wetted with 1 mL concentration H_2SO_4, swirled, and allowed to stand for 5 min. Then, 5 mL MilliQ water and 5 mL 40 percent NaOH were also added down the flask wall and swirled and cooled. The absorbance of the yellow color was measured at 410 nm immediately after the solution solutions were cooled.

Indices of Groundwater Quality Parameters for Irrigation Purposes

Residual Sodium Carbonate (RSC)

The RSC exists when the content of the (CO_3^- and HCO_3^-) exceeds the (Ca^{++} and Mg^{++}) content of the irrigation water. RSC was defined as follows [14] (Equation (A1)).

$$RSC = (CO_3^- + HCO_3^-) - (Ca^{++} + Mg^{++}) \quad (A1)$$

The safe limit of RSC for irrigation purpose is <1.24 meq. L^{-1} (Table S2). Irrigation water having RSC values higher than safe limits pose hazards to the crop development (or growth) and leads to accumulation of Na in the soils. It also exerts higher EC and excess salinity in soils.

Sodium Adsorption Ratio (SAR)

The SAR is the relative concentration of Na to the combined concentration of (Ca^{++} and Mg^{++}). Usually, it is used to predict the hazards of Na^{+} in soils. As such, it is a measure of the sodicity/alkaline hazard of irrigation water [14], calculated (Equation (A2)) using the concentration of ions in millimol (+)/1.

$$\text{SAR} = \frac{\text{Na}^{+}}{\sqrt{\frac{\text{Ca}^{++}+\text{Mg}^{++}}{2}}} \tag{A2}$$

Irrigation water is considered to be hazardous with a SAR value >1.0 [14].

References

1. Abdel-Satar, A.M.; Al-Khabbas, M.H.; Alahmad, W.R.; Yousef, W.M.; Alsomadi, R.H.; Iqbal, T. Quality assessment of well water and agricultural soil in Hail region, Saudi Arabia. *Egypt. J. Aquat. Res.* **2017**, *43*, 55–64. [CrossRef]
2. MacDonald, A.M.; Bonsor, H.C.; Ahmed, K.M.; Burgess, W.G.; Basharat, M.; Calow, R.C.; Dixit, A.; Foster, S.S.D.; Gopal, K.; Lapworth, D.J.; et al. Groundwater quality and depletion in the Indo-Gangetic Basin mapped from in situ observations. *Nat. Geosci.* **2016**, *9*, 762–766. [CrossRef]
3. Malan, A.; Sharma, H.R. Groundwater quality in open-defecation-free villages (NIRMAL grams) of Kurukshetra district, Haryana, India. *Environ. Monit. Assess.* **2018**, *190*, 472. [CrossRef] [PubMed]
4. Sahoo, P.K.; Kim, K.; Powell, M.A. Managing well water nitrate contamination from livestock farms: Implication for nitrate management guidelines. *Current Pollution Reports.* **2016**, *2*, 178–187. [CrossRef]
5. Laegreid, M.; Bøckman, O.C.; Kaarstad, O. *Agriculture, Fertilizers and the Environment*; Norsk Hydro ASA: Porsgrunn, Norway, 1999.
6. Thorburn, P.J.; Biggs, J.S.; Weier, K.L.; Keating, B.A. Nitrate in well water of intensive agricultural areas in coastal Northeastern Australia. *Agric. Ecosyst. Environ.* **2003**, *94*, 49–58. [CrossRef]
7. Cui, Z.; Chen, X.; Zhang, F. Current nitrogen management status and measures to improve the intensive wheat–maize system in China. *AMBIO* **2010**, *39*, 376–384. [CrossRef] [PubMed]
8. Van der Ploeg, R.R.; Horton, R.; Kirkham, D. Steady flow to drains and wells. In *Agricultural Drainage*; Agronomy Series No. 38, ASA–CSSA–SSSA; Skaggs, R.W., van Schilfgaarde, J., Eds.; Wiley Publishers: Madison, WI, USA, 1999; pp. 213–263.
9. Beaudoin, N.; Saad, J.K.; Van Laethem, C.; Machet, J.M.; Maucorps, J.; Mary, B. Nitrate leaching in intensive agriculture in Northern France: Effect of farming practices, soils and crop rotations. *Agric. Ecosyst. Environ.* **2005**, *111*, 292–310. [CrossRef]
10. Sihi, D.; Sharma, D.K.; Pathak, H.; Singh, Y.V.; Sharma, O.P.; Nain, L.; Chaudhary, A.; Dari, B. Effect of organic farming on productivity and quality of basmati rice. *Oryza Int. J. Rice* **2012**, *49*, 24–29.
11. Sihi, D.; Dari, B.; Sharma, D.K.; Pathak, H.; Nain, L.; Sharma, O.P. Evaluation of soil health in organic vs. conventional farming of basmati rice in North India. *J. Plant Nutr. Soil Sci.* **2017**, *180*, 389–406. [CrossRef]
12. SESD. *U.S. Environmental Protection Agency-Science and Ecosystem Division, Operating Procedure for Well Water Sampling*; SESDPROC-301-R3; 2013. Available online: https://www.epa.gov/quality/surface-water-sampling (accessed on 16 December 2016).
13. APHA; AWWA. *Standard Methods for the Examination of Water and Wastewater*; American Public Health Association/American Water Works Association/Water Environment Federation: Washington, DC, USA, 1995.
14. Richards, L.A. *Diagnosis and Improvement of Saline and Alkali Soils*; Government Printing Office: Washington, DC, USA, 1954.
15. Chowdary, V.M.; Rao, N.H.; Sharma, P.B.S. Decision support framework for assessment of non-point-source pollution of well water in large irrigation projects. *Agric. Water Manag.* **2005**, *75*, 194–225. [CrossRef]
16. Vendrell, P.F.; Zupancic, J. Determination of soil nitrate by transnitration of salicylic acid. *Commun. Soil Sci. Plant Anal.* **1990**, *21*, 1705–1713. [CrossRef]
17. JMP. *Statistical Software 2010 (Version 10.0)*; SAS Institute. Inc.: Cary, NC, USA, 2010.

18. Environmental Protection Agency (EPA) (Ed.) *Drinking Water Standards and Health Advisories*; Oxford Academics: Washington, DC, USA, 2012.
19. World Health Organization. *Guidelines for Drinking Water Quality. Incorporating the First and Second Addenda, Volume 1, Recommendations*, 3rd ed.; WHO Chronicle: Geneva, Switzerland, 2008.
20. Lenka, S.; Singh, A.K.; Lenka, N.K. Soil water and nitrogen interaction effect on residual soil nitrate and crop nitrogen recovery under maize–wheat cropping system in the semi-arid region of northern India. *Agric. Ecosyst. Environ.* **2013**, *179*, 108–115. [CrossRef]
21. Ibrikci, H.; Cetin, M.; Karnez, E.; Flügel, W.A.; Tilkici, B.; Bulbul, Y.; Ryan, J. Irrigation-induced nitrate losses assessed in a Mediterranean irrigation district. *Agric. Water Manag.* **2015**, *148*, 223–231. [CrossRef]
22. Drinkwater, L.E.; Wagoner, P.; Sarrantonio, M. Legume-based cropping systems have reduced carbon and nitrogen losses. *Nature* **1998**, *396*, 262–265. [CrossRef]
23. Poonia, S.R.; Mehta, S.C.; Pal, R. Calcium-sodium, magnesium exchange equilibria in relation to organic matter in soils. In *International Symposium on Salt Affected Soils*; Central Soil Salinity Research Institute: Karnal, India, 1980; pp. 134–142.
24. Suhag, R. *Overview of Ground Water in India*; PRS Legislative Research Standing Committee on Water Resources: Delhi, India, 2016.
25. Sharma, A. *Groundwater Information Booklet Kaithal District, Haryono*; Central Groundwater Board: Chandigarh, India, 2008.

Article

Study on the Ecological Operation and Watershed Management of Urban Rivers in Northern China

Guangyi Deng [1], Xiaohan Yao [1], Haibo Jiang [1,*], Yingyue Cao [1], Yang Wen [2], Wenjia Wang [1], She Zhao [1] and Chunguang He [1,*]

1 State Environmental Protection Key Laboratory of Wetland Ecology and Vegetation Restoration, Key Laboratory for Vegetation Ecology, Ministry of Education, Jilin Provincial Science and Technology Innovation Center of Wetland Restoration and Function Development, Northeast Normal University, Changchun 130117, China; denggy409@nenu.edu.cn (G.D.); yaoxh511@nenu.edu.cn (X.Y.); yczoro3104@gmail.com (Y.C.); wangwenjia007@live.com (W.W.); zhaos@nenu.edu.cn (S.Z.)

2 Key Laboratory of Environmental Materials and Pollution Control, the Education Department of Jilin Province, School of Environmental Science and Engineering, Jilin Normal University, Siping 136000, China; weny501@126.com

* Correspondence: jianghb625@nenu.edu.cn (H.J.); he-cg@nenu.edu.cn (C.H.); Tel.: +86-431-891-65611 (C.H.)

Received: 25 February 2020; Accepted: 20 March 2020; Published: 24 March 2020

Abstract: Small- and medium-sized rivers are facing a serious degradation of ecological function in water resource-scarce regions of Northern China. Reservoir ecological operation can restore the damaged river ecological environment. Research on reservoir ecological operation and watershed management of urban rivers is limited in cold regions of middle and high latitudes. In this paper, the urban section of the Yitong River was selected as the research object in Changchun, Northern China. The total ecological water demand and reservoir operation water (79.35×10^6 m^3 and 15.52×10^6 m^3, respectively) were calculated by the ecological water demand method, and a reservoir operation scheme was established to restore the ecological function of the urban section of the river. To examine the scientific basis and rationality of the operation scheme, the water quality of the river and physical habitat after carrying out the scheme were simulated by the MIKE 11 one-dimensional hydrodynamic-water quality model and the Physical Habitat Simulation Model (PHABSIM). The results indicate that the implementation of the operation scheme can improve the ecological environment of the urban section of the Yitong River. A reform scheme was proposed for the management of the Yitong River Basin based on the problems in the process of carrying out the operation schemes, including clarifying department responsibility, improving laws and regulations, strengthening service management, and enhancing public participation.

Keywords: ecological water demand; reservoir ecological operation; MIKE 11 model; PHABSIM model; watershed management

1. Introduction

With the rapid development of urbanization and industrialization, the ecological environment of rivers is severely affected by human activities [1]. Research on river ecological restoration has been carried out based on the damaged ecological function of rivers [2,3]. Reservoir ecological operation plays an important role in river ecological restoration [4–6]. Research on river ecological restoration is mainly focused on the scheme and management of reservoir ecological operation [7].

River ecological protection is an important component of reservoir operation decisions [8]. Reservoirs can operate based on specific conditions [9,10]. Since the 1980s, in order to eliminate the adverse impact of dams on the ecological health of downstream rivers, the mode of reservoir operation has been adjusted from two aspects by the management and decision departments of some developed countries. First, operation modes are optimized, and supporting technical facilities are studied; second, the impact of technical methods on the ecological environment is evaluated. When the ecosystem is not obviously disturbed, the improved effects of rivers are ensured. Studies on the theory and method of ecological operation have been carried out in the Yangtze River and Yellow River in China, including water operations, sediment operations, and water quality operations [11–13]. However, most research on reservoir ecological operation is focused on large rivers with abundant water in Southern China [14]. Studies on river ecological operation are limited in Northern China. Supplying water is the main function of reservoirs in the water resource-scarce region of Northern China, resulting in neglecting the requirement of water resources for downstream river ecosystems [15]. The improper operation of reservoirs with limited water resources will pose a serious hazard to river health and regional ecological safety [16]. Therefore, to supply water for downstream rivers, the ecological operation mode of northern water supply reservoirs should be improved.

The river ecological water demand needs to be calculated, and the water supply of basins should be considered in the development of reservoir ecological operation schemes [17–19]. The purpose is to determine whether river ecological restoration needs are met. Therefore, watershed management is crucial. Mature watershed management includes the establishment of watershed management institutions, the implementation of coordinated power, and the establishment of a water rights trading market [20–23]. Current research on watershed management is still in the theory stage in China [24].

In this study, the urban section of the Yitong River was selected as the research object in Changchun. The objectives of our study are as follows: (1) to establish an ecological operation scheme of the Xinlicheng Reservoir in the upstream Yitong River; (2) to examine the scientific basis and rationality of the reservoir ecological operation scheme; and (3) to propose construction and provide advice for watershed management of the Yitong River.

2. Materials and Methods

2.1. Study Area

The Yitong River is the second tributary of the Songhua River, with a total length of 343.5 km and a basin area of 8440 km^2. Most of the Yitong River is located in Changchun, Jilin Province, China. The Xinlicheng Reservoir is 20 km south of Changchun and was established in the upstream Yitong River. The Xinlicheng Reservoir is a water supply reservoir with a total storage of 5.92×10^8 m^3. The annual average water storage of the Xinlicheng Reservoir is 0.61×10^8 m^3. There is no clear ecological water supply purpose except supplying water for human activities.

The downstream Xinlicheng Reservoir is a natural watercourse without human disturbance, with a length of 9.77 km. After entering the urban area of Changchun, the watercourse was artificially expanded and hardened. The study area was located from downstream of the Xinlicheng Reservoir to the Yangjiawaizi Bridge of the Yitong River (Figure 1).

2.2. Ecological Water Demand Calculation

River ecological water demand includes the watercourse base flow, evaporation water demand, leakage water demand, self-purification water demand, river outside water demand, and sediment transport water demand [25,26]. In the study area, sewage from tributaries, open trenches, and sluices must be treated by sewage treatment plants, further purified by constructed wetlands, and then discharged into the Yitong River after reaching the Class V standard of surface water. The major pollution load of the basin has been effectively resolved. Therefore, self-purification water demand can be neglected in the process of calculating the river ecological water demand. Because the sediment concentration of the study area is low, the sediment transport water demand can also be neglected. The rivers of Northern China have a nonfrozen period from April to October and a frozen period from November to March in the next year. Therefore, ecological water demand is calculated from April to October.

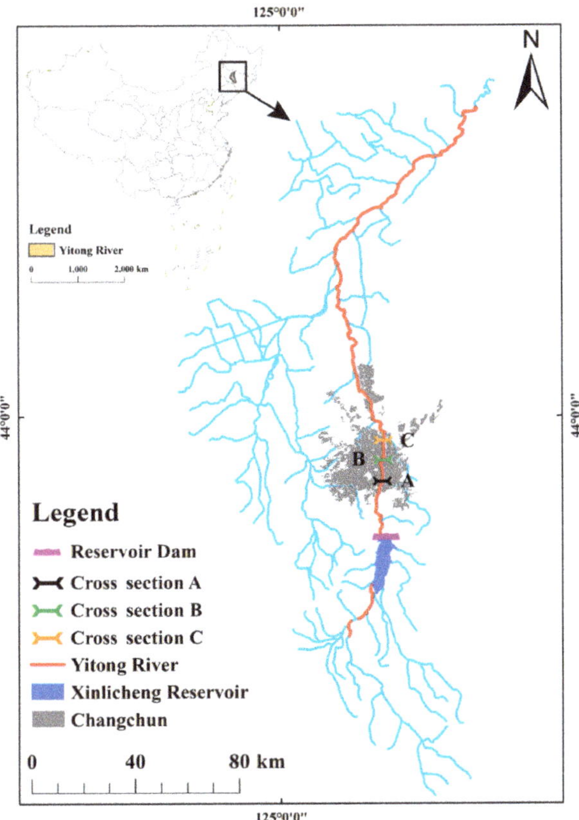

Figure 1. Location of the study area.

Watercourse base water demand (W_h, m^3) can be calculated by the Tennant method. Table 1 shows the standard of flow classification [27]. Because the Yitong River is located in the water resource scarce region of Northern China, general or poor was selected as the base flow standard. The minimum ecological flows from October to March and from April to September are 10% and 30% of the average flow, respectively.

Table 1. Tennant method for the recommended base flow standard.

Qualitative Description of Flow Values and Corresponding Habitat	Recommended Base Flow Standard (The Percentage of Average Flow)	
	General Water Period (October–March)	Fish Spawning Breeding Season (April–September)
Maximum	200	200
Optimum range	60–100	60–100
Very good	40	60
Good	30	50
Better	20	40
General or poor	10	30
Poor or minimum	10	10
Severe degradation	<10	<10

Evaporation water demand is the water needed to maintain the normal environmental function of river ecosystems. When precipitation is lower than evaporation, the lost water needs to be supplied from outside the watercourse. Evaporation water demand can be calculated by the following equation.

$$W_z = \begin{cases} A(E-P) & E > P \\ 0 & E < P \end{cases} \tag{1}$$

where W_z is the evaporation water demand (m^3); A is the average water surface area (m^2); E is the average evaporation (mm); and P is the average precipitation (mm).

Leakage water demand is the sum of water supplementation from rivers and groundwater. Leakage water demand can be calculated by the following equation.

$$W_l = K \times A \tag{2}$$

where W_l is the leakage water demand (m^3); K is the empirical coefficient; and A is the leakage area (m^2).

River outside water demand is mainly the vegetation water demand outside of the river, and it can be calculated by the following equation.

$$Q_e = F_f \times Z_f + F_u \times Z_u \tag{3}$$

where Q_e is the river outside water demand (m^3); F_f is the farmland area (hm^2); Z_f is the water quota of the farmland ecological environment (m^3/hm^2); F_u is the urban greenbelt area (hm^2); and Z_u is the water quota of the urban greenbelt ecological environment (m^3/hm^2).

River ecological water demand (W_s, m^3) is the sum of the watercourse base flow, evaporation water demand, leakage water demand, and river outside water demand. River ecological water demand can be calculated by the following equation.

$$W_s = W_h + W_z + W_l + Q_e \tag{4}$$

2.3. Water Supplement Scheme Examination

2.3.1. MIKE 11 One-Dimensional Hydraulics-Water Quality Model

MIKE 11 is a one-dimensional hydraulics-water quality model that is mainly used for the simulation, design, and management of river basin systems. The basic model was established by four types of data files, including river network generalization files, riverbed section files, watercourse boundary condition files, and initial condition files. To ensure the accuracy and reliability of the results, the model was further corrected for hydrodynamics and water quality. A one-dimensional hydrodynamic-water quality model was established for the study area.

The model includes an HD hydrodynamic module, an SO controllable building module, and an AD convection diffusion module in this study. The HD hydrodynamic module is the basic module, and it can be calculated by the following equation.

$$\begin{cases} \frac{\partial A}{\partial t} + \frac{\partial Q}{\partial x} = q \\ \frac{\partial Q}{\partial t} + \frac{\partial}{\partial x}\left(\frac{Q^2}{A}\right) + gA\frac{\partial h}{\partial x} + g\frac{Q|Q|}{C^2AR} = 0 \end{cases} \tag{5}$$

where A is the cross-sectional area (m^2); t is the time (s); Q is the flow (m^3/s); x is the distance coordinate (m); q is the inflow of the side (m^3/s); h is the water level (m); C is the Chezy coefficient; R is the hydraulic radius (m); and g is the gravitational acceleration.

The SO controllable building module can simulate the operation mode of hydraulic buildings and set different operation schemes and operation orders. If a suitable control point and target point are selected, the simulated capability of the MIKE 11 model will be improved.

The AD convection diffusion module is on the basis of the HD hydrodynamic module and mainly used to establish a river water quality model. Water quality should be simulated based on the hydraulics model. The convection diffusion process of pollutants in water can be simulated by setting the attenuation coefficient and diffusion coefficient. It can be calculated by the following equation.

$$\frac{\partial c}{\partial t} + v\frac{\partial c}{\partial x} = \frac{\partial}{\partial x}\left(E_x\frac{\partial c}{\partial x}\right) - kc \tag{6}$$

where c is the pollutant concentration (mg/L); E_x is the convection diffusion coefficient; v is the river flow speed (m/s); k is the attenuation coefficient; x is the spatial displacement (m); and t is the time (s).

2.3.2. Physical Habitat Simulation Model (PHABSIM)

The habitat conditions required by aquatic organisms for growth and reproduction can be simulated by PHABSIM based on the calculation of hydraulics combined with the theory of physical habitat simulation and according to the specific flow determined from the different hydraulic conditions in the different life cycles of representative aquatic organisms. In general, fish are the dominant species in river ecosystems for measuring the ecological health of rivers [28,29]. The dominant fish species in the Yitong River is *Cypriniformes*. Therefore, carp was selected as the representative species to simulate habitat.

The calculation process of PHABSIM includes a hydraulic simulation and habitat simulation. The relationship between the water level and flow can be predicted by the MANSQ method of the hydraulic simulation. Furthermore, the relationship between the flow and weighted available area of the study area can be estimated based on the hydraulic simulation result and accommodation curve of the carp habitat.

3. Results and Discussion

3.1. Ecological Water Demand

3.1.1. Watercourse Base Water Demand

The watercourse base water demand of the study area can be calculated from the monthly average flow of the Yitong River (Table 2). The total watercourse base water demand of the study area was 12.56×10^6 m^3 from April to October.

Table 2. Basic flow and water demand of the study area from April to October.

Month	April	May	June	July	August	September	October	Average/Total
Monthly average flow (m^3/s)	0.62	1.23	1.99	4.88	5.69	1.37	0.92	2.39
Minimum ecological flow (m^3/s)	0.19	0.37	0.60	1.46	1.71	0.41	0.09	0.69
Monthly water demand (10^6 m^3)	0.48	0.97	1.52	3.83	4.48	1.04	0.24	12.56

3.1.2. Evaporation Water Demand

The average water surface area in the study area was 3.2×10^6 m^2. The evaporation water demand of the study area from April to October can be calculated by Equation (1) (Table 3).

Table 3. Evaporation water demand of the study area from April to October (10^6 m^3).

Month	April	May	June	July	August	September	October	Total
Evaporation water demand	0.63	0.88	0.51	0.06	0.11	0.33	0.30	2.82

3.1.3. Leakage Water Demand

The empirical coefficients of the natural and artificial sections of the Yitong River were 1.0 m and 0.5 m, respectively. The leakage water demand of the study area from April to October can be calculated by Equation (2), with a total leakage water demand of 0.17×10^6 m^3.

3.1.4. River Outside Water Demand

The main crop in the study area is corn. The water quota of corn is 4275 m^3/hm^2 [30,31]. The water supplement area of study is approximately 3300 hm^2. The quota water of the greenbelt in the urban section of the river was obtained from "Water Quota (Local Standard of Jilin Province)" (DB/T389–2010). The river outside water demand of the study area can be calculated by Equation (3) (Table 4).

Table 4. River outside water demand of the study area.

Section	Area (hm^2)	Water Quota (m^3/hm^2)	Total Water Demand (10^6 m^3)
Natural	3300	4275	14.10
Urban	12426	4000	49.70
Total	15726	-	63.80

3.1.5. River Ecological Water Demand

The total river ecological water demand of the study area from April to October can be calculated by Equation (4) (Table 5).

Table 5. Ecological water demand of the study area (10^6 m^3).

Water Demand	Watercourse Base	Evaporation	Leakage	River Outside	Total
Amount	12.56	2.82	0.17	63.80	79.35

The river ecological water demand of the study area from April to October is shown in Table 6. The largest ecological water demand was in August because the aquatic biodiversity and biomass were the largest in the river, and the plants were abundantly growing.

Table 6. River ecological water demand of the study area from April to October (10^8 m^3).

Month	April	May	June	July	August	September	October	Total
Total	0.08	0.10	0.11	0.16	0.18	0.09	0.07	0.79

3.2. Ecological Water Supplement and Ecological Operation of the Xinlicheng Reservoir

The southeast sewage treatment plant of Changchun, which has a designed daily treatment capacity of 15×10^4 m^3, is located between natural and urban sections of the Yitong River downstream of the Xinlicheng Reservoir. The daily treated water is approximately 7×10^4 m^3 and is discharged into the Yitong River. Therefore, the water from the sewage treatment plant in the study area is 14.70×10^6 m^3 from April to October, which can meet the watercourse base water demand of the urban section of the Yitong River.

The annual precipitation and effective precipitation of the study area were 575 mm and 85.50 × 10^6 m^3, respectively, based on meteorological data from Changchun from 1980 to 2011. Therefore, precipitation can meet the river outside water demand of the study area.

As mentioned above, the ecological water supplement of the Xinlicheng Reservoir for the study area includes the watercourse base water demand of the natural section, evaporation water demand, and leakage water demand. The ecological water supplement from April to October can be calculated from the monthly river ecological water demand, the watercourse base water demand of the urban section and the river outside water demand of the study area (Table 7).

Table 7. Ecological water supplement and daily release of the Xinlicheng Reservoir from April to October.

Month	April	May	June	July	August	September	October	Total
Ecological water supplement (10^6 m^3)	1.13	1.87	2.05	3.91	4.61	1.39	0.56	15.52
Daily release (m^3/s)	0.44	0.70	0.79	1.46	1.72	0.54	0.21	—

3.3. Model Simulation

3.3.1. Simulation of the MIKE 11 One-Dimensional Hydrodynamic-Water Quality Model

The simulation points are located at typical gate dams in the study area. The MIKE 11 one-dimensional hydrodynamic model was estimated from the measured meteorological data, boundary water level, and reservoir release in 2010. Figures 2–4 show that the predicted water level by MIKE 11 and the measured water level fit well. The Nash-Sutcliffe coefficients (R^2) used in evaluation of the hydrological model for Cross Sections A, B, and C are 0.96, 0.95, and 0.97, respectively. The simulated results show that the MIKE 11 model has high reliability and accuracy.

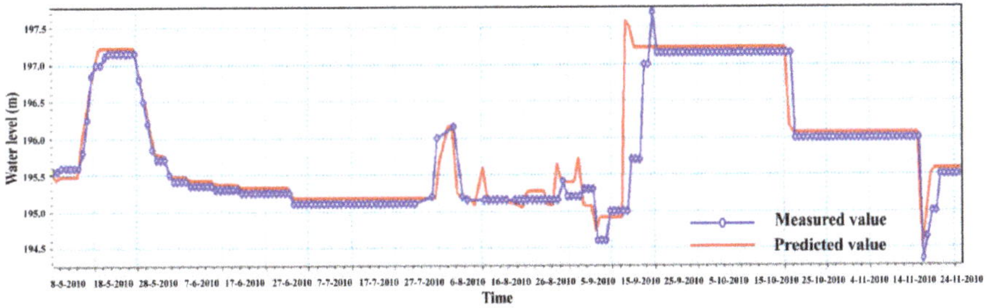

Figure 2. Simulated result of the water level at Cross Section A.

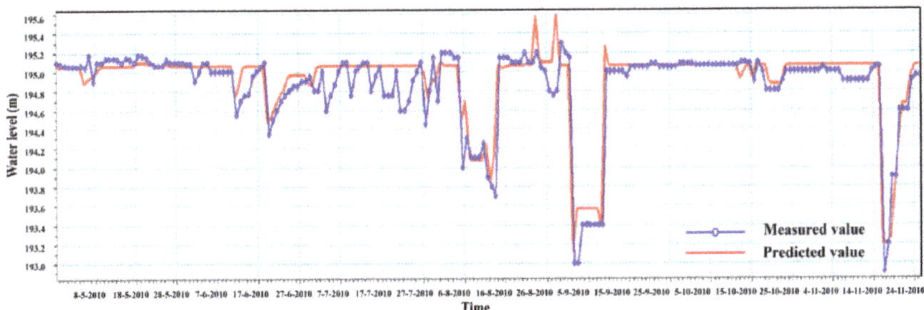

Figure 3. Simulated result of the water level at Cross Section B.

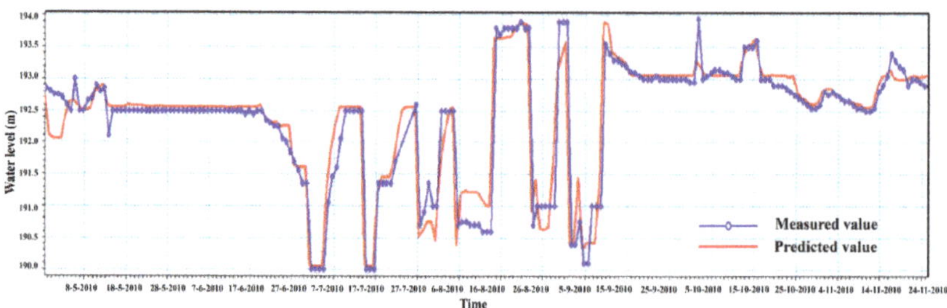

Figure 4. Simulated result of the water level at Cross Section C.

Chemical oxygen demand (COD) and ammonia nitrogen (NH_3-N) are used as simulated objects in the water quality model. In the simulated process, the COD and NH_3-N of the study area were calculated in 2010. Figures 5 and 6 show that the predicted variations in water quality are consistent with the measured variation trends. The R^2 of COD and NH_3-N in the model are 0.82 and 0.84, respectively. This indicates that the simulation effect of the model is good based on the evaluation criterion of the goodness of fit. Therefore, the MIKE 11 one-dimensional hydrodynamic-water quality model can be used in the research of the study area.

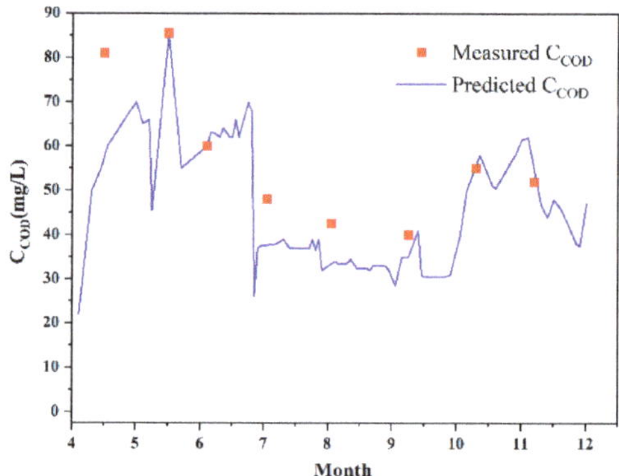

Figure 5. Simulated results of the COD concentration in 2010.

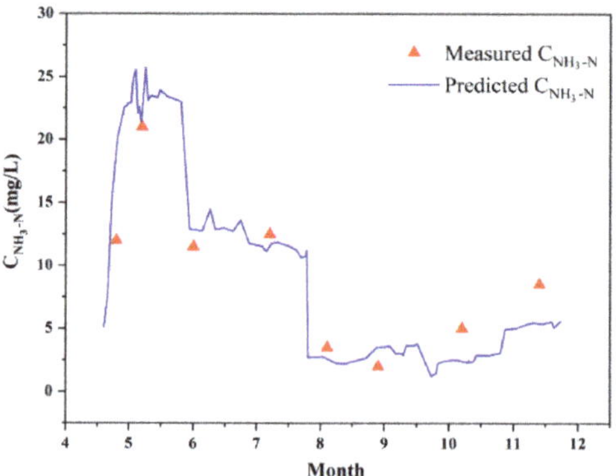

Figure 6. Simulated results of the NH_3-N concentration in 2010.

The water quality of the Xinlicheng Reservoir before and after carrying out the ecological operation scheme was simulated by the established model from April to October. After ecological operation, the COD concentration decreased significantly in the downstream river, especially from June to July (Figure 7). However, the effect of the ecological operation on the NH_3-N concentration is not significant. The simulated result shows that the NH_3-N concentration is 3–4 mg/L after the ecological operation, which has not yet reached the Class V standard of surface water (Figure 8). Therefore, when ecological operation is carried out, the water quality of the Xinlicheng Reservoir should also be improved. To reduce the NH_3-N concentration in the river, the nonpoint source pollution of the Xinlicheng Reservoir and the Yitong River should be reduced, and the purification effect of water quality in the riverside wetlands should be promoted.

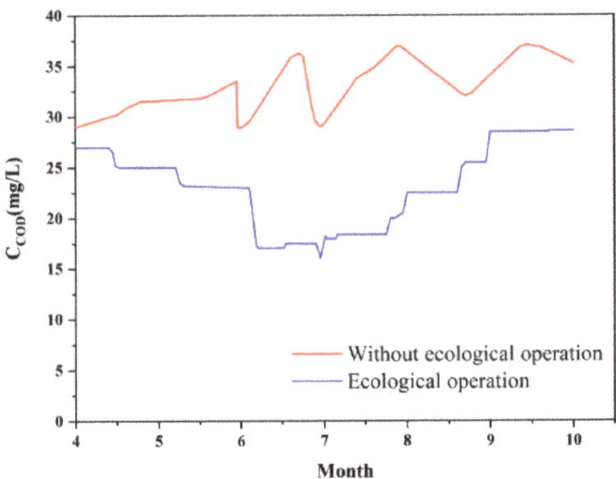

Figure 7. Simulated COD concentration before and after the ecological operation scheme.

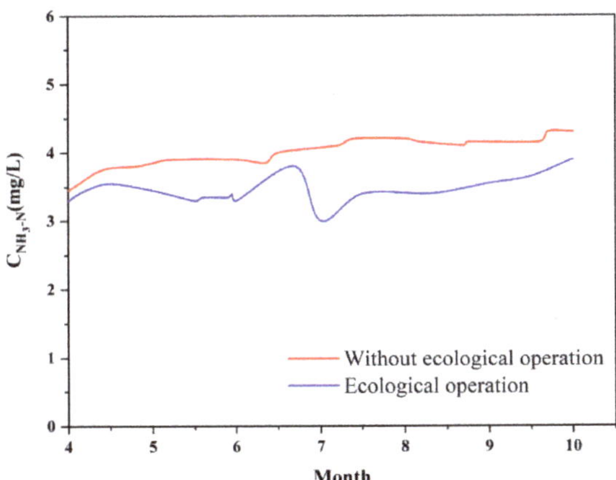

Figure 8. Simulated NH_3-N concentration before and after the ecological operation scheme.

3.3.2. Simulation of the PHABSIM Model

The relationship between the flow and weighted available area of the study area is shown in Figure 9. When the flow of the study area is less than 1.58 m^3/s, the weighted available area of carp physical habitat increases with increasing flow. When the flow of the study area is 1.58–3.00 m^3/s, the weighted available area decreases with increasing flow. When the flow of the study area is greater than 3.00 m^3/s, the weighted available area reaches a steady state.

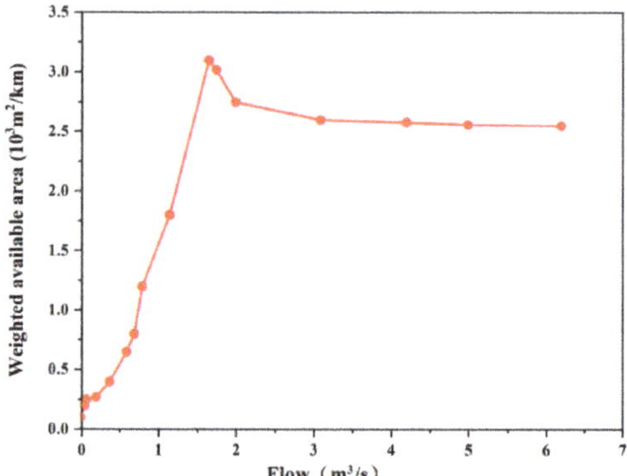

Figure 9. Weighted available area of the study area vs. flow.

The Xinlicheng Reservoir stores water in a brief rain period and discharges floods in an abundant rain period. The flow of the river is extremely small, and the phenomenon of intermittent flow occurs in the brief rain period. However, the flow of the river can reach 50–60 m^3/s in abundant rain periods, which cannot meet the need for a suitable ecological environment for carp. After ecological operation, the flow of the river can be maintained at 0.44–1.72 m^3/s, and the physical habitat area of carp can remain from 0.60×10^3 to 3.10×10^3 m^2/km from April to September. Maintaining physical habitat areas can improve the ecological environment of the river.

The simulated results of the MIKE 11 one-dimensional hydrodynamic-water quality model and PHABSIM model indicate that the ecological operation scheme can better restore the ecological environment of the Yitong River urban section. However, implementation of the water supplement scheme requires sufficient water resources from the Xinlicheng Reservoir. According to the storage capacity and hydrological statistics of the Xinlicheng Reservoir, the annual water storage of the reservoir is approximately between 44.3×10^6 and 134×10^6 m^3 with a guarantee rate of 50%–98%. The annual average water storage capacity is approximately 61×10^6 m^3. The Xinlicheng Reservoir needs to supply 88×10^6 m^3 of domestic water to Changchun every year. The supplemental water for ecological operation is approximately 15.52×10^6 m^3. Therefore, the water storage of the Xinlicheng Reservoir cannot guarantee the realization of ecological operation under the condition of meeting the domestic water supply. To achieve ecological operation, the inflow of the Xinlicheng Reservoir should be scientifically managed.

3.4. Watershed Management Reform and Suggestion for the Yitong River

The Xinlicheng Reservoir is a water supply reservoir in the Yitong River Basin. To solve the water shortage problem of the Xinlicheng Reservoir, the water distribution and management of the Yitong River Basin should be reformed.

3.4.1. Main Problems of Watershed Management

(1) Lack of watershed management institution

The "River Chief" was proposed to take charge of the protection and management of rivers and lakes in December 2016 and was assumed by the principles of the party and government. Jilin Province further instituted a "River Sheriff" and established specifications, task goals, and responsibilities. However, a unified management department is lacking for achieving the cooperation of public security, environmental protection, water conservancy, and animal husbandry. Each department works independently, causing conflict in water supply and affecting ecological operation.

(2) Incomplete legal framework of the basin

At present, there is only a national water law for watershed management of the Yitong River. Although river protection ordinances have been proposed, there is no law applied to basins. Law enforcement of water administration is mainly based on the "Water Law" (revised in 2016) and the "Water and Soil Conservation Law" (revised in 2010). The pertinence of the current legal framework is not strong, and the cost of violating the law is low, resulting in wasted water resources.

(3) Unscientific water distribution

The water service of the Yitong River is managed by Jilin Province and the local water service institution. Every water service institution can draw water based on requirements, resulting in an imbalance of water distribution and a reduction in reservoir inflow in the Xinlicheng Reservoir. Ecological operations cannot be carried out.

3.4.2. Countermeasures and Suggestions for Watershed Management

The following reforms and suggestions are proposed based on the main problems of water resource management in the Yitong River Basin combined with the feasibility of the ecological operation of the Xinlicheng Reservoir.

(1) Clarify department responsibility

The "Yitong River Basin Comprehensive Management and Protection Committee" should be established. This committee can further optimize water resource management, strengthen the communication and cooperation of departments, reduce the conflict between supply and demand, and maximize the utilization of water resources based on the overall Yitong River Basin.

(2) Improve law and regulation

A law is recommended to establish a basin, such as the "Yitong River Basin Management Regulation". This law can clarify the responsibilities of departments related to water resource management and establish the legal status for the "Yitong River Basin Comprehensive Management and Protection Committee" to carry out the management work of supervision, overall planning, and coordination. The cost of violating laws can be increased, leading to the avoidance of wasting water resources.

(3) Strengthen Water Service Management

The water service group of the Yitong River Basin is proposed to manage the related work of water supply, drainage, and water pollution treatment. Competition and performance mechanisms are introduced to promote development. However, the premise of carrying out water service reform is that there is a complete and feasible legal system and management model. The reasonable and efficient allocation of water resources is still a long and complicated process.

(4) Enhance Public Participation

Public and stakeholder participation is more beneficial to manage water resources. A publicly administered institution can be introduced to give real rights for publics. Their participation can get more rights for themselves and contribute to the sustainable development of watershed. Through public participation, watershed management builds a sense of community, helps reduce conflicts, increases commitment to the actions necessary to meet environmental goals, and ultimately improves the likelihood of success for the watershed management scheme.

4. Conclusions

The calculated river ecological water demand was 79.35×10^6 m^3 for the urban section of the Yitong River. An ecological operation scheme was proposed for the Xinlicheng Reservoir, including a water supplement of 15.52×10^6 m^3 and daily release from April to October. The MIKE 11 one-dimensional hydrodynamic-water quality model and the PHABSIM model were used to simulate the water quality and physical habitat of the river urban section. The simulated results indicate that the estimated models have higher accuracy and reliability. The water quality and physical habitat were simulated by the models after carrying out ecological operation. The results show that the COD of the river can be improved, but the improvement effect of NH_3-N was not obvious. Further water pollution control of the basin should be carried out. The increase in the physical habitat area and the ecological environment of the river can be improved after ecological operation. To realize the ecological operation of the Xinlicheng Reservoir, the "Yitong River Basin Comprehensive Management and Protection Committee" is proposed to establish the basin. Establishing "Yitong River Basin Management Regulations" should be suitable for the local characteristics. A water service investment group can be established to realize comprehensive reform and improvement of the water resources in the Yitong River Basin. Public participation should be enhanced to improve the likelihood of success for the watershed management scheme. This research can serve as a reference for reservoir ecological operation and watershed management in small and medium cities of the same latitude or type.

Author Contributions: Data curation: Y.C.; formal analysis: S.Z.; investigation: X.Y.; methodology: H.J.; project administration: H.J. and C.H.; resources: H.J.; software: W.W.; supervision: C.H.; validation: H.J.; writing—original draft: G.D. and H.J.; writing—review & editing: G.D. and Y.W. All authors have read and agreed to the published version of the manuscript.

Funding: This research was founded by the National Natural Science Foundation of China, grant numbers 41901116, the Foundation of Jilin Scientific and Technological Development Project, grant numbers 20190103137JH and 20190701048GH, and the Foundation of Jilin Educational Committee, grant numbers JJKH20180792KJ.

Conflicts of Interest: The authors declare that there is no conflict of interest.

References

1. Dunham, J.B.; Angermeier, P.L.; Crausbay, S.D.; Cravens, A.E.; Gosnell, H.; McEvoy, J.; Moritz, M.A.; Raheem, N.; Sanford, T. Rivers are social–ecological systems: Time to integrate human dimensions into riverscape ecology and management. *Water* **2018**, *5*, e1291. [CrossRef]
2. Rinaldi, M.; Gurnell, A.M.; del Tánago, M.G.; Bussettini, M.; Hendriks, D. Classification of river morphology and hydrology to support management and restoration. *Aquat. Sci.* **2016**, *78*, 17–33. [CrossRef]
3. Roy, S.G.; Uchida, E.; Souza, S.P.; Blachly, B.; Fox, E.; Gardner, K.; Gold, A.J.; Jansujwicz, J.; Klein, S.; McGreavy, B. A multiscale approach to balance trade-offs among dam infrastructure, river restoration, and cost. *Proc. Natl. Acad. Sci. USA* **2018**, *115*, 12069–12074. [CrossRef] [PubMed]
4. Yu, Y.; Wang, P.F.; Wang, C.; Wang, X. Optimal reservoir operation using multi-objective evolutionary algorithms for potential estuarine eutrophication control. *J. Environ. Manag.* **2018**, *223*, 758–770. [CrossRef]
5. Yang, Z.; Yang, K.; Hu, H.; Su, L.W. The cascade reservoirs multi-objective ecological operation optimization considering different ecological flow demand. *Water Resour. Manag.* **2019**, *33*, 207–228. [CrossRef]
6. Suen, J.-P.; Eheart, J.W. Reservoir management to balance ecosystem and human needs: Incorporating the paradigm of the ecological flow regime. *Water Resour. Res.* **2006**, *42*, W03417. [CrossRef]
7. Meißner, T.; Schütt, M.; Sures, B.; Feld, C.K. Riverine regime shifts through reservoir dams reveal options for ecological management. *Ecol. Appl.* **2018**, *28*, 1897–1908. [CrossRef]
8. Lu, S.B.; Shang, Y.Z.; Li, W.; Wu, X.H.; Zhang, H.B. Basic theories and methods of watershed ecological regulation and control system. *J. Water Clim. Chang.* **2018**, *9*, 293–306. [CrossRef]
9. Ma, L.J.; Zhang, X.N.; Huan, W.; Qi, C.J. Characteristics and practices of ecological flow in rivers with flow reductions due to water storage and hydropower projects in China. *Water* **2018**, *10*, 1091. [CrossRef]
10. He, S.; Yin, X.A.; Yu, C.X.; Xu, Z.H.; Yang, Z.F. Quantifying parameter uncertainty in reservoir operation associated with environmental flow management. *J. Clean. Prod.* **2018**, *176*, 1271–1282. [CrossRef]
11. Jia, W.H.; Dong, Z.C.; Duan, C.G.; Ni, X.K.; Zhu, Z.Y. Ecological reservoir operation based on DFM and improved PA-DDS algorithm: A case study in Jinsha river, China. *Hum. Ecol. Risk Assess. Int. J.* **2019**, 1–19. [CrossRef]

12. Wang, Y.Z.; Fan, Y.B.; Bu, F.; Zhou, D.M. Quantifying effects of water and sediment regulation scheme on the sand bar in the yellow river estuary in 2014. *Ecohydrol. Hydrobiol.* **2019**. Available online: https://doi.org/10.1016/j.ecohyd.2019.10.004 (accessed on 20 March 2020). [CrossRef]
13. Kong, D.X.; Miao, C.Y.; Wu, J.W.; Duan, Q.Y.; Sun, Q.H.; Ye, A.Z.; Di, Z.H.; Gong, W. The hydro-environmental response on the lower Yellow River to the water–sediment regulation scheme. *Ecol. Eng.* **2015**, *79*, 69–79. [CrossRef]
14. Dudgeon, D. River regulation in Southern China: Ecological implications, conservation and environmental management. *Regul. Rivers Res. Manag.* **1995**, *11*, 35–54. [CrossRef]
15. Mi, Y.J.; He, C.G.; Bian, H.F.; Cai, Y.P.; Sheng, L.S.; Ma, L. Ecological engineering restoration of a non-point source polluted river in Northern China. *Ecol. Eng.* **2015**, *76*, 142–150. [CrossRef]
16. Lv, J.; Xu, J.L.; Wang, H.X.; Li, W.; Liu, X.J.; Yao, D.F.; Lu, Y.; Zheng, X. X Study on ecological protection and rehabilitation technology of a reservoir-type water source in the northeastern region of China. *Hum. Ecol. Risk Assess. Int. J.* **2019**, *25*, 1802–1815. [CrossRef]
17. Suen, J.-P. Determining the ecological flow regime for existing reservoir operation. *Water Resour. Manag.* **2011**, *25*, 817–835. [CrossRef]
18. Kingsford, R.T. Ecological impacts of dams, water diversions and river management on floodplain wetlands in Australia. *Austral Ecol.* **2000**, *25*, 109–127. [CrossRef]
19. Daldorph, P.W.G. A reservoir in management-induced transition between ecological states. *Ecol. Bases Lake Reserv. Manag.* **1999**, *395–396*, 325–333.
20. Thompson, R.M.; Bond, N.; Poff, N.L.; Byron, N. Towards a systems approach for river basin management—Lessons from Australia's largest river. *River Res. Appl.* **2019**, *35*, 466–475. [CrossRef]
21. den Brandeler, F.; Gupta, J.; Hordijk, M. Megacities and rivers: Scalar mismatches between urban water management and river basin management. *J. Hydrol.* **2019**, *573*, 1067–1074. [CrossRef]
22. Taha, R.; Dietrich, J.; Dehnhardt, A.; Hirschfeld, J. Scaling effects in spatial multi-criteria decision aggregation in Integrated River Basin Management. *Water* **2019**, *11*, 355. [CrossRef]
23. Delipınar, Ş.; Karpuzcu, M. Policy, legislative and institutional assessments for integrated river basin management in Turkey. *Environ. Sci. Policy* **2017**, *72*, 20–29. [CrossRef]
24. Xue, L.Q.; Wang, J.; Zhang, L.C.; Wei, G.H.; Zhu, B.L. Spatiotemporal analysis of ecological vulnerability and management in the Tarim River Basin, China. *Sci. Total Environ.* **2019**, *649*, 876–888. [CrossRef] [PubMed]
25. Yang, Z.F.; Cui, B.S.; Liu, J.L.; Wang, X.Q.; Liu, C.M. *Theory, Methods and Practices of Ecological Environment Water Demand*; Science Press: Beijing, China, 2003.
26. Jiang, H.B.; He, C.G.; Luo, W.B.; Yang, H.J.; Sheng, L.X.; Bian, H.F.; Zou, C.L. Hydrological Restoration and Water Resource Management of Siberian Crane (*Grus leucogeranus*) Stopover Wetlands. *Water* **2018**, *10*, 1714. [CrossRef]
27. Men, B.H.; Yu, T.; Kong, F.L.; Yin, H. Study on the minimum and appropriate instream ecological flow in Yitong River based on Tennant method. *Nat. Environ. Pollut. Technol.* **2014**, *13*, 541–546.
28. Herman, M.R.; Nejadhashemi, A.P. A review of macroinvertebrate- and fish-based stream health indices. *Ecohydrol. Hydrobiol.* **2015**, *15*, 53–67. [CrossRef]
29. O'Brien, A.; Townsend, K.; Hale, R.; Sharley, D.; Pettigrove, V. How is ecosystem health defined and measured? A critical review of freshwater and estuarine studies. *Ecol. Indic.* **2016**, *69*, 722–729. [CrossRef]
30. Han, N.N.; Wang, Y.R.; Zhou, Q.Y.; Li, S.M.; Ye, L.T.; Jin, J.H. Analysis on Spatial and Temporal Variations of Maize Irrigation Water Requirement in Jilin Province. *IOP Ser. Earth Environ. Sci.* **2018**, *128*, 012024. [CrossRef]
31. Tan, M.H.; Zheng, L.Q. Different Irrigation water requirements of seed corn and field corn in the Heihe River Basin. *Water* **2017**, *9*, 606. [CrossRef]

Article

Projected Changes in the Water Budget for Eastern Colombia Due to Climate Change

Oscar Molina *, Thi Thanh Luong and Christian Bernhofer

Institute of Environmental Sciences, University TU Dresden, 01062 Dresden, Germany; thanh_thi.luong@tu-dresden.de (T.T.L.); christian.bernhofer@tu-dresden.de (C.B.)

* Correspondence: oscar_david.molina_rincon1@tu-dresden.de

Received: 22 October 2019; Accepted: 12 December 2019; Published: 23 December 2019

Abstract: There is a lack of information about the effect of climate change on the water budget for the eastern side of Colombia, which is currently experiencing an increased pressure on its water resources due to the demand for food, industrial use, and human demand for drinking and hygiene. In this study, the lumped model BROOK90 was utilized with input based on the available historical and projected meteorological data, as well as land use and soil information. With this data, we were able to determine the changes in the water balance components in four different regions, representing four different water districts in Eastern Colombia. These four regions reflect four different sets of climate and geographic conditions. The projected data were obtained using the Statistical Downscaling Model (SDSM), in which two global climate models were used in addition to two different climate scenarios from each. These are the Representative Concentration Pathways (RCP) RCP 2.6 and RCP 8.5. Results showed that the temporal and spatial distribution of water balance components were considerably affected by the changing climate. A reduction in the generated streamflow for all of the studied regions is shown and changes in the evapotranspiration and stored water were varied for each region according to both the climate scenario as well as the characteristics of soil and land use for each area. The results of spatial change of the water balance components showed a direct link to the geography of each region. Soil moisture was reduced considerably in the next decades, and the percentage of decrease varied for each scenario.

Keywords: climate change; water budget; general circulation model; modeling; stream flow changes; soil water; RCP

---

1. Introduction

According to the Intergovernmental Panel on Climate Change [1], the amount of greenhouse gases released into the atmosphere might have an high influence on the global warming of Earth's surface over the next decades. These changes in climate can have long-term implications on social, economic, and ecological processes, while also affecting the natural development of ecosystems [2]. Thanks to advances in the modelling field as well as the physical understanding of the climate system-processes, more regional climate change projections have been developed for several regions of the world, throughout the last years. It is expected that the increment of average annual and seasonal temperatures in the tropics and subtropics will be higher than that in the mid-latitudes. On the other hand, average annual precipitation is expected to decrease in many regions located in mid-latitudes and subtropics. It is estimated that for each degree produced from global warming, there will be a reduction of at least 20% in hydric resources for approximately 7% of the global population [3].

Water budgets are a useful method for evaluating availability and sustainability of a water supply—it shows the balance between the water stored in an area and the water that flows into and out of the area. Observed changes in water budgets of an area over time can be used to assess the effects of climate variability and human activities on water resources. Furthermore, they provide a basis for assessing how natural or human-induced changes in one part of the hydrologic cycle might affect other aspects of the cycle [4]. A large number of published articles show the important impacts of climate change on water resources—some of which are related to the hydrological cycle, e.g., [5–9], while other studies are related to groundwater, recharge, changes in the vegetation cover, and the impact on ecosystems. Alterations in the climate will produce changes in the hydrological cycle, including an increase of evaporation due to higher temperatures, as well as an increase in global and regional evapotranspiration, which will be directly related to precipitation levels, spatiotemporal changes in rain distribution, vapor pressure deficit, and wind speed [10–12]. These features could have a negative influence on water sources.

Soil moisture variation is caused by rising temperatures and other climate variations; soil moisture affects agricultural productivity and has a negative influence on the land's ability to store carbon. Moreover, soil moisture information is valuable to a wide range of government agencies and private companies concerned with issues of weather and climate, runoff potential and flood control, soil erosion and slope failure, reservoir management, geotechnical engineering, and water quality. Soil moisture is a key variable in controlling the exchange of water and heat energy between the land surface and the atmosphere, through evaporation and plant transpiration.

The eastern region of Colombia is highly vulnerable to the effects of climate change, especially with regards to its high diversity of fauna and flora, the expansion of the agricultural frontier, in addition to pressure on water resources for industrial activities. When a dry climatic condition occurs in the country, the water yield reduces significantly, as compared to normal conditions; thus, the natural supply of the hydric resource in an average year and a dry year have regional differences that are important to consider [13]. This hydric shortage could affect areas including both the agriculture an energy sector. There is a lack of data and detailed climate studies throughout this region; therefore, research on a water budget approach is necessary to determine on a regional scale the possible change in the availability of water for future decades. This research can, thus, contribute valuable information to development planners, decision makers, researchers, and other stakeholders as to when to plan and implement appropriate management strategies for adapting to climate change in this area.

This study aims to determine the change of the water balance components under climate change scenarios (RCP 2.6 and RCP 8.5) in two periods of time (2021–2050 and 2071–2100), based on the modeling of climatic and hydrologic parameters on four representative regions characterized as individual water districts on the east side and in the middle of Colombia—regions with very different geographic and climatic conditions. Included complementary to this study, are descriptions of the soil moisture in the projected climate scenarios.

2. Study Area

The areas analyzed in this study comprises four representative areas characterized as individual water sectors on the east side and in the middle of Colombia with different geographic and climatic conditions. These regions lie between 74°56′13″ and 66°82′29″ west longitude, and between 12°24′40″ north and 2°18′25″ south latitudes. These regions are named Alta Guajira, Bajo Meta, Rio Catatumbo and Sabana de Bogota. They were selected due to each region's variability of conditions and the sufficient availability of data for analysis.

Colombia is located in the northwestern corner of South America, exhibiting complex geographical, environmental, and hydroecological features. Colombia is crossed by three rugged parallel ranges of the Andes Mountains—namely, the Eastern, Central, and Western Cordilleras. Precipitation along the country is highly influenced by the Inter Tropical Convergence Zone (ITCZ); however, the climate is also conditioned by local particularities like those caused by mountain barriers to the atmospheric circulation. On seasonal time scales, the displacement of the ITCZ exerts a strong control on the annual cycle of Colombia's hydroclimatology [14–16]. Some regions of the country experience a bimodal annual cycle of precipitation with distinct rainy seasons and dry seasons, while others experience a unimodal annual cycle, which result from the different passages of the ITCZ over those regions.

Moisture transported from the Amazon basin encounters the orographic barrier of the Andes, thus, focusing and enhancing deep convection and rainfall in the eastern flank of the Cordillera, with the maximum rainfall occurring during June–August. The interannual variability of the diurnal cycle is dominated by the effects of both phases of El Niño Southern Oscillation (ENSO).

The east side of Colombia is hot in most of its extension, with a range of medium temperature from 12 to 34 °C. The eastern side of the country borders Venezuela, the Amazon is to the south, extensive valleys and the Andean mountains are on the mid-eastern side, and coastal plains towards the higher north. These plains are tropical grasslands that undergo seasonal flooding; they are suitable for livestock grazing and in some areas for the cultivation of crops. Additionally, major petroleum discoveries have been made in the eastern region.

3. Materials and Methods

3.1. Meteorological Data

Historical daily data of precipitation, maximum temperature, minimum temperature, and relative humidity from 153 hydrometeorological stations along the four studied regions was provided by the Institute of Hydrology, Meteorology and Environmental Studies of Colombia (IDEAM). From this, only datasets with less than 30% of missing values for the time range of 1980–2015 were considered for the analysis. This was used in concordance with the minimum extension of 30 year-records, which is recommended by the World Meteorological Organization [17] in order to obtain reliable statistics. Figure 1 and Table 1 show the location and description of the 4 analyzed regions or water districts; the climate characterization refers to Lang's Index (I = Pr/Tm), where Pr is the mean annual precipitation amount and Tm is the mean annual temperature. In some of these water districts, several climate conditions coexist.

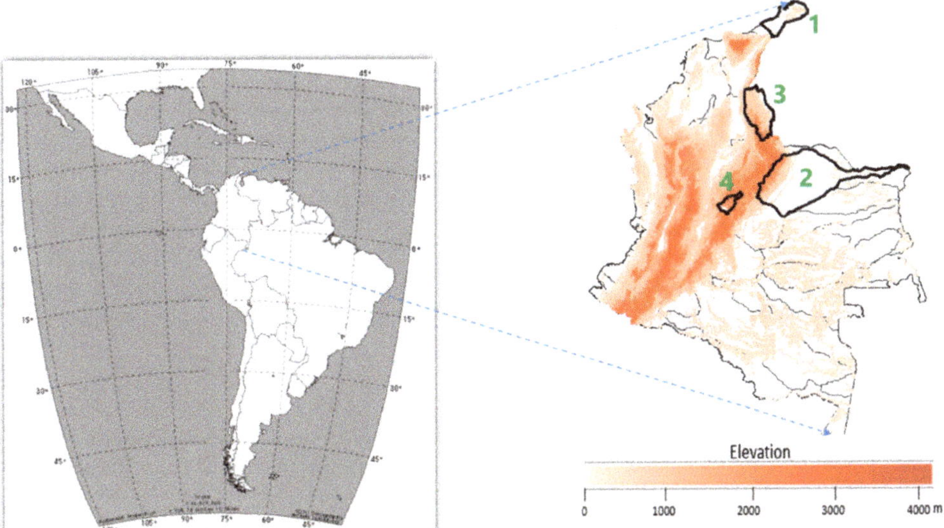

Figure 1. Location of the studied areas with elevation.

Table 1. Characteristics of the four analyzed regions.

	Region/Water District	Climate	Area (km^2)	N° of Stations (Precipitation)	Min. Elevation (m.a.s.l.)	Max. Elevation (m.a.s.l.)
1	Alta Guajira	arid, desertic	12,348	25	1	390
2	Bajo meta	semihumid	42,655	42	45	3520
3	Rio Catatumbo	humid	17,960	47	83	1740
4	Sabana de Bogota	semihumid, semiarid	2245	39	2540	3800

Projected daily data sets for the same variables and for the future periods 2021–2050 and 2071–2100 were created from a regional downscaling procedure [18], using the statistical downscaling model SDSM and datasets from two Global Climate Models (GCM), which are part of the CMIP5-project, the Global Climate Model CanESM2 developed by Canadian Centre for Climate Modelling and Analysis and the model IPSL-CM5A-MR developed by The Institut Pierre Simon Laplace. Both GCM included in the study (as most GCM used to date) used fundamental physical laws, which were then subjected to physical approximations like equations of Geophysical Fluid-Dynamics that are appropriate for describing the atmosphere and the ocean at large enough scales. The Representative Concentration Pathways—RCP 2.6 and RCP 8.5 were considered for both models, representing two different possible future emission trajectories and radiative forcings. The RCP 8.5 combines assumptions about high population and relatively slow income growth with modest rates of technological change and energy intensity improvements, leading to a high energy demands and greenhouse gas (GHG) emissions in the long term, in the absence of climate change policies. The RCP 2.6 might be described as the best case for limiting anthropogenic climate change. In this scenario, Global CO2 emissions peak by 2020 and decline to around zero by 2080. The concentrations in the atmosphere peak at around 440 ppm in midcentury and then slowly start declining.

3.2. Soil and Land Cover

The values of the different canopy and vegetation variables were taken from available local studies, and maps provided by IDEAM. Physical characteristics of the different types of soils for each region were obtained from regional studies in the areas. The overall available information about the general soil characteristics of the whole extension of the study sites was relatively low, most of the data used for the analysis corresponded to the maps provided by the IDEAM and the local authorities. Three of the studied regions, not including Alta guajira, were relatively similar in terms of soil properties but did show significant differences in terms of land cover. In these regions, numerous wetlands can be found along with some urban areas; however, grassland and tropical forest are the main land-cover types for all of regions, even though the portions differ.

3.3. Model

The model BROOK90 [19] was used in this study for the water balance assessment in the historical and the different projected scenarios. BROOK90 is a deterministic, process-oriented, lumped parameter hydrologic model that can be used to simulate the water balance in most land surfaces at a daily time-step, year-round. The model has a strong physically based description that simulates the above and below liquid phases of the precipitation–evaporation–streamflow–ground water flow part of the hydrological cycle for a point-scale stand at a daily time-step [20]. The BROOK90 model calculates evaporation through the Shuttleworth–Wallace approach [21], as well as an improvement of the Penman–Monteith equation. The characteristics of the soil water were determined using a modified approach of the Brooks and Corey [22], and Saxton et al. [23]. The water movement through the soil was simulated using the Darcy–Richards equation. To calculate streamflow, the model used a simplified process—storm flow by source area flow or subsurface pipe-flow and delayed flow, from vertical or downslope soil drainage and first-order groundwater storage. A general water balance equation can be represented as follows:

$$PREC = EVAPOT + FLOW + STORAGE \tag{1}$$

where *PREC* represents precipitation (mm), *EVAPOT* is the evapotranspiration (mm), *FLOW* is the corresponding simulated total streamflow (mm) derived from surface flow and groundwater flow, and *STORAGE* is the deep seepage loss from groundwater (mm). Applications of the BROOK90 model have been demonstrated in grasslands, temperate evergreen and deciduous forests [19], and cultivated lands [24], among various vegetations with satisfactory performance. The model is also applicable in the tropics after adjusting the parameters to local conditions.

3.4. Data-Grid and Interpolation

The study intends to show the change in water budget caused by climate change over the studied regions on a bigger scale than a watershed scale. For this reason, the study was focused primarily on water sectors that covered a much bigger extension of an area. This would give a wider overview and understanding of the availability of water for several cities and settlements located in and around these regions, as well as the productive activities developed in the area. With this purpose, the data from the stations (including historical and future data) were interpolated and converted into a 10 km × 10 km grid of datasets. This is an appropriate approach considering the irregular distribution of the stations and the highly variable geography of the areas. It further enables the possibility of conducting a water balance calculation in areas where no historical data are available, while also being located at different elevations from the station point. The data were interpolated using the Thin Plate Spline Method (TPS). This is a spline-based technique for data interpolation and smoothing and it has been proven to perform a good interpolation for precipitation data (Tait et al., 2006). For interpolation of scattered $z(x,y)$ data, the TPS is just a special case of Radial Basis Function (RBF) interpolation:

$$z(x,y) = p(x,y) + \sum_i l_i \, \phi(r) \tag{2}$$

where $p(x,y)$ is a polynomial function and ϕ is an RBF. In the case of TPS, $\phi = r^2 \ln(r)$. The water balance would be calculated for each of the grid points using BROOK90 and all available data. The results for the baseline historical period of 1981–2010 would be compared with the results obtained for the different projected scenarios, considering the two Representative Concentration Pathways RCP 2.6 and RCP 8.5, the two GCM, and the two projected periods 2021–2050 and 2071–2100. The comparison would allow one to visualize the expected differences in the availability of water for the future decades, due to the effects of climate change.

4. Results

The graphs provided in Appendix A show an overview of monthly average results for the 4 studied areas. Here, one can compare the three periods of time (historical baseline 1981–2010, future projections 2021–2050, and 2071–2100) for each area, each GCM, and each Representative Concentration Pathway. As an example, in Figure A1a of Appendix A, the results for the three periods of time in the Alta Guajira region and the scenario with the model CanESM2 and Representative Concentration Pathway RCP 2.6 can be observed. The results are expressed in terms of the water balance components given by Equation (1). Table 2 shows the relative increment or decrease for each projected component of the water balance in the different climate scenarios, compared to the reference period of 1981–2010.

Projected soil moisture for the 4 studied regions as an averaged monthly value can be seen in Figure 2, the results for the climate scenarios are then compared with the baseline period of 1981–2010.

Table 2. Percentual change of the water-balance components in the projected scenarios.

	Precipitation (mm)		Streamflow (mm)		Evapotranspiration (mm)		Storage (mm)	
	2021–2050	2071–2100	2021–2050	2071–2100	2021–2050	2071–2100	2021–2050	2071–2100
Alta Guajira								
CanESM2 (RCP 2.6)	9.12	3.68	17.73	35.07	13.73	−10.36	−22.34	−21.03
CanESM2 (RCP 8.5)	−0.88	−24.15	66	−22.75	−22.24	−26.74	−44.64	25.34
IPSL-CM5A-MR (RCP 2.6)	−35.22	−26.59	−1.97	−15.73	−4.37	−25.27	−28.88	14.41
IPSL-CM5A-MR (RCP 8.5)	−35.13	−22.07	−22.8	−52.77	−18.3	0.4	5.97	30.3
Bajo Meta								
CanESM2 (RCP 2.6)	−11.41	−11.58	−12.48	−12.81	−6.67	−5.29	7.74	6.52
CanESM2 (RCP 8.5)	−19.33	−20.73	−20.45	−22.52	−15.18	−13.06	16.3	14.85
IPSL-CM5A-MR (RCP 2.6)	−1.5	−6.91	−10.58	−15.85	18.12	27.03	−9.04	−18.09
IPSL-CM5A-MR (RCP 8.5)	−9.81	−17.67	−10.47	−19.93	7.67	8.43	−7.01	−6.17
Rio Catatubo								
CanESM2 (RCP 2.6)	−3.64	−2.44	−24.92	−23.93	17.15	18.45	4.13	3.04
CanESM2 (RCP 8.5)	−8.9	−10.57	−30.73	−39.7	12.41	17.78	9.42	11.35
IPSL-CM5A-MR (RCP 2.6)	−6.25	−5.92	−30.64	−30.53	17.57	18.02	6.82	6.59
IPSL-CM5A-MR (RCP 8.5)	−14.17	−13.68	−35.08	−40.32	6.24	12.24	14.67	14.4
Sabana de Bogota								
CanESM2 (RCP 2.6)	10.33	10.53	−16.18	−18.17	30.11	32.33	−3.6	−3.63
CanESM2 (RCP 8.5)	17.84	16.78	−24.59	−22.39	50.25	46.41	−7.82	−7.24
IPSL-CM5A-MR (RCP 2.6)	−2.57	−1.77	−15.27	−16.16	7.07	8.99	5.63	5.4
IPSL-CM5A-MR (RCP 8.5)	12.54	20.72	−20.8	−11.09	38.1	44.67	−4.76	−12.86

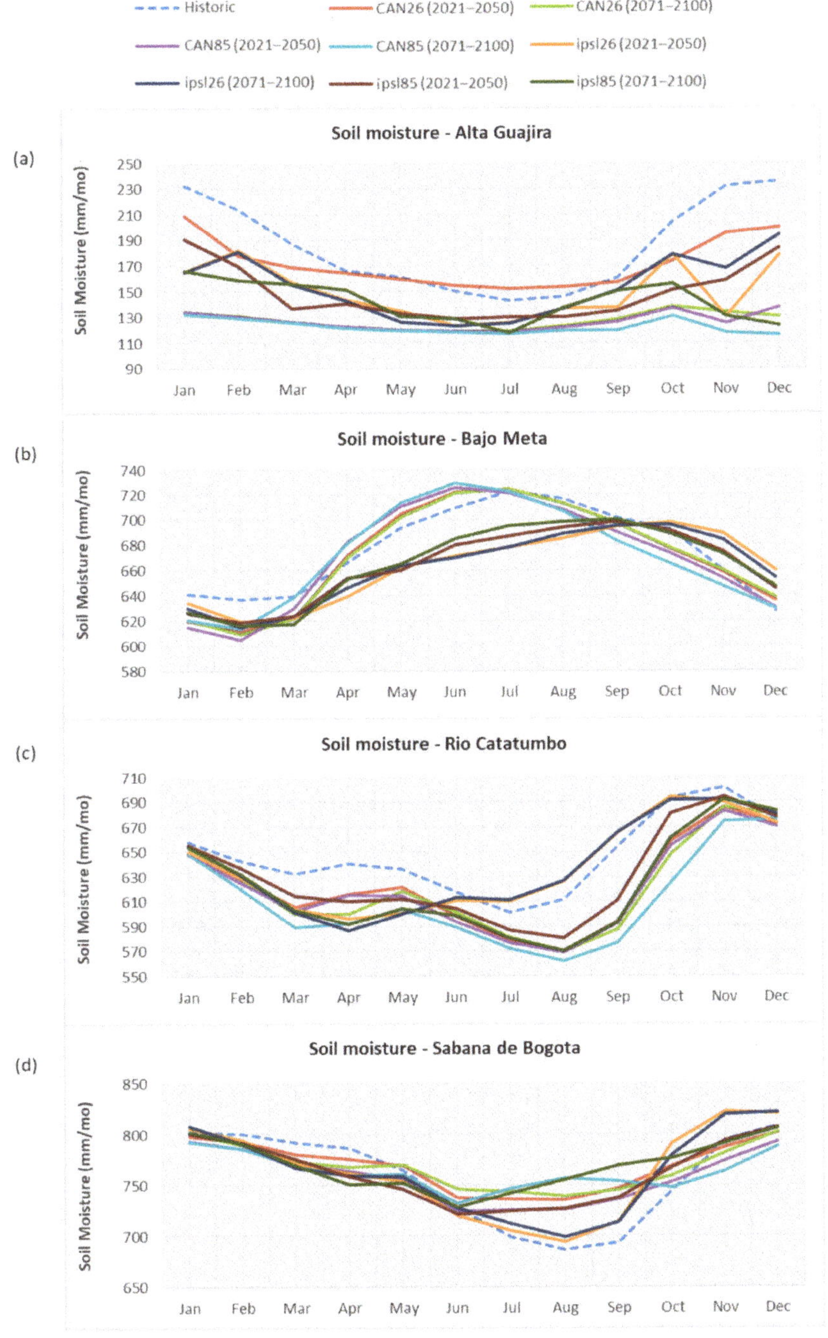

Figure 2. Monthly averaged soil moisture in (**a**) Alta Guajira; (**b**) Bajo Meta; (**c**) Rio Catatumbo; (**d**) Sabana de Bogota.

5. Discussion

In Appendices A and B, one can appreciate the notorious difference between the climate conditions and the water availability between the four studied regions, with regards to the historical period as well as to the projections. In general terms, the water balance components in the different regions showed different patterns magnitudes due to variability in precipitation along the Colombian territory. The region of Sabana de Bogota (Appendix A, m–p) showed a clear bimodal precipitation regime and the region Meta (Appendix A, e–h) showed a clear monomodal regime; the other two regions presented a not-so-clearly defined bimodal regime—these precipitation conditions were obedient to the displacement of the ITCZ over the regions.

The region of Alta Guajira being an arid/desert region shows very low levels of precipitation for most of the year. It reaches a peak of about 105 mm by the month of October, generates low levels of streamflow, since almost 85% of the total precipitation in the year is converted into evapotranspiration due to the high temperatures. In the first three months of the year, evapotranspiration can be almost 8 times larger than precipitation. During this time, the storage water produced as a consequence of the rainy season in the last months of the year is constantly being evaporated. The projections for Alta Guajira showed a decrease in precipitation, in general terms, and, therefore, a decrease in the other components of the water balance. This was the case for both projected periods of time of 2021–2050 and 2071–2100. Only the model CanESM2 with scenario RCP 2.6 showed a slight increment of precipitation in the short and long term.

From the four analyzed areas in this study, Bajo Meta presented the highest amount of rain on a yearly basis. In this region, the results showed an almost "normal" distribution of precipitation throughout the year for the historical records—presenting the highest values in the months of June and July with a peak of 430 mm/month and a non-rainy season at the end and beginning of the year. Most of the precipitation in Bajo Meta was converted into streamflow during the year (74.3%). This might be due to the characteristics of the soil (a predominant silty loam type for a big part of this region), which does not allow a big rate of infiltration. The projections for this region showed a decrease in precipitation, which led to a directly proportional decrease in streamflow; evaporation showed variable results depending on the model; CanESM2 indicated a slight decrease for both RCP scenarios, while IPSL-CM5A indicated the opposite. These results are reasonable considering that for the first model the decrease in precipitation was much bigger in magnitude than the second model.

The historical period indicates that in the region of Rio Catatumbo, half of the precipitation in the year was evaporated (50.7%) and a similar level was converted into streamflow (45.1%). In general terms, the projections for the future showed a slight decrease of the precipitation regimes, with around 6% from the model CanESM2 and around 9% from the model IPSL-CM5A but with a much higher decrease in the levels of streamflow produced by this precipitation. Evaporation in both models was projected to increase at levels of around 20% for the next decades. This was linked to the projected increase in the temperature for the region, which was close to 4 °C for the end of the century with the scenario RCP 8.5, according to the results from the regional downscaling procedure [18].

In the region of Sabana de Bogota, the historical period showed that 58% of the precipitation was evaporated while only 41% converted into streamflow. The typical clay type of soil predominant in a big part of the region was reflected in the low levels of stored water. Sabana de Bogota was the only one from the analyzed four regions where an increase of precipitation was projected for the next decades; Table 2 shows that the biggest increment was in the period of 2071–2100, with the model IPSL-CM5A and the scenario RCP 8.5. These projections showed that the levels of evaporation would increase in a considerable rate compared with the historical period. It was clear from Appendix A (m–p) that a bigger percentage of the projected precipitation would evaporate, compared with the historical baseline.

It is important to consider that the results obtained in this study and shown in Figure 2 and Appendix A are the averaged-out product of the results of each station in the region. This was made with the purpose of obtaining results in a macro-scale, to have a notion of the projected scenarios for water districts, where productive and social activities are planned in accordance with the available water in that area. An analysis for an individual station or an area of a much higher resolution could also show variable results, depending on the elevation of the studied area. The rasters produced in Appendix B are a useful representation of the spatial variability of the historical and projected results for two of the studied regions, when considering the geographical variation presented in each of them. As was said before, the spatial distribution density of meteorological stations at the other two regions was too low to allow a proper interpolation process to create a figure that reliably showed a representation of spatial variability of the results throughout the regions.

Figure 2 shows that, in general, soil moisture would be reduced considerably in the next decades, the percentage of decrease could vary for each scenario; the only exception was the region of Sabana de Bogota, where precipitation is projected to increase, which would result in an increase in the soil moisture. These results are important in relation to agricultural activities and the planning/use of soil for the next decades.

The lack of available historical records of discharge in the studied areas as well as the wide extent of these areas, made the performance of a respective process of calibration and validation of the water balance components that comprises all extensions of the studied areas unsuitable, however, an intensive review of the input data to the model was carried out to ensure an appropriate parameterization. This included a detailed selection of values regarding the information of soils in the studied areas, through technical information from the public and private sector. Information regarding vegetation comes mostly from Governmental institutes, as well as the detailed land use information that was also obtained from regional, territorial development plans. In the same way, a review of the results of the model and a comparison with results of other studies nearby, in tropical or similar areas was made [25–31], to verify the veracity of the results and ensure that they are within a correct range of magnitudes.

The main differences of both GCM that were used for a regional downscaling as a source of the projected climate data used in this study are their spatial resolution; CanESM2 with 2.79° latitude × 2.81° longitude and IPSL-CM5A-MR with 1.26° latitude × 2.5° longitude, as well as the model-components with which they were coupled; CanESM2 consists of a physical atmosphere–ocean model coupled to a terrestrial carbon model and an ocean carbon model, while the IPSL-CM5A-MR model couples four components of the Earth system, atmospheric dynamics and physics, ocean dynamics, sea ice dynamics and thermodynamics, and land surface. Moreover, every single GCM differs in the parametrization of the physical modeled processes, for this reason they offer varied results that might be more successfully correlated with real measurements in some areas than others. The model IPSL-CM5A, in spite of its slightly higher spatial resolution has shown a better performance than other models to identify extreme events in South America and other regions [32], but a lower performance to appropriately reproduce precipitation historical records in comparison with CanESM2 and other models, in Colombia and South America [33,34]. This agreed with the results of this study, where extreme events and seasonal precipitation was more clearly identifiable for the model IPSL-CM5A, especially in regions of lower elevations, where the model seemed to overestimate the projected change for the different variables.

Rainfall seasonality and its interannual variability have been observed to change in magnitude, timing, and duration, in the tropics [35]. As mentioned before, the climate in Colombia is conditioned by local particularities like those caused by mountain barriers to the atmospheric circulation but the annual cycle of Colombia's hydroclimatology is mostly influenced by the displacement of the ITCZ. The different passages of the ITCZ over the regions determine either a bimodal annual cycle of precipitation with distinct rainy seasons and dry seasons, while others experience a unimodal annual cycle that result from the moisture transported from the Amazon basin when it encounters the orographic barrier of the Andes. The seasonality of hydrological elements in the different water districts shows larger variability due to their different conditions of topography, hydrogeology, and vegetation. Higher regions like Sabana de Bogota or parts of Rio Catatumbo are highly dependent on altitude, and since there is no snow formation in any of the regions, there is no considerable time lag between the precipitation event, the stream flow, and the soil moisture, which can be observed in Figure 2 and Appendix A. A prolonged positive soil humidity in the humid regions seen in Figure 2 linked directly with rain events explained by a permeable soil and temperature that was not high enough to increase evapotranspiration for several months. Climate seasonality is a defining feature of many ecosystems, often characterized in the tropics by a distinct non-uniformity in their timing of annual rainfall. This results in one or two wet seasons during which most of the annual rainfall occurs, separated by prolonged dry periods. In regions like Bajo Meta or Rio Catatumbo, under conditions of relative water abundance, long-term evapotranspiration becomes limited by the potential evapotranspiration, while in arid regions like Alta Guajira, where the energy supply is high, precipitation is the main constraint to evapotranspiration. In the former case, water supply exceeds demand, while in the latter case water supply is outstripped by the demand [36]. In Alta Guajira a projected increase in mean temperature would likely lead to increase in the frequency and the intensity of seasonal droughts [37].

It is important to consider that although the RCP 2.6 scenario might be described as the best case for limiting anthropogenic GHG emissions, their atmospheric concentrations will continue to increase even after emissions slow down and then will eventually start to decrease [38]. Carbon dioxide accumulates in the atmosphere and stays there for decades. Even if emissions start reducing in 2020, the concentration continues increasing and starts falling very slowly, only after 2050. This might explain why in some of the results of RCP 2.6 in Table 2, a bigger percentual change is observed for the period 2020–2050 than for the period of 2070–2100. However, as expected, the results obtained for the projections in the scenario RCP 8.5 showed a higher projected change than those in the scenario 2.6, this is the case for the analyzed regions except in Alta Guajira with model CanESM2, where the results showed a slight increase in precipitation under scenario 2.6, and a negative change under scenario 8.5 and model IPSL-CM5A_MR. This could have been caused by the incapacity of the model to accurately predict changes of precipitation in very arid areas that are characterized by little but highly variable and unpredictable rainfall and has been shown in other studies like Zhao [39] who analyzed the performance of the GCM models used in the CMIP5-project in several arid regions of the world, or by Mingxia [40] which found similar results across dryland areas.

The inherent existence of uncertainty in every water budget approach must be taken into consideration. Uncertainty related to hydrological modeling is affected by the input data, validation data, model structure, and model parameters. In order to reduce the uncertainties as much as possible in the hydrological modeling in this study, a detailed parametrization of the model was intended for each of the grids cells that the regions were divided into. For this, soil, vegetation, land use, and topography data were taken from local private studies and maps provided by governmental institutions and the parameters were individually defined for each of the cell grids where the model was run individually.

The hydro-climatic model chain typically consists of the components—emission scenario, GCM, regional climate model or statistical downscaling, and hydrological model [41]. This study represents the last step of that chain, but all of these components constitute a potential uncertainty source for the results. The uncertainty associated with the individual components of this chain has been investigated by an increasing number of studies. In some of them, the GCM structure is identified as the dominant source of uncertainty, e.g., [42–44]. A common finding for other studies is that in the hydrological model, uncertainty is less important than other sources but cannot be ignored [45–47]. Ideally, the analysis of hydrologic change in future studies should comprehend the full suite of uncertainties associated with global climate modeling, climate downscaling, hydrologic modeling, and natural climate variability. In this manner, the water resources planning and management community can make more informed decisions. Parameter uncertainty estimation is one of the major challenges in hydrological modeling and analysis of future change for the water sector is an interdisciplinary endeavor [48]. Ongoing parallel efforts to monitor and verify water budget components would help to improve accuracy. Posterior analysis could be done in an effort to determine the magnitude of the uncertainty of the hydrological response to climate change. Due to the uncertainties associated with the study of climate change and the limitations of models in representing climate and hydrological response, the most trustworthy indicator is still the trends observed at the measuring stations while the predictions of models in a big scale like that in this study might only be used to have a general notion of trends for the studied variables under different potential climate conditions.

The main objective of this study was to provide an overview of water budget response to climate change for a region where no other studies have been performed, and where not so much information was available since the few existing studies in Colombia have been aimed at regions with a bigger density of available data. In the analyzed regions in this study and especially in Alta Guajira and Bajo Meta, there is a lack of observed data with sufficient detail and quality. We encourage the future improvement of collection and testing of reliable data in a range of spatial and temporal scales in these regions, since it is critical to improve our understanding of hydrological processes [49].

6. Conclusions

The model BROOK90, historical data, as well as projected meteorological data were used to determine the changes in the water balance components in four different regions, which represent the four water districts in Eastern Colombia. The four regions reflect four different climatic and geographic conditions. The projected data were obtained from a statistical regional downscaling procedure, where two GCMs (CanESM2 and IPSL-CM5A-MR) and two different climate scenarios from RCP 2.6 and RCP 8.5 were used.

Results have shown a potential reduction in the generated streamflow for all studied regions. The temporal distribution of water balance components was considerably affected by the changing climate, which moreover, might have a profound impact on the hydrological regimes in these regions. Changes in evapotranspiration and stored water could vary from each region, according to the climate scenario, and the characteristics of soil and land use for each area. Results of spatial change of the water balance components have shown a direct link to the geography of each region and how the values differed accordingly, at different elevations. Soil moisture would be reduced considerably in the next decades and the percentage of decrease could vary for each scenario. Only in the region of Sabana de Bogota did the results show the opposite—this agreed with the precipitation projections that are to increase and, therefore, also the soil moisture.

Application of the model BROOK90 proved to be valuable for water cycle analysis and for the purpose of this study in offering a general overview to the change of water balance components, throughout the east side of Colombia, due to future climate change. Prediction of the impact of climate change on water budget components is a transcendent, practical, and theoretical problem to which each country and its institutions should dedicate more resources—especially for countries and regions that are more vulnerable to climate change.

Uncertainties associated with the GCMs, hydrological models, and the approaches used in this study have a direct effect on the outcome, and they have to be considered and evaluated for the use of these results, in addition to their uses for future works. The results obtained in this study should be considered as indicative of the expected trend in water resources of the studied regions, as a result of climate change. These results might serve as a baseline information for creating mitigating measures. However, future work using other models and other techniques for the analysis of water resources throughout these areas is encouraged.

Author Contributions: Conceptualization, methodology, software, analysis, writing, O.M.; Software, review, T.T.L.; Supervision, review and guidance, C.B. All authors have read and agreed to the published version of the manuscript.

Funding: This research received no external funding. The DAAD (German Academic Exchange Service) is the provider of the scholarship in which this research project took place.

Acknowledgments: We acknowledge the support given by the Open Access Publication Funds of the SLUB/TU Dresden. We thank IDEAM for providing the available historical records in the area.

Conflicts of Interest: The authors declare no conflict of interest.

Appendix A

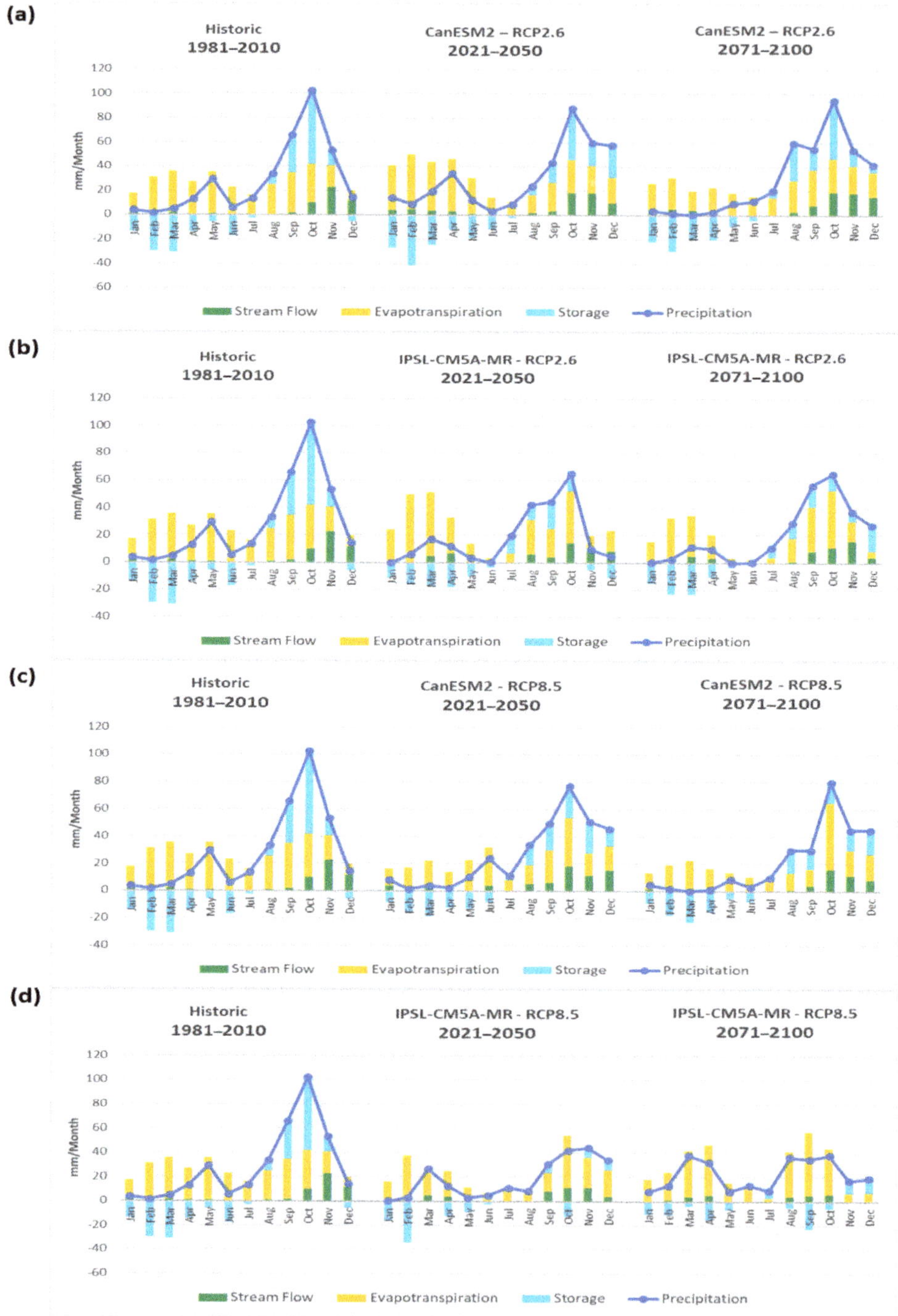

Figure A1. Alta Guajira.

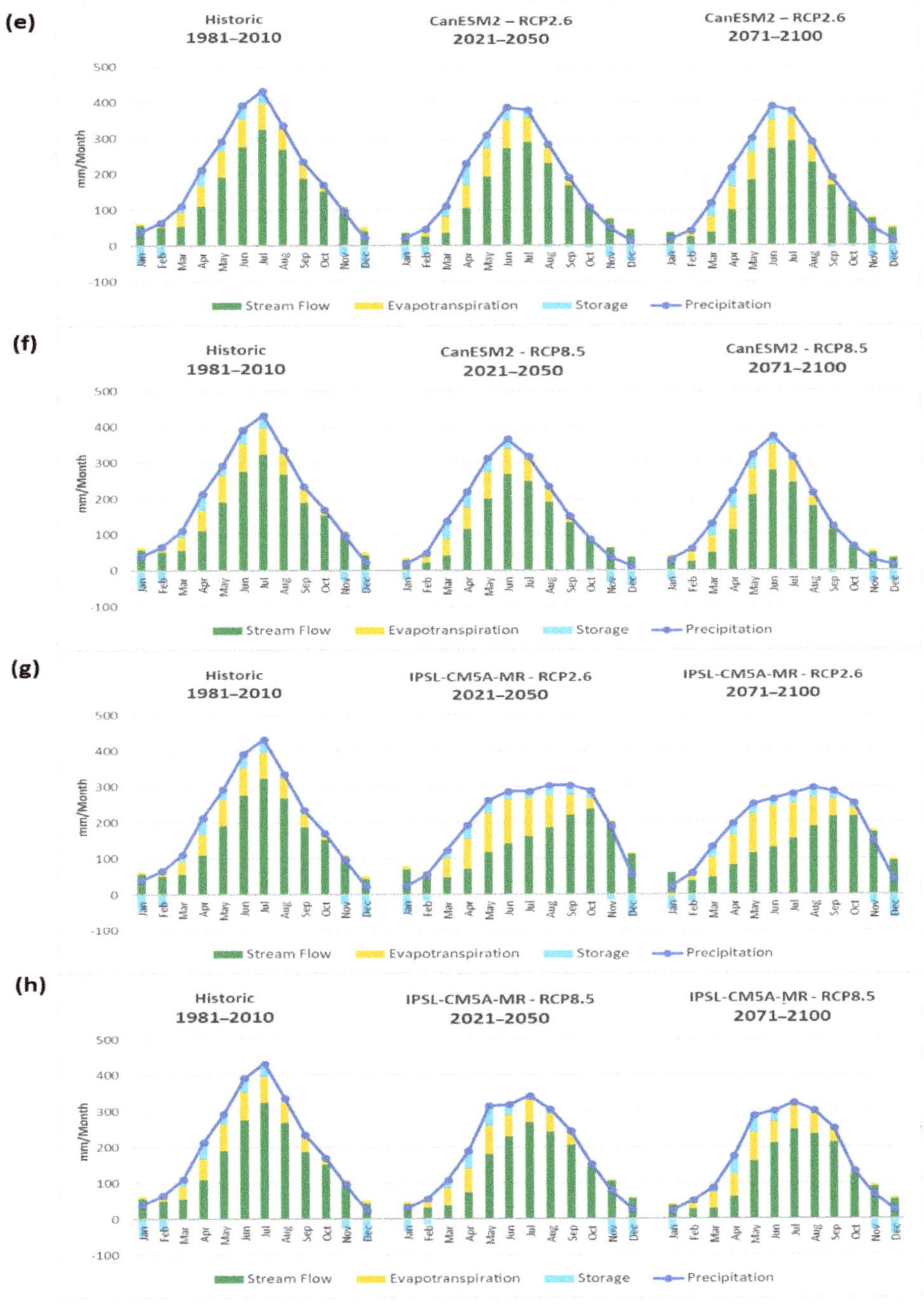

Figure A2. Bajo Meta.

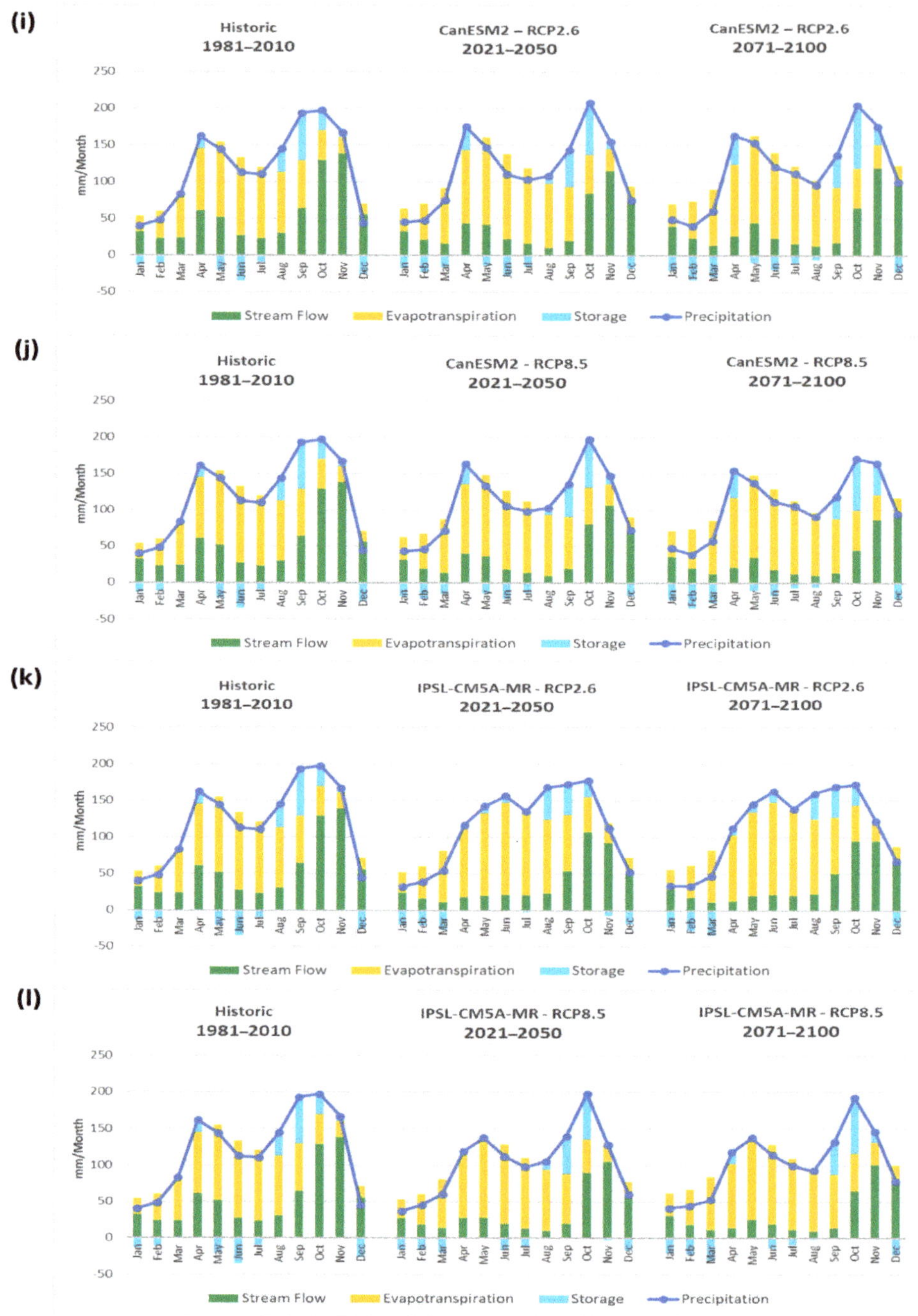

Figure A3. Rio Catatumbo.

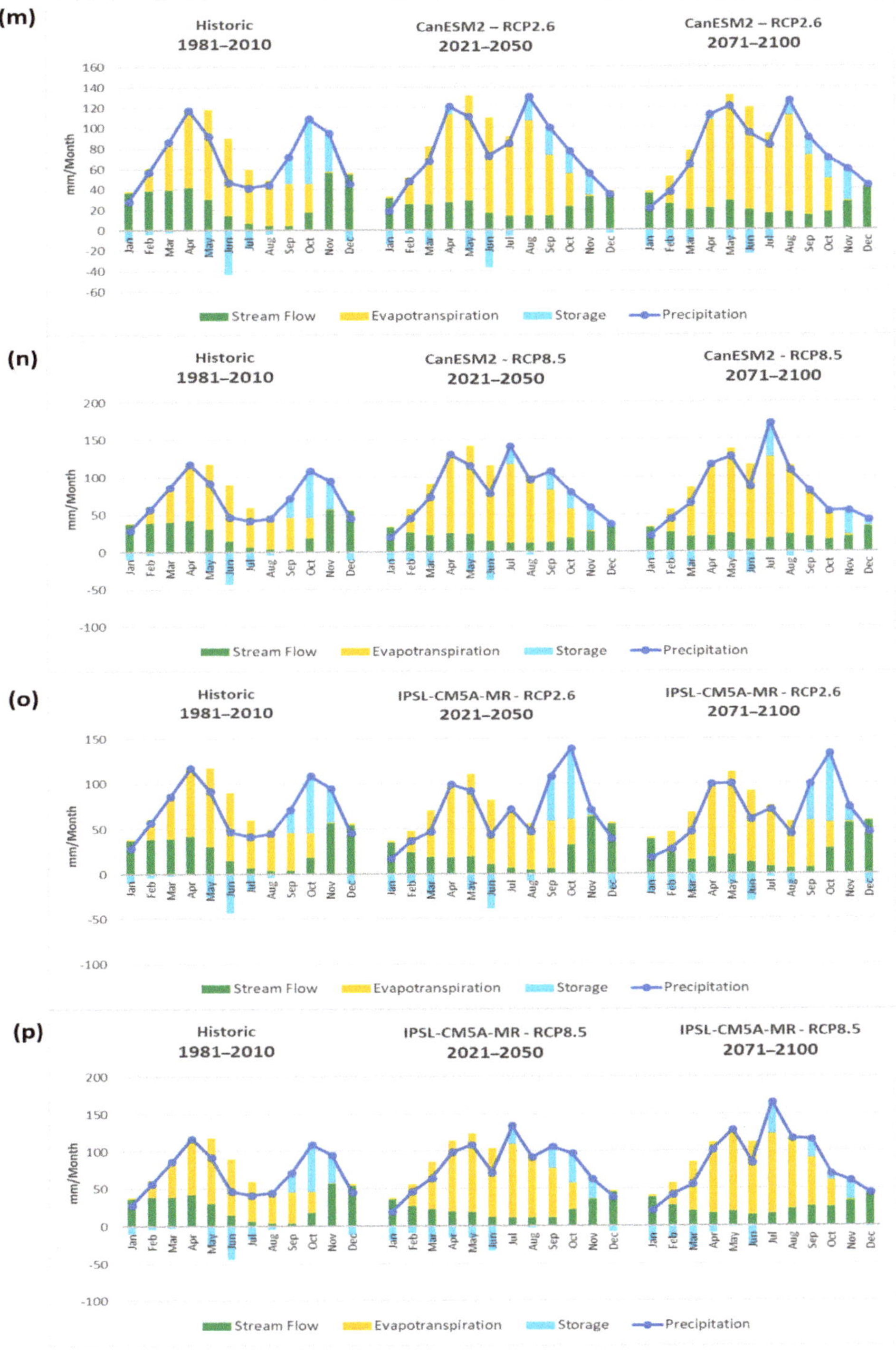

Figure A4. Sabana de Bogota.

Appendix B

Appendix B.1 Rio Catatumbo

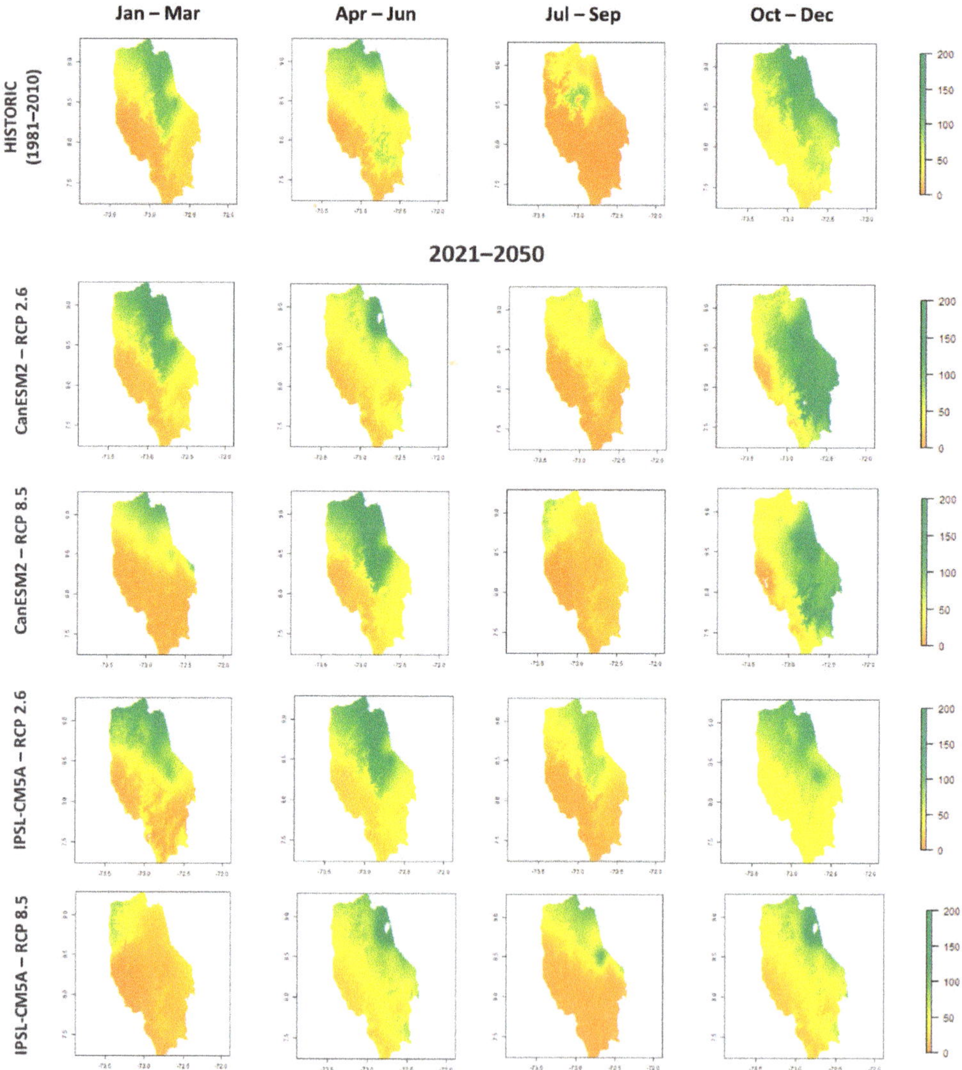

Figure A5. Stream Flow 2021–2050.

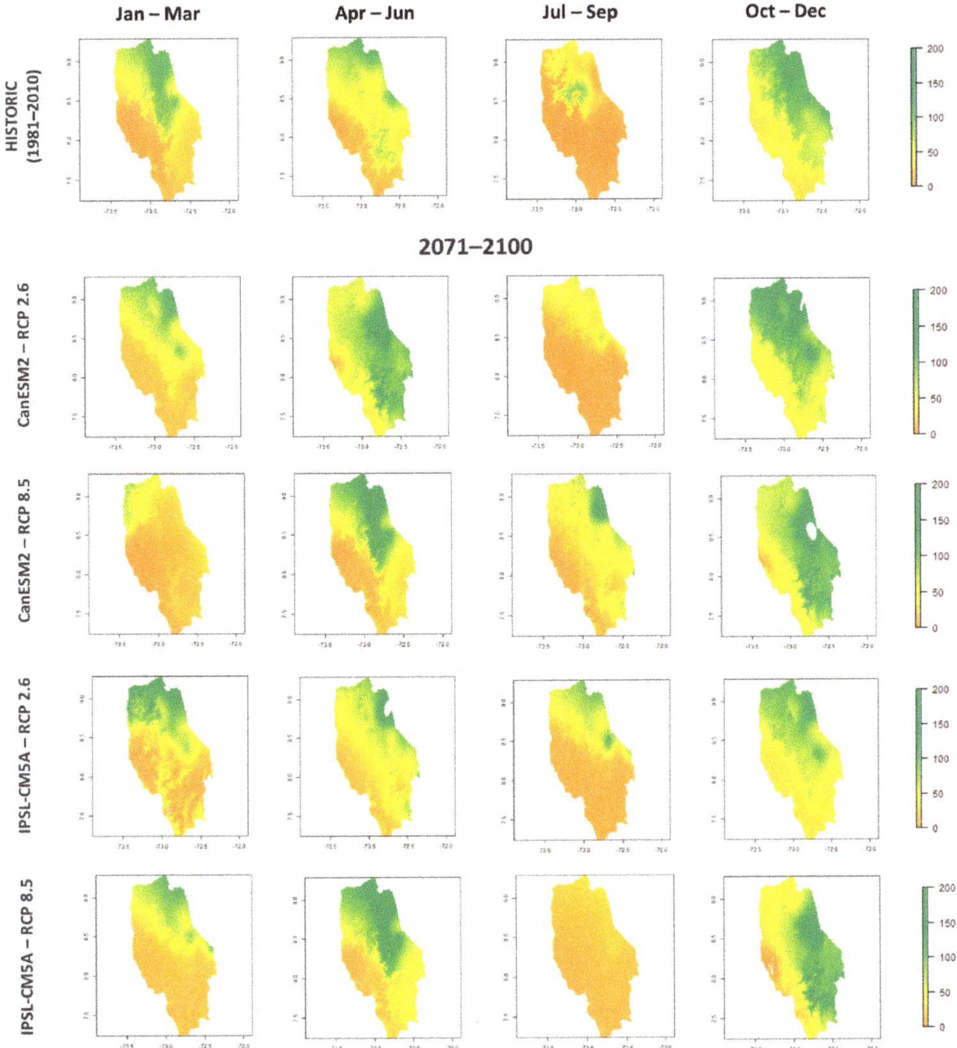

Figure A6. Stream Flow 2071–2100.

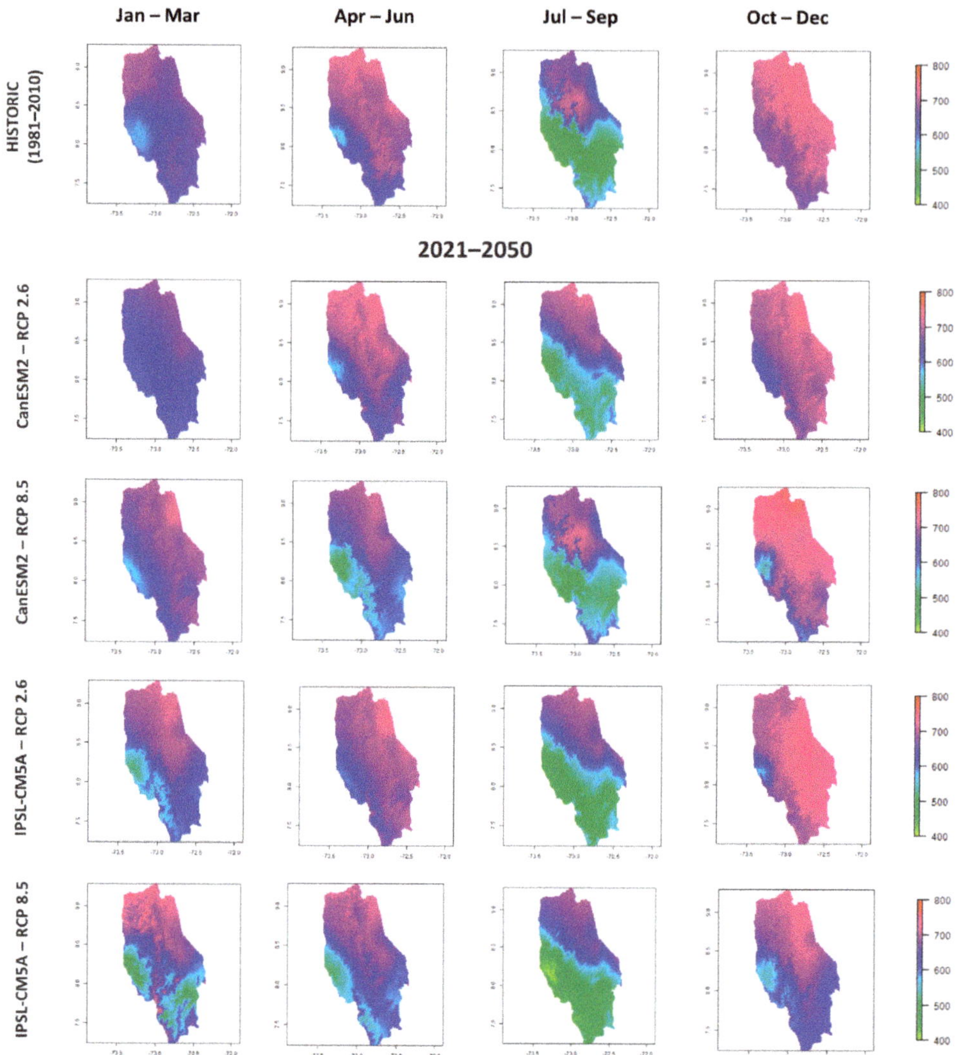

Figure A7. Soil Moisture 2021–2050.

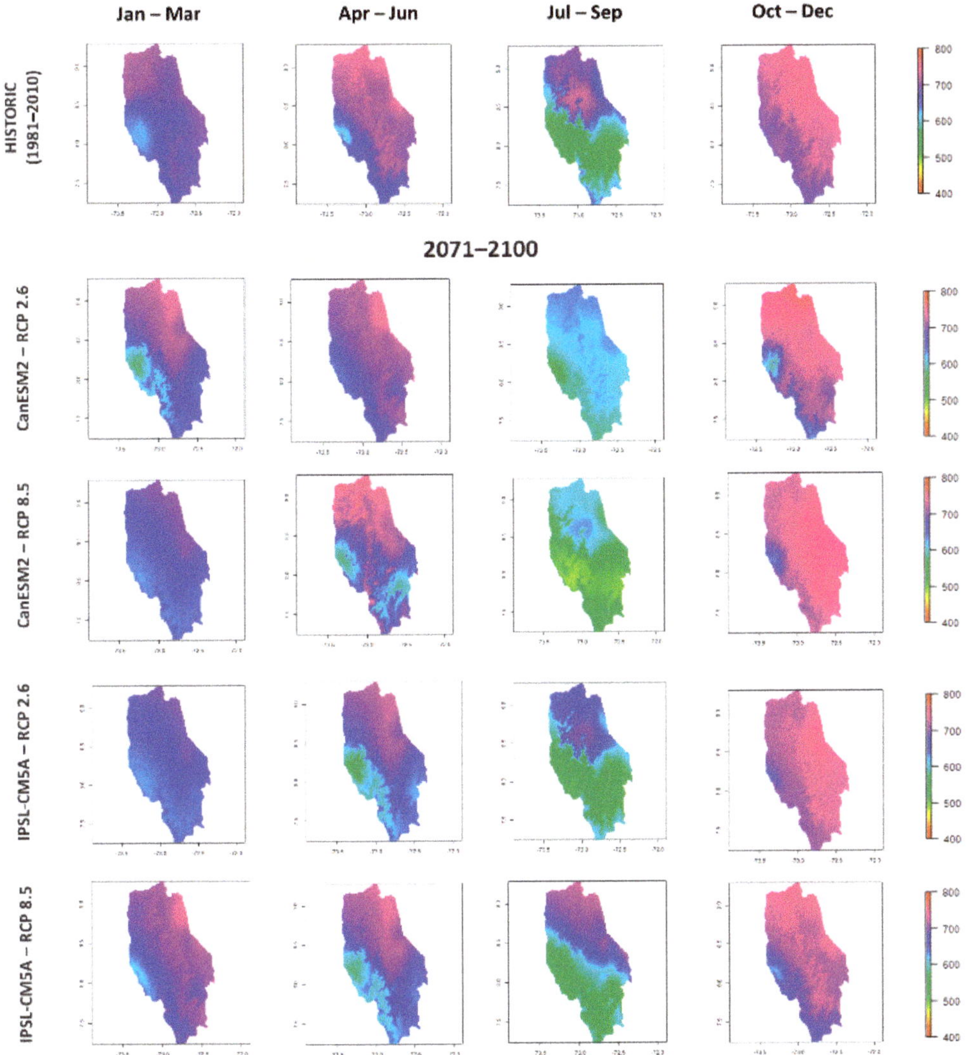

Figure A8. Soil Moisture 2071–2100.

Appendix B.2 Sabana de Bogota

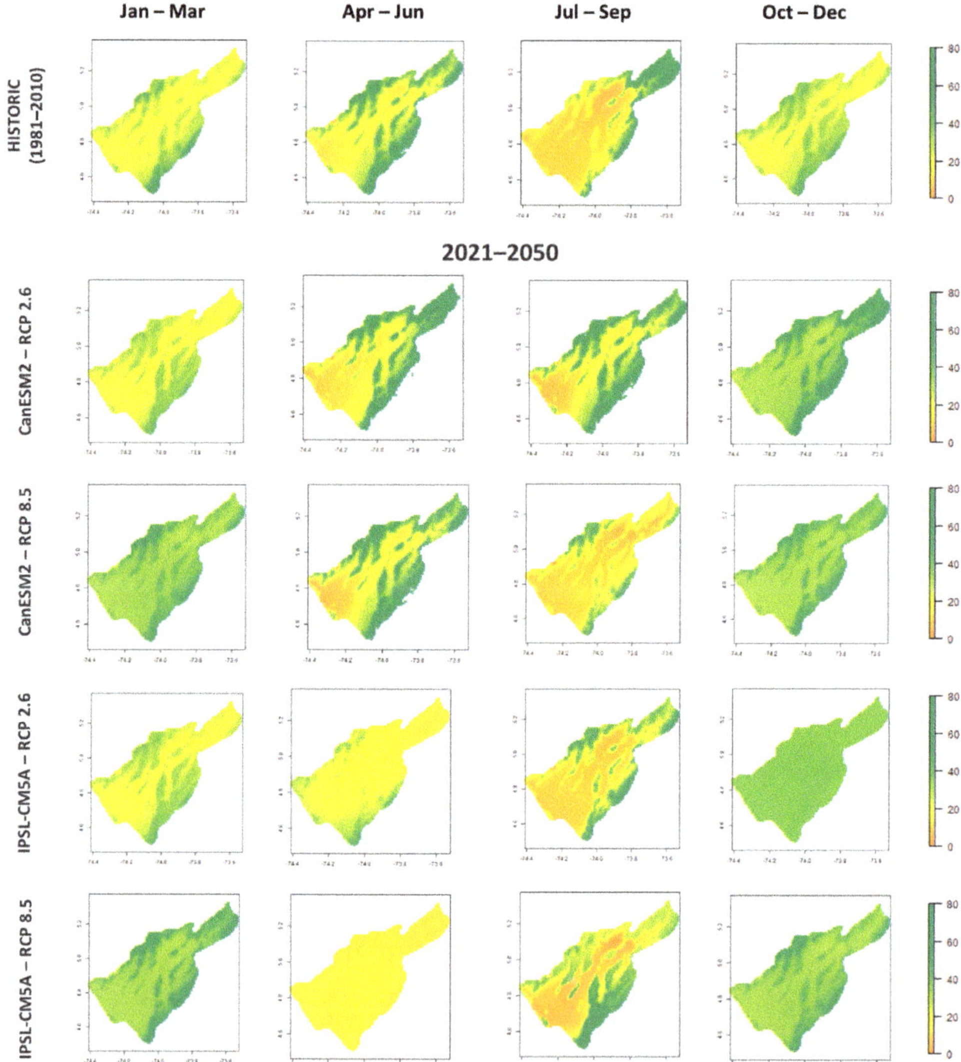

Figure A9. Stream Flow 2021–2050.

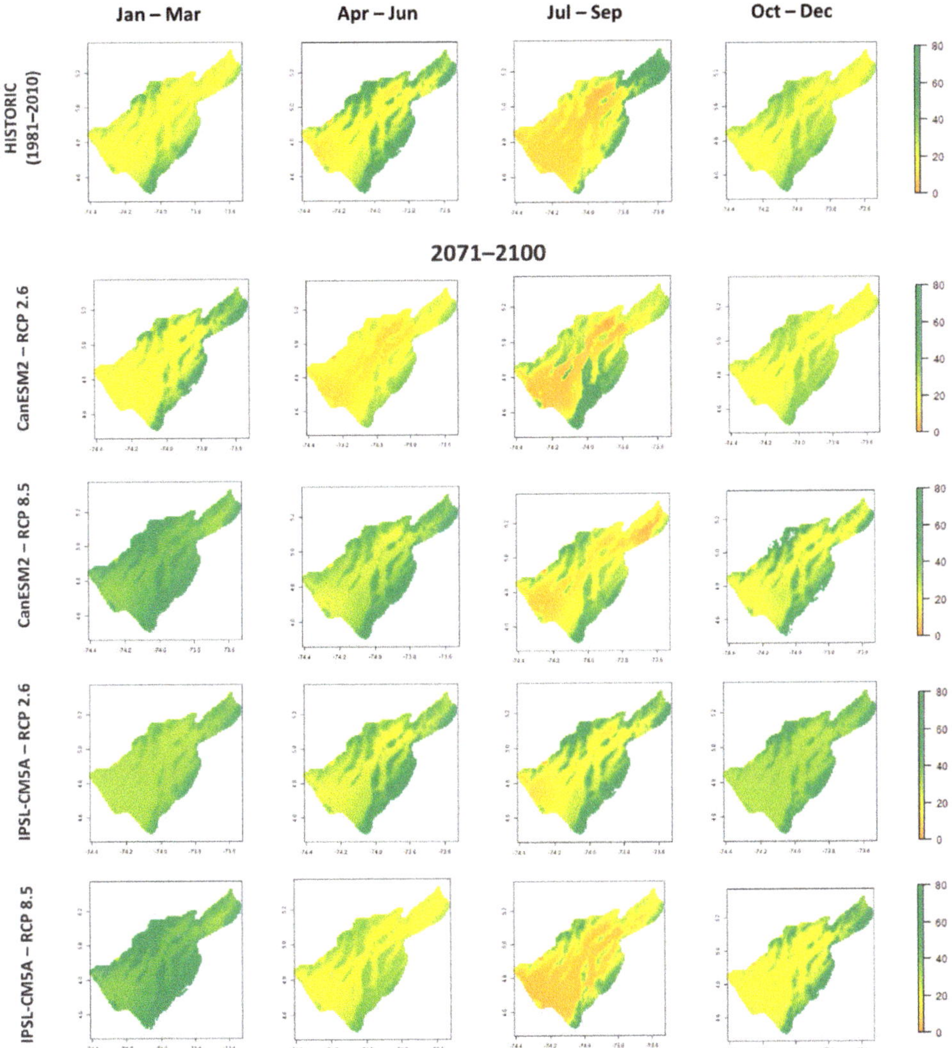

Figure A10. Stream Flow 2071–2100.

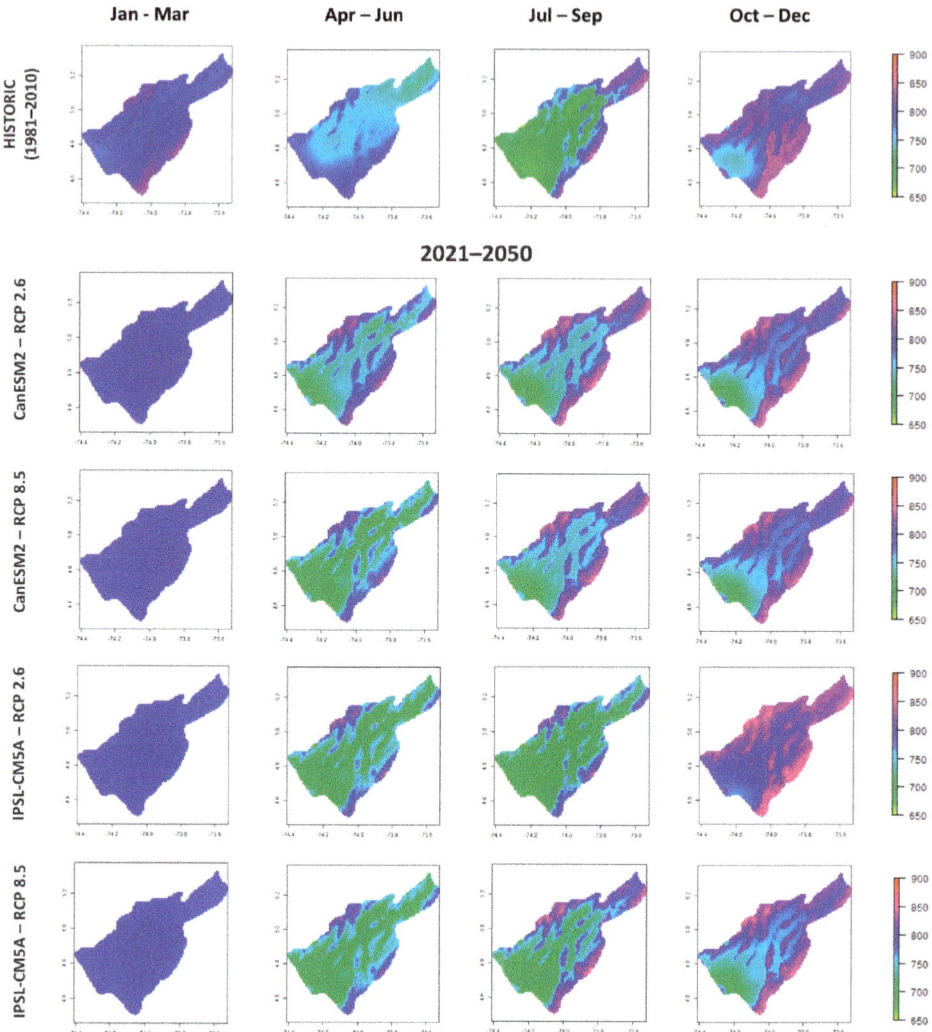

Figure A11. Soil Moisture 2021–2050.

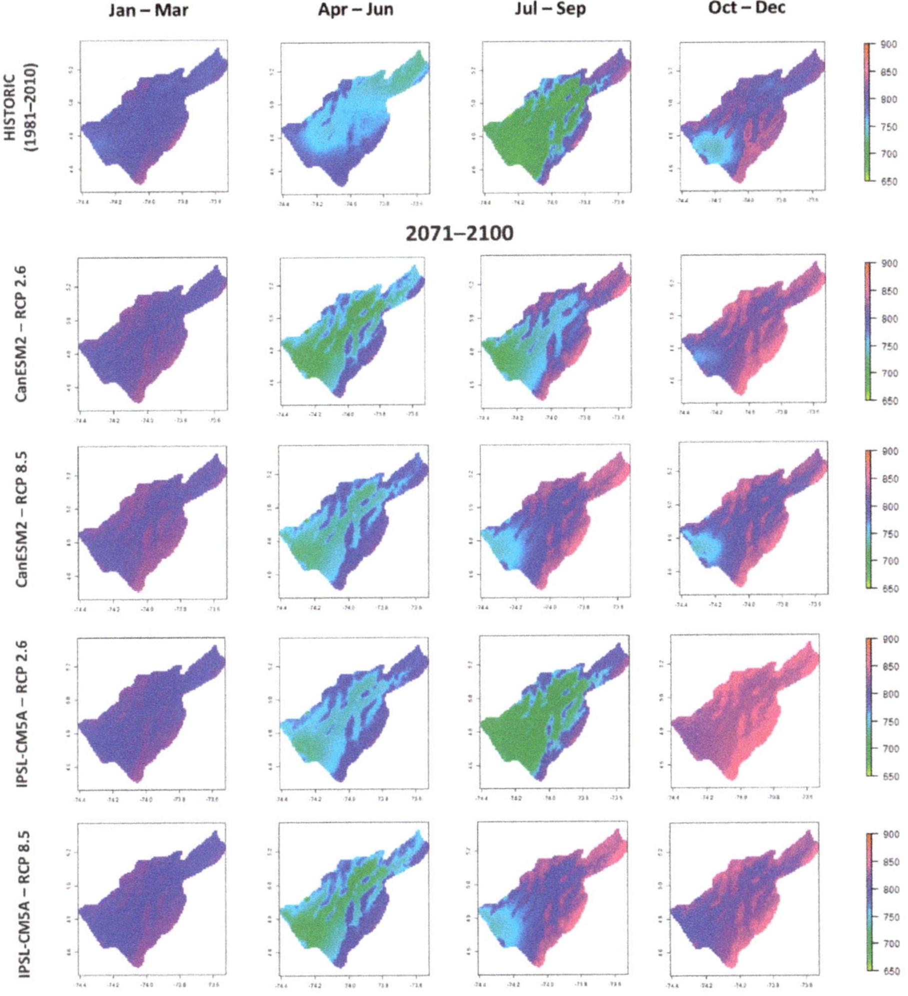

Figure A12. Soil Moisture 2071–2100.

References

1. IPCC. *Climate Change 2014: Synthesis Report. Contribution of Working Groups I, II and III to the Fifth Assessment Report of the Intergovernmental Panel on Climate Change*; Pachauri, R.K., Meyer, L.A., Eds.; IPCC: Geneva, Switzerland, 2014; 151p.
2. Band, L.; Mackay, D.; Creed, I.; Semkin, R.; Jeffries, D. Ecosystem processes at the watershed scale: Sensitivity to potential climate change. *Limnol. Oceanogr.* **1996**, *5*, 928–938. [CrossRef]
3. Jimenez Cisneros, B.E.; Oki, T.; Arnell, N.W.; Benito, G.; Cogley, J.G.; Döll, P.; Jiang, T.; Mwakalila, S.S. *Impacts, Adaptation and Vulnerability. Part A: Global and Sectorial Aspects. Contribution of Working Group II to the Fifth Assessment Report of the Intergovernmental Panel on Climate Change 2014*; Cambridge University Press: Cambridge, UK, 2014; pp. 229–269.
4. Healy, R.W.; Winter, T.C.; LaBaugh, J.W.; Franke, O.L. *Water Budgets: Foundations for Effective Water-Resources and Environmental Management*; U.S. Geological Survey Circular: Reston, VA, USA, 2007; Volume 1308, 90p.

5. Burns, D.A.; Klaus, J.; McHale, M.R. Recent climate trends and implications for water resources in the Catskill Mountain region, New York, USA. *J. Hydrol.* **2007**, *336*, 155–170. [CrossRef]
6. Candela, L.; Elorza, F.J.; Jiménez-Martínez, J.; von Igel, W. Global change and agricultural management options for groundwater sustainability. *Comput. Electron. Agric.* **2012**, *86*, 120–130. [CrossRef]
7. Hagg, W.; Braun, L.N.; Kuhn, M.; Nesgaard, T.I. Modelling of hydrological response to climate change in glacierized Central Asian catchments. *J. Hydrol.* **2007**, *332*, 40–53. [CrossRef]
8. Ruth, M.; Coelho, D. Understanding and managing the complexity of urban systems under climate change. *Clim. Policy* **2007**, *7*, 317–336. [CrossRef]
9. Werritty, A. Living with uncertainty: Climate change, river flows and water resource management in Scotland. *Sci. Total Environ.* **2002**, *294*, 29–40. [CrossRef]
10. Bates, B.C.; Kundzewicz, Z.W.; Wu, S.; Palutikof, J.P. (Eds.) *Climate Change and Water*; Technical Paper; Interguvernmental Panel on Climate Change: Geneva, Switzerland, 2008; 210p.
11. Fu, G.; Charles, S.P.; Yu, J. A critical overview of pan evaporation trends over the last 50 years. *Clim. Chang.* **2009**, *97*, 193–214. [CrossRef]
12. Miralles, D.G.; Holmes, T.R.H.; de Jeu, R.A.M.; Gash, J.H.; Meesters, A.G.C.A.; Dolman, A.J. Global land-surface evaporation estimated from satellite-based observations. *Hydrol. Earth Syst. Sci.* **2011**, *15*, 453–469. [CrossRef]
13. IDEAM; PNUD; MADS; CANCILLERÍA; DNP. New Climate Change Scenarios for Colombia 2011–2100. Scientific Tools for National-Regional Level Decision-Making. Available online: http://documentacion.ideam.gov.co/openbiblio/bvirtual/022964/documento_nacional_departamental.pdf (accessed on 11 September 2019).
14. Snow, J.W. *The Climate of Northern South America. Climates of Central and South America*; Schwerdtfeger, W., Ed.; Elsevier: Amsterdam, The Netherlands, 1976; pp. 295–403.
15. Mejía, J.F.; Mesa, O.J.; Poveda, G.; Vélez, J.I.; Hoyos, C.D.; Mantilla, R.; Barco, J.; Cuartas, A.; Montoya, M.; Botero, B. Spatial distribution, annual and semi-annual cycles of precipitation in Colombia. *DYNA* **1999**, *127*, 7–26. (In Spanish)
16. León, G.E.; Zea, J.A.; Eslava, J.A. General circulation and the intertropical convergence zone in Colombia. *Meteor. Colomb.* **2000**, *1*, 31–38. (In Spanish)
17. WMO. *WMO Guidelines on the Calculation of Climate Normals*; WMO-No. 1203; Chairperson, Publications Board: Geneva, Switzerland, 2017; Available online: https://library.wmo.int/doc_num.php?explnum_id=4166 (accessed on 21 June 2019).
18. Molina, O.D.; Bernhofer, C. Projected climate changes in four different regions in Colombia. *Environ. Syst. Res.* **2019**, *8*, 33. [CrossRef]
19. Federer, C.A. BROOK 90: A Simulation Model for Evaporation, Soil Water, and Streamflow. 2002. Available online: http://www.ecoshift.net/brook/brook90.htm (accessed on 21 June 2019).
20. Combalicer, E.A.; Lee, S.H.; Ahn, S.; Kim, D.Y.; Im, S. Modeling water balance for the small-forested watershed in Korea. *KSCE J. Civ. Eng.* **2008**, *12*, 339–348. [CrossRef]
21. Shuttleworth, W.J.; Wallace, J.S. Evaporation from sparse crops—An energy combination theory. *Q. J. R. Meteorol. Soc.* **1985**, *111*, 839–855. [CrossRef]
22. Brooks, R.H.; Corey, A.T. Hydraulic properties of porous media. *Hydrol. Pap.* **1964**, *3*, 1–27.
23. Saxton, K.E.; Rawls, W.J.; Romberger, J.S.; Papendick, R.I. Estimating generalized soil water characteristics from texture. *Trans. Am. Soc. Agric. Eng.* **1986**, *50*, 1031–1035. [CrossRef]
24. Wahren, A.; Schwärzel, K.; Feger, K.H.; Münch, A.; Dittrich, I. Identification and model based assessment of the potential water retention caused by land-use changes. *Adv. Geosci. Eur. Geosci. Union* **2007**, *11*, 49–56. [CrossRef]
25. Bastidas Osejo, B.; Betancur Vargas, T.; Alejandro Martinez, J. Spatial distribution of precipitation and evapotranspiration estimates from Worldclim and Chelsa datasets: Improving long-term water balance at the watershed-scale in the Urabá region of Colombia. *Int. J. Sustain. Dev. Plan.* **2019**, *14*, 105–117. [CrossRef]
26. Leta, O.T.; El-Kadi, A.I.; Dulai, H.; Ghazal, K.A. Assessment of climate change impacts on water balance components of Heeia watershed in Hawaii. *J. Hydrol. Reg. Stud.* **2016**, *8*, 182–197. [CrossRef]
27. Louzada, F.L.R.; de, O.; Xavier, A.C.; Pezzopane, J.E.M. Climatological water balance with data estimated by tropical rainfall measuring mission for the Doce river basin. *Eng. Agric.* **2018**, *38*, 376–386. [CrossRef]

28. Silva, A.L.; Roveratti, R.; Reichardt, K.; Bacchi, O.O.; Timm, L.C.; Bruno, I.P.; Oliveira, J.C.; Dourado Neto, D. Variability of water balance components in a coffee crop in Brazil. *Sci. Agric.* **2006**, *63*, 105–114. [CrossRef]
29. Almeida, A.Q.; Ribeiro, A.; Leite, F.P.; Souza, R.; Gonzaga, M.S.; Santos, W.A. Water Balance in a Tropical Eucalyptus plantations in the Doce River Basin, Eastern Brazil. *JAS J. Agric. Sci.* **2019**, *11*, 209–217. [CrossRef]
30. Schwerdtfeger, J.; Weiler, M.; Johnson, M.S.; Couto, E.G. Estimating water balance components of tropical wetland lakes in the Pantanal dry season, Brazil. *Hydrol. Sci. J.* **2014**, *59*, 2158–2172. [CrossRef]
31. Escurra, J.J.; Vazquez, V.; Cestti, R.; De Nys, E.; Srinivasan, R. Climate change impact on countrywide water balance in Bolivia. *Reg. Environ. Chang.* **2014**, *14*, 727–742. [CrossRef]
32. McSweeney, C.F.; Jones, R.G.; Lee, R.W.; Rowell, D.P. Selecting CMIP5 GCMs for downscaling over multiple regions. *Clim. Dyn.* **2015**, *44*, 3237. [CrossRef]
33. Bonilla-Ovallos, C.A.; Mesa, O.J. Validación de la precipitación estimada por modelos climáticos acoplados del proyecto de intercomparación CMIP5 en Colombia. *Rev. De La Acad. Colomb. De Cienc. Exactas Físicas Y Nat.* **2017**, *41*, 107. [CrossRef]
34. Yin, L.; Fu, R.; Shevliakova, E.; Dickinson, R.; Dickinson, R.E. How well can CMIP5 simulate precipitation and its controlling processes over tropical South America? *Clim. Dyn.* **2012**, *41*, 3127–3143. [CrossRef]
35. Feng, X.; Porporato, A.; Rodriguez-Iturbe, I. Changes in rainfall seasonality in the tropics. *Nat. Clim. Chang.* **2013**, *3*, 811–815. [CrossRef]
36. Feng, X.; Vico, G.; Porporato, A. On the effects of seasonality on soil water balance and plant growth. *Water Resour. Res.* **2012**, *48*, W05543. [CrossRef]
37. Hartmann, D.L.; Tank, A.M.; Rusticucci, M.; Alexander, L.V.; Brönnimann, S.; Charabi, Y.A.; Dentener, F.J.; Dlugokencky, E.J.; Easterling, D.R.; Kaplan, A.; et al. Observations: Atmosphere and surface. In *Climate Change 2013: The Physical Science Bases. Contribution of Working Group I to the Fifth Assessment Report of the Intergovernmental Panel on Climate Change*; Stocker, T.F., Ed.; Cambridge University Press: Cambridge, UK, 2013; pp. 159–254. Available online: http://www.climatechange2013.org/report/full-report/ (accessed on 7 August 2019).
38. MacDougall, A.H.; Eby, M.; Weaver, A.J. If anthropogenic CO2 emissions cease, will atmospheric CO_2 concentration continue to increase? *J. Clim.* **2013**, *26*, 9563–9576. [CrossRef]
39. Zhao, T.; Chen, L.; Ma, Z. Simulation of historical and projected climate change in arid and semiarid areas by CMIP5 models. *Chin. Sci. Bull.* **2014**, *59*, 412–429. [CrossRef]
40. Ji, M.; Huang, J.; Xie, Y.; Liu, J. Comparison of dryland climate change in observations and CMIP5 simulations. *Adv. Atmos. Sci.* **2015**, *32*, 1565–1574. [CrossRef]
41. Muerth, M.J.; St-Denis, G.; Ricard, B.; Velázquez, S.; Schmid, J.A.; Minville, M.; Caya, D.; Chaumont, D.; Ludwig, R.; Turcotte, R. On the need for bias correction in regional climate scenarios to assess climate change impacts on river runoff. *Hydrol. Earth Syst. Sci.* **2013**, *17*, 1189–1204. [CrossRef]
42. Prudhomme, C.; Davies, H. Assessing uncertainties in climate change impact analyses on the river flow regimes in the UK. Part 2: Future climate. *Clim. Chang.* **2009**, *93*, 197–222. [CrossRef]
43. Hagemann, S.; Chen, C.; Haerter, J.O.; Heinke, J.; Gerten, D.; Piani, C. Impact of a statistical bias correction on the projected hydrological changes obtained from three GCMs and two hydrology models. *J. Hydrometeorol.* **2011**, *12*, 556–578. [CrossRef]
44. Dobler, C.; Hagemann, S.; Wilby, R.L.; Stötter, J. Quantifying different sources of uncertainty in hydrological projections in an Alpine watershed, Hydrol. *Earth Syst. Sci.* **2012**, *16*, 4343–4360. [CrossRef]
45. Thompson, J.R.; Green, A.J.; Kingston, D.G.; Gosling, S.N. Assessment of uncertainty in river flow projections for the Mekong River using multiple GCMs and hydrological models. *J. Hydrol.* **2013**, *486*, 1–30. [CrossRef]
46. Velázquez, J.A.; Schmid, J.; Ricard, S.; Muerth, M.J.; Gauvin St- Denis, B.; Minville, M.; Chaumont, D.; Caya, D.; Ludwig, R.; Turcotte, R. An ensemble approach to assess hydrological models' contribution to uncertainties in the analysis of climate change impact on water resources. *Hydrol. Earth Syst. Sci.* **2013**, *17*, 565–578. [CrossRef]
47. Jobst, A.M.; Kingston, D.G.; Cullen, N.J.; Schmid, J. Intercomparison of different uncertainty sources in hydrological climate change projections for an alpine catchment (upper Clutha River, New Zealand). *Hydrol. Earth Syst. Sci.* **2018**, *22*, 3125–3142. [CrossRef]

48. Clark, M.P.; Wilby, R.L.; Gutmann, E.D.; Vano, J.A.; Gangopadhyay, S.; Wood, A.W.; Fowler, H.J.; Prudhomme, C.; Arnold, J.R.; Brekke, L.D. Characterizing uncertainty of the hydrologic impacts of climate change. *Clim. Chang. Rep.* **2016**, *2*, 55–64. [CrossRef]
49. Xu, C.; Widén, E.; Halldin, S. Modelling hydrological consequences of climate change—Progress and challenges. *Adv. Atmos. Sci.* **2005**, *22*, 789–797. [CrossRef]

MDPI
St. Alban-Anlage 66
4052 Basel
Switzerland
Tel. +41 61 683 77 34
Fax +41 61 302 89 18
www.mdpi.com

Water Editorial Office
E-mail: water@mdpi.com
www.mdpi.com/journal/water

www.ingramcontent.com/pod-product-compliance
Lightning Source LLC
LaVergne TN
LVHW070202170726
843515LV00007B/1982

* 9 7 8 3 0 3 9 4 3 3 0 6 3 *